ADVANCES IN MOLECULAR AND CELL BIOLOGY

Volume 23A • 1998

ION PUMPS

ADVANCES IN MOLECULAR AND CELL BIOLOGY

ION PUMPS

Series Editor: E. EDWARD BITTAR
Department of Physiology
University of Wisconsin
Madison, Wisconsin

Guest Editor: JENS PETER ANDERSEN
Institute of Physiology
University of Aarhus
Aarhus, Denmark

VOLUME 23A • 1998

JAI PRESS INC.

Greenwich, Connecticut *London, England*

55 Old Post Road No. 2
Greenwich, Connecticut 06836

JAI PRESS LTD.
38 Tavistock Street
Covent Garden
London WC2E 7PB
England

ISBN: 0-7623-0287-9

Printed and bound in Great Britain by
CPI Antony Rowe, Chippenham and Eastbourne

CONTENTS (Volume 23A)

CONTENTS (VOLUME 23B)

LIST OF CONTRIBUTORS

Robert Aggeler	Institute of Molecular Biology University of Oregon Eugene, Oregon
Krister Bamberg	Wadsworth Veterans Administration Medical Center Los Angeles, California
Denis Bayle	Wadsworth Veterans Administration Medical Center Los Angeles, California
Roderick A. Capaldi	Institute of Molecular Biology University of Oregon Eugene, Oregon
Leopoldo de Meis	Departamento de Bioquimica Médica Universidade Federal do Rio de Janeiro Rio de Janeiro, Brazil
Peter Dimroth	Mikrobiologisches Institut Eidgenössische Technische Hochschule Zürich, Switzerland
Jan Eggermont	Laboratorium voor Fysiologie Katholieke Universiteit Leuven Leuven, Belgium
Simone Engelender	Departamento de Bioquimica Médica Universidade Federal do Rio de Janeiro Rio de Janeiro, Brazil

Agnes Enyedi	National Institute of Haematology, Blood Transfusion and Immunology Budapest, Hungary
Michael Forgac	Department of Cellular and Molecular Physiology Tufts University Boston, Massachusetts
Käthi Geering	Institute of Pharmacology and Toxicology University of Lausanne Lausanne, Switzerland
Ursula Gerike	School of Biology and Biochemistry University of Bath Bath, England
James E. Haber	Rosenstiel Basic Medical Sciences Research Center and Department of Biology Brandeis University Waltham, Massachusetts
Parjit Kaur	Department of Biology Georgia State Univesity Atlanta, Georgia
Daniel Khananshvili	Department of Physiology and Pharmacology Sackler School of Medicine Tel-Aviv University Tel-Aviv, Israel
David B. McIntosh	MRC Biomembrane Research Unit and Department of Chemical Pathology University of Cape Town Medical School Cape Town, South Africa
Luc Mertens	Laboratorium voor Fysiologie Katholieke Universiteit Leuven Leuven, Belgium
John T. Penniston	Department of Biochemistry and Molecular Biology Mayo Clinic/Foundation Rochester, Minnesota

David S. Perlin — Public Health Research Institute
New York, New York

Luc Raeymaekers — Laboratorium voor Fysiologie
Katholieke Universiteit Leuven
Leuven, Belgium

George Sachs — Wadsworth Veterans Administration
Medical Center
Los Angeles, California

Jai Moo Shin — Wadsworth Veterans Administration
Medical Center
Los Angeles, California

Marc Solioz — Department of Clinical Pharmacology
University of Berne
Berne, Switzerland

Ludo Van Den Bosch — Laboratorium voor Fysiologie
Katholieke Universiteit Leuven
Leuven, Belgium

Hilde Verboomen — Laboratorium voor Fysiologie
Katholieke Universiteit Leuven
Leuven, Belgium

Herman Wolosker — Departamento de Bioquimica Médica
Universidade Federal do Rio de Janeiro
Rio de Janeiro, Brazil

Frank Wuytack — Laboratorium voor Fysiologie
Katholieke Universiteit Leuven
Leuven, Belgium

PREFACE

Both eukaryotic and prokaryotic cells depend strongly on the function of ion pumps present in their membranes. The term ion pump, synonymous with active ion-transport system, refers to a membrane-associated protein that translocates ions uphill against an electrochemical potential gradient. Primary ion pumps utilize energy derived from chemical reactions or from the absorption of light, while secondary ion pumps derive the energy for uphill movement of one ionic species from the downhill movement of another species.

In the present volume, various aspects of ion pump structure, mechanism, and regulation are treated using mostly the ion-transporting ATPases as examples. One chapter has been devoted to a secondary ion pump, the Na^+-Ca^{2+} exchanger, not only because of the vital role played by this transport system in regulation of cardiac contractility, but also because it exemplifies the interesting mechanistic and structural similarities between primary and secondary pumps.

The cation-transporting ATPases fall into two major categories. The P-type ATPases (treated in eight chapters) are characterized by the formation during the catalytic cycle of an aspartyl phosphorylated intermediate, and by their rather simple subunit composition (one catalytic α-chain, supplemented in Na^+,K^+ and H^+,K^+ pumps with a β-chain that has a crucial role in expression of functional pumps at the cell surface). The F/V-type ATPases (treated in three chapters) do not form a recognizable phosphoenzyme intermediate and are composed of multiple different subunits whose individual roles are now beginning to be understood. Similarly, some of the anion-transporting ATPases (treated in a single chapter) seem to be composed of multiple subunits with separate roles in ATP hydrolysis and ion translocation.

Although there are too many structural and functional differences between the various ion transport ATPase families to warrant speculations about a common evolutionary origin of all families, they do seem to resemble each other in some fundamental aspects. Thus, irrespective of whether an ATP-driven pump consists of one or several subunits, it has a large hydrophilic "pump head" protruding from the membrane, which is responsible for the ATP hydrolysis/synthesis reactions. The vectorial transport process is carried out by a more slender membrane-buried part forming a channel-like structure. Energy coupling seems to occur by long-range communication between the membrane-buried and extramembranous domains/subunits, mediated by conformational changes. A major challenge is to elucidate the exact nature of the coupling process and to define the ion-translocation pathway through the membrane-buried part of the protein. As discussed in some of the chapters in this book, the distinction between ion pumps and ion channels now appears less sharp than previously, since several transporters can exhibit both types of function. It appears reasonable to consider an ion pump as a channel equipped with locks and gate controls.

The three-dimensional structure of the extramembranous catalytic F_1 part of F-type ATPases has recently been determined, and this has helped to provide a firm basis for speculations on the energy-coupling mechanism. Similarly, although a high-resolution crystal structure is still a distant goal in the case of P-type ATPases, analysis of the amino acid sequences of these pumps, in conjunction with various site-directed approaches such as mutagenesis and affinity labeling, has paved the way for realistic modeling of the structure of the ATP binding domain and its coupling with the ion transport domain.

The topological features of the membranous part of P-type ATPases have long been a matter of conjecture, but several pieces of evidence now support a 10-transmembrane segment structure in Ca^{2+}-, H^+,K^+-, and Na^+,K^+-ATPases. A core region constituted by six of these transmembrane segments seems to have been retained also during evolution of P-type ATPases involved in heavy-metal-ion transport.

The determinants of the specific ion selectivities of the pumps are poorly understood, but some interesting clues have been provided in studies of hybrids between H^+- and Na^+-transporting F-type ATPases and by comparing heavy-metal-transporting P-type ATPases with other P-type ATPases.

Many ion pumps are known to occur as multiple isoenzymes differing with respect to tissue distribution and functional and regulatory characteristics. Regulation occurs both at the level of gene expression and through changes in kinetic parameters that determine the molecular pump rates. In health as well as in disease, control is exerted by the concentrations of ions to be pumped and by hormonal and developmental influences. Although the molecular details of the regulatory events are now beginnning to be understood, as described in several of the chapters in this volume, there seems to be much more to learn.

Jens Peter Andersen
Guest Editor

REACTION MECHANISM OF THE SARCOPLASMIC RETICULUM Ca^{2+}-ATPase

Herman Wolosker, Simone Engelender,

and Leopoldo de Meis

Advances in Molecular and Cell Biology
Volume 23A, pages 1-31.

ISBN: 0-7623-0287-9

I. THE RELAXING FACTOR

The discovery of the Ca^{2+}-ATPase is intimately associated with the understanding of the mechanism of muscle relaxation. The first report of the Ca^{2+}-ATPase was published by Kielley and Meyerhof (1948). Who found a Mg^{2+}-dependent ATPase activity separate from actomyosin that was associated with the microsomal fraction of skeletal muscle (1952). At that time, the function of this new ATPase was unknown. Marsh showed that an aqueous muscle extract was able to induce swelling of muscle homogenate in the presence of ATP, suggesting a relaxation process of the contractile proteins. Bendall (1952) further showed that the "Marsh" factor was also able to relax glycerinated fibers. Shortly after, it was recognized that the fraction containing the Kielley-Meyerhof enzyme was identical to the "Marsh" relaxing factor (Kumagai et al., 1955) and corresponded to fragments of sarcoplasmic reticulum (Portzehl, 1957; Ebashi, 1958; Nagai et al., 1960; Muscatello et al., 1962). After these discoveries, different approaches were employed to study the mechanism of action of the relaxing factor, but few of them focused on the role of Ca^{2+} ion in the process of muscle contraction and relaxation. Bendall had shown that the relaxing activity of the factor was inactivated by the addition of low concentrations of Ca^{2+} (Bendall, 1953). Furthermore, EDTA, a divalent cation chelating agent, mimicked the relaxing factor by inducing relaxation of glycerinated fibers (Bozler, 1954). Weber (1959) later showed that the contraction and ATPase activity of myofibrils required low concentrations of Ca^{2+}. Aware of these findings and working at the laboratory of Fritz Lipmann, Setsuro Ebashi (1960) found a relationship between Ca^{2+} binding capacities of various chelating agents and the ability to promote relaxation of glycerinated fibers. He soon realized that the relaxing factor is able to promote muscle relaxation by removing Ca^{2+} from the contractile system . Ebashi (1961) also showed that the relaxing factor was able to bind Ca^{2+} in the presence of ATP (Ebashi, 1961). Since the amount of Ca^{2+} associated with the relaxing factor was low and it was not possible to detect any Ca^{2+}-dependent ATPase activity, Ebashi thought that the Ca^{2+} ions bind to the membrane of sarcoplasmic vesicles, rather than being accumulated in an active process inside the vesicles (Ebashi, 1961, Ebashi and Lipmann, 1962). Excellent reviews of the discovery of relaxing factor can be found in Ebashi and Endo (1968) and Ebashi, (1985).

II. THE CALCIUM PUMP

The first evidence for an active Ca^{2+} pump in the relaxing factor was reported by Hasselbach and Makinose (1961) who investigated the Ca^{2+}-binding capacity of different muscle components in the presence of oxalate, a Ca^{2+}-binding anion. They found that Ca^{2+} was accumulated in sarcoplasmic reticulum vesicles in a process coupled with the hydrolysis of ATP (Hasselbach and Makinose, 1961,1962). This discovery was made possible by the inclusion of oxalate in the assay medium. Oxalate diffuses through the vesicle membrane and allows the accumulation of large quantities of Ca^{2+} inside the vesicles as Ca^{2+}-oxalate precipitates. These were later identified with the use of electron microscopy (Hasselbach, 1964). Addition of micromolar Ca^{2+} concentrations led to a striking increase of the ATPase activity of sarcoplasmic reticulum vesicles, which was called ATP "extra"-splitting. This ATPase activity was coupled with the Ca^{2+} uptake by sarcoplasmic reticulum vesicles (Figure 1). The stoichiometry of two Ca^{2+} ions accumulated by the vesicles for each molecule of ATP hydrolyzed was determined by measuring the initial velocities of both Ca^{2+} uptake and ATP hydrolysis (Hasselbach and Makinose, 1963).

Subsequently, Makinose and Hasselbach (1965) observed that the Ca^{2+} gradient formed across the sarcoplasmic reticulum membrane regulates Ca^{2+} transport. When the Ca^{2+} concentration inside the sarcoplasmic reticulum vesicles was increased to the millimolar range, the rates of both ATP hydrolysis and Ca^{2+} transport decreased, providing evidence for a regulatory role of the Ca^{2+} gradient that was called "back inhibition" (Makinose and Hasselbach; Weber et al., 1966; Weber 1971a, 1971b; Yamada and Tonomura, 1972).

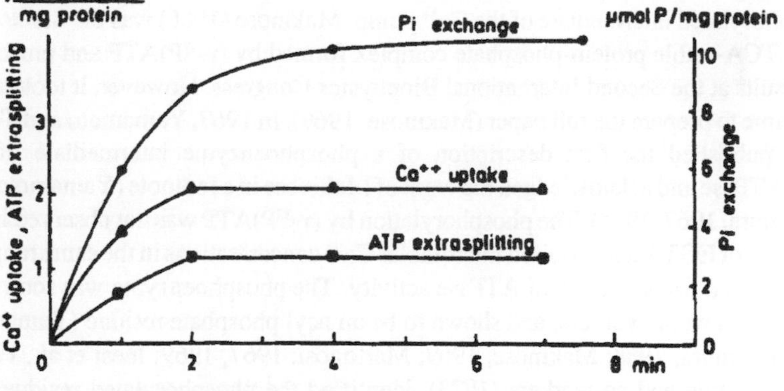

Figure 1. Simultaneous kinetic measurements of Ca^{2+} uptake, ATP hydrolysis and Ca^{2+}-dependent ADP/ ATP exchange in the presence of 5 mM oxalate (Hasselbach and Makinose, 1963).

III. PHOSPHOENZYME OF HIGH ENERGY

Hasselbach and Makinose (1962) and Ebashi and Lipmann (1962), independently found that the microsomal preparation catalyzed a phosphate exchange between ATP and ADP. This was the first indication that the reaction of ATP hydrolysis could be reversed. The existence of a high-energy phosphorylated intermediate was proposed to account for the transfer of phosphate from ATP to the molecule of ADP (Hasselbach and Makinose, 1962; Ebashi and Lipmann, 1962). At that time, it was not known which component of the membrane was phosphorylated by ATP; that is in the membrane lipids or the enzyme molecule itself. Hasselbach and Makinose (1962) proposed the participation of a phosphorylated intermediate in the Ca^{2+}-transport process as follows:

> The unknown substance mentioned becomes a carrier by phosphorylation on the outer surface of the granules which increases its affinity for calcium. The calcium complex of the phosphorylated carrier diffuses to the inner surface of the membrane of the granule. There the phosphate group is split off from the carrier whereby the calcium affinity of the carrier is strongly diminished. Thus the bound calcium is liberated in spite of the comparatively high concentration of calcium 'inside' (Hasselbach and Makinose, 1962).

In spite of the influences derived from the ATP,ADP exchange experiments, the first direct measurement of phosphoenzyme was not obtained with Ca^{2+}-ATPase, but rather with Na^+-K^+-ATPase discovered by Skou (1957). In 1965, Post was able to show that the Na^+-K^+-ATPase of kidney membranes was phosphorylated by (γ-^{32}P)ATP (Post et al., 1965) and then the phosphoenzyme was acid stable and participated in the hydrolytic cycle of the Na^+-K^+-ATPase. Both Na^+ and Mg^{2+} stimulated its formation while the hydrolysis of the phosphoenzyme was stimulated by K^+. Following these observations with Na^+-K^+-ATPase, attempts were made to identify a phosphorylated intermediate of the Ca^{2+} pump. Makinose (1966) was the first to isolate a TCA-stable protein-phosphate complex formed by (γ-^{32}P)ATP and presented his results at the Second International Biophysics Congress. However, it took him a long time to prepare the full paper (Makinose, 1969). In 1967, Yamamoto and Tonomura published the first description of a phosphoenzyme intermediate of the Ca^{2+}-ATPase and acknowledged the work of Makinose in a footnote (Yamamoto and Tonomura, 1967, 1968). The phosphorylation by (γ-^{32}P)ATP was not observed in the presence of EGTA and required micromolar Ca^{2+} concentrations in the same range as that required for activation of ATPase activity. The phosphoenzyme was soon confirmed by several workers, and shown to be an acyl phosphate residue (Yamamoto and Tonomura, 1968; Makinose, 1969; Martonosi, 1967, 1969; Inesi et al., 1970). Later, Bastide and co-workers (1973), identified the phosphorylated residue as a phosphoaspartate at the catalytic site of the Ca^{2+}-ATPase.

The formation and hydrolysis of the phosphoenzyme intermediate were described by two coupled reactions:

$$ATP + E \rightleftharpoons E{\sim}P + ADP \text{ (Reaction 1)}$$

$$E{\sim}P \rightarrow E + P_i \text{ (Reaction 2)}$$

The enzyme would be phosphorylated by ATP, with the generation of a phosphoenzyme intermediate having a high-energy phosphate bond (Reaction 1). Reaction 1 would be fully reversible and have an equilibrium constant of one, in order to account for the phosphate exchange between ATP and ADP (Hasselbach and Makinose, 1962; Ebashi and Lipmann, 1962). Reversal of this reaction would be possible only if the phosphoenzyme were an acyl phosphate residue, which had a high phosphate "group transfer potential" (Lipmann, 1941). If the phosphoenzyme were of a phosphoester type, then it was thought that the hydrolytic reaction would not be reversible, since a phosphoester has a low phosphate "group transfer potential" and could not transfer its phosphate bond to ADP. At that time it was believed that the hydrolysis of the phosphoenzyme would provide energy for the transport of Ca^{2+} across the membrane (Reaction 2), so that the Ca^{2+} ion would be transported only after the hydrolysis of the acylphosphate. The changes in protein tertiary structure that would be required for Ca^{2+} transport would be minimal, and little energy would be spent by the conformational changes of the protein per se. From this sequence, it was inferred that work would be performed in a part of the enzyme molecule where energy is released, for example, in the immediate vicinity of the catalytic site.

IV. THE USE OF CHEMIOSMOTIC ENERGY AND THE REVERSAL OF TRANSPORT ATPases

A new concept regarding the mechanism of energy transduction in biological systems was put forward by Mitchell (1961). Studying the mechanism of ATP synthesis by the F_0F_1 ATPase, Mitchell observed that during cellular respiration an electrochemical proton gradient was formed across the mitochondrial membrane, and that the energy from this gradient would be used by the F_0F_1 to synthesize ATP from ADP and P_i. This was called the "protonmotive force" (Mitchell, 1961, 1966). The protonmotive force was defined as the sum of the electrical potential and proton gradient across the membrane. As H^+ flows through the F_0F_1-ATPase, the enzyme would synthesize ATP from ADP and P_i by converting the osmotic energy into chemical energy. Since the F_0F_1-ATPase does not form a phosphorylated intermediate during the process of ATP synthesis, it was not clear whether the chemiosmotic hypothesis could be applied to other ATPases such as the Ca^{2+}-ATPase and the Na^+-K^+-ATPase that form an acyl phosphate intermediate during the catalytic cycle.

Garrahan and Glynn (1967) showed that the Na^+-K^+-ATPase could use the energy derived from the gradients of Na^+ and K^+ across the membrane to synthesize

ATP. These workers prepared red blood cells containing high K^+ inside and low K^+ in a medium containing only Na^+. They found that ATP was synthesized from ADP and P_i in a process involving inward Na^+ movement and outward K^+ transport through the Na^+-K^+-ATPase (Garrahan and Glynn, 1967).

Shortly after, it was shown that the entire process of Ca^{2+} transport could also be reversed by establishing a Ca^{2+} concentration gradient across the sarcoplasmic reticulum vesicles membrane (Barlogie et al., 1971; Makinose and Hasselbach, 1971; Makinose, 1971, 1972). In these experiments the vesicles were previously loaded with Ca^{2+} and incubated in a medium containing EGTA. In this condition Ca^{2+} flows out of the vesicles at a slow rate due to the low Ca^{2+} permeability of the membrane. The Ca^{2+} efflux increased sharply when ADP⇌P_i and Mg^{2+} were added to the incubation medium. The fast efflux of Ca^{2+} was found to be coupled with the synthesis of ATP from ADP and P_i (Figure 2). For every two Ca^{2+} ions released from the vesicles, one molecule of ATP was synthesized. Another approach used to demonstrate the reversal of the Ca^{2+} pump involved the ATP⇌P_i exchange reaction. Makinose (1971) showed that after steady-state Ca^{2+} accumulation was reached, the Ca^{2+}-ATPase catalyzed an exchange between (^{32}P)P_i and the gamma phosphate of ATP (Figure 3). This exchange indicated that the Ca^{2+} pump can operate simultaneously in both directions, for example, hydrolyzing and synthesizing ATP (Makinose, 1971). The disruption of the vesicle integrity either with phospholipase A or diethyl ether abolished the synthesis of ATP, which indicated that the calcium gradient was essential to the synthesis of ATP. From these data, it was inferred that

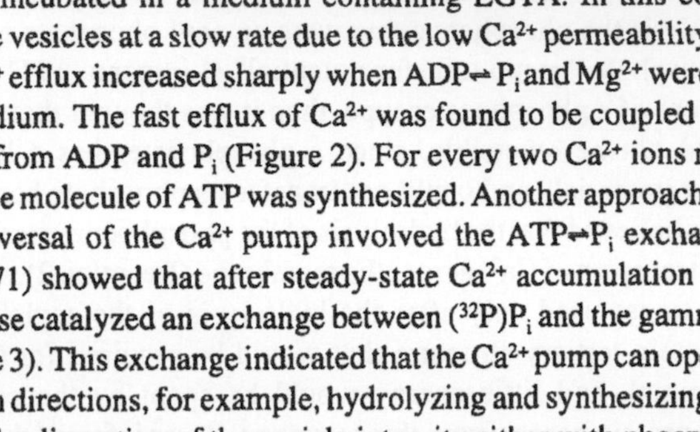

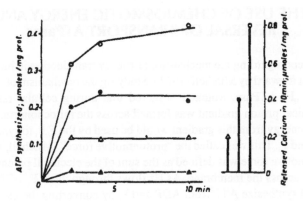

Figure 2. Ca^{2+} release and net ATP synthesis. During an incubation period of fifteen minutes in media containing (mM) 2 acetyl phosphate, 7 $MgCl_2$, 40 KCl, 0.1 glucose, 20 Na(^{32}P) orthophosphate, and 0.5 mg/ml vesicular protein at pH 7.0, the vesicles (1 mg protein) were loaded with different amounts of calcium (nmoles): (○) 800, (●) 400, (-) 200, and (▲) no calcium. Subsequently, the Ca^{2+} release was started by addition of 2 mM ADP, hexokinase (0.02 mg/ml), and 1 to 2 mM EGTA at time 0. For experimental details, see Makinose and Hasselbach (1971).

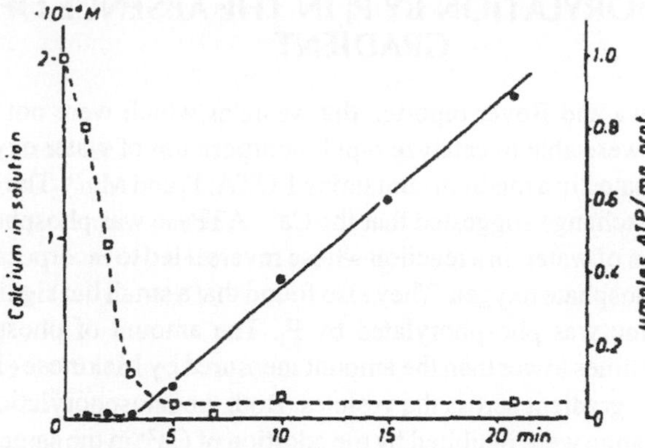

Figure 3. Calcium uptake and ATP formation. The reactions were started by addition of the vesicles at time 0. The composition of the medium is given in the text. *Solid line:* ATP formation. The reaction was stopped by addition of 5 percent perchloric acid. After filtration and neutralization with KOH, hexokinase and glucose were added to the supernatant. From the ^{32}P-radioactivity in the glucose-6-phosphate fraction (isolated by paper chromatography), the amount of ATP formed was calculated (right ordinate). *Dotted line:* Decrease in the calcium concentration in the assay resulted from calcium uptake of the vesicles. The ^{45}Ca level was determined in aliquots filtered through Millipore filter (450 n)(left ordinate). For details, see Makinose (1971).

when the steady state between Ca^{2+} influx and Ca^{2+} efflux was reached, the energy derived from the hydrolysis of ATP was used to maintain the Ca^{2+} gradient, and the energy derived from the gradient was used for the synthesis of ATP (Makinose, 1971; de Meis and Carvalho, 1974; Carvalho et al., 1976).

If the synthesis of ATP was indeed the direct result of reversal of the Ca^{2+} pump, then it would be expected that the enzyme should be phosphorylated by P_i as follows:

$$E + P_i \rightarrow EP$$

$$E{\sim}P + ADP \rightarrow E + ATP$$

A phosphoenzyme intermediate formed from (^{32}P)P_i was in fact detected by Makinose (1972) when a Ca^{2+} gradient was formed across the vesicle membrane. This phosphorylated intermediate could transfer its high-energy phosphate bond to ADP, resulting in ATP synthesis. At that time it was thought that the energy derived from the gradient would be trapped by the enzyme and used for the formation of the high-energy bond phosphoenzyme.

V. PHOSPHORYLATION BY Pi IN THE ABSENCE OF A Ca^{2+} GRADIENT

In 1973, Kanazawa and Boyer reported that vesicles which were not previously loaded with Ca^{2+} were able to catalyze rapid incorporation of water oxygen atoms into P_i when incubated in a medium containing EGTA, P_i and Mg^{2+}. The occurrence of the $P_i \rightharpoonup$ HOH exchange suggested that the Ca^{2+}-ATPase was phosphorylated by P_i with elimination of water, in a reaction whose reversal led to incorporation of water oxygen into phosphate oxygen. They also found that a small but significant fraction of the enzyme was phosphorylated by P_i. The amount of phosphoenzyme measured was 20 times lower than the amount measured by Makinose (1972) in the presence of a Ca^{2+} gradient across the vesicles. Both the phosphorylation by P_i and the P_i HOH exchange were inhibited by the addition of Ca^{2+} in the same concentration range as that needed for the activation of ATP hydrolysis. Addition of the detergent triton X-100, which renders the vesicles leaky to Ca^{2+}, abolished the small phosphorylation by P_i and the $P_i \rightharpoonup$ HOH exchange. It was already known that sarcoplasmic reticulum vesicles contain a small amount of endogenous Ca^{2+} (Duggan and Martonosi, 1970; Chevalier and Butow, 1971). On the basis of the low amount of phosphoenzyme obtained and its inhibition by triton X-100, these workers concluded that phosphorylation by P_i and the $P_i \rightharpoonup$ HOH exchange was promoted by the small transmembrane Ca^{2+} gradient, formed between the contaminant Ca^{2+} inside the vesicles and EGTA in the medium.

Shortly afterwards, Masuda and de Meis demonstrated the phosphorylation of the enzyme by $(^{32}P)P_i$ in the absence of a transmembrane Ca^{2+} gradient (Masuda and de Meis, 1973; de Meis and Masuda, 1974). In these reports, the vesicles were rendered leaky with the use of diethyl ether instead of triton X-100, and the chosen pH was 6.0 and not 7.0, as in the Kanazawa and Boyer experiments. At pH 6.0, Masuda and de Meis found high levels of phosphoenzyme in the leaky vesicles which were similar to those obtained in the presence of a Ca^{2+} gradient. The phosphoenzyme was hydroxylamine-sensitive and unstable at alkaline pH. This provided definitive evidence that an acyl phosphate could be formed spontaneously from P_i in the absence of a transmembrane Ca^{2+} gradient. The affinity of the enzyme to P_i increases several fold when a Ca^{2+} gradient is formed across the membrane, or when the pH is decreased to 6.0 (Masuda and de Meis, 1973; de Meis and Masuda, 1974; de Meis, 1976). Later, Kanazawa (1975) showed that the Ca^{2+}-ATPase was unstable in the presence of triton X-100, thus, explaining the earlier findings of Kanazawa and Boyer. After the addition of triton X-100, a large phosphorylation by P_i could only be observed in the first seconds of the reaction in the absence of a Ca^{2+}.

The formation of an acyl phosphate in the Ca^{2+}-ATPase without a Ca^{2+} gradient across the membrane was a puzzling finding, since at that time it was thought that an acyl phosphate, possessing a high-energy bond, could not be formed spontaneously without input of energy in to the system. That the

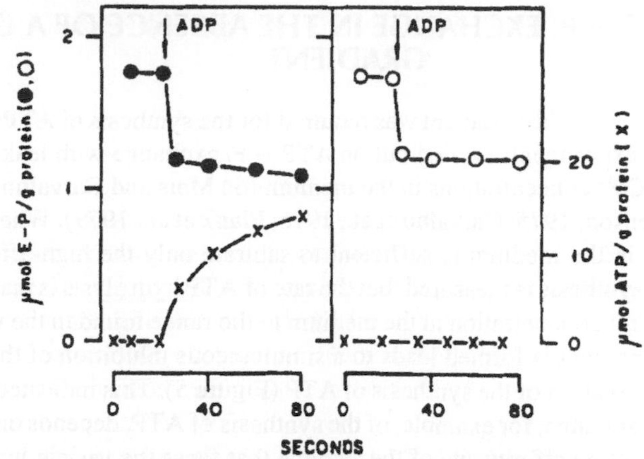

Figure 4. Phosphorylation by P_i and ATP synthesis. The assay medium composition was Tris maleate buffer (pH 6.5), 30 mM $MgCl_2$, 15 mM EGTA, and 2 mM (*left*) or 5 mM (*right*) $^{32}P_i$. The reaction was started by addition of vesicles which were loaded with calcium phosphate (*left*) or empty and leaky (*right*). After different incubation intervals at 30°C, samples were taken to determine phosphoenzyme (E-P; ● and ○) and γ^{32}ATP (×). The arrow indicates time of addition of ADP, to a final concentration of 0.05 mM in the assay medium. For further details see de Meis (1976).

Ca^{2+}-ATPase phosphorylated by P_i in the absence of a Ca^{2+} gradient was not able to transfer the phosphate from the phosphoenzyme to ADP was shown by de Meis (1976). After phosphorylation by P_i, the addition of ADP to the medium led to a decrease in the level of the phosphoenzyme from either leaky vesicles or vesicles previously loaded with Ca^{2+}. However, only the vesicles pre-loaded with Ca^{2+} were able to synthesize ATP (Figure 4). Therefore, the phosphoenzyme formed in the absence of a Ca^{2+} gradient was regarded as being of "low energy" since it was spontaneously formed from P_i but unable to transfer its phosphate bond to ADP. Conversely, the phosphoenzyme formed in the presence of a Ca^{2+} gradient was referred as "high energy" because it was able to transfer its phosphate bond to ADP and hence, synthesize ATP (de Meis, 1976; de Meis and Vianna, 1979).

The phosphorylation of leaky vesicles by P_i is inhibited by the same Ca^{2+} concentrations required for activation of the phosphorylation of the enzyme by ATP. This finding led to the conclusion that ATP and P_i should interact with a common catalytic site of the enzyme, which undergoes a conformational change depending on the binding of Ca^{2+} in the outer surface of the vesicles. In one conformation the enzyme would be phosphorylated only by P_i, and in the other conformation the enzyme would be phosphorylated only by ATP. Therefore, the binding of Ca^{2+} would determine the choice of substrate for the phosphorylation reaction (Masuda and de Meis, 1973; de Meis and Masuda, 1974).

VI. ATP ⇌ Pi EXCHANGE IN THE ABSENCE OF A Ca^{2+} GRADIENT

The concept that the Ca^{2+} gradient was required for the synthesis of ATP was challenged by the experiments carried out on ATP ⇌ P_i exchange with leaky vesicles and different Ca^{2+} concentrations in the medium (de Meis and Carvalho, 1974; de Meis and Sorenson, 1975; Carvalho et al., 1976; Plank et al., 1979). When the Ca^{2+} concentration in the medium is sufficient to saturate only the high-affinity Ca^{2+} sites, no ATP synthesis is measured, but the rate of ATP hydrolysis is maximal. Increasing the Ca^{2+} concentration in the medium to the range found in the vesicle lumen when a gradient is formed leads to a simultaneous inhibition of the ATPase activity and activation of the synthesis of ATP (Figure 5). This indicated that activation of the exchange, for example, of the synthesis of ATP, depends on the binding of Ca^{2+} to a low affinity site of the enzyme that faces the vesicle lumen.

The rate of ATP ⇌ P_i exchange varies with the P_i concentration in the medium both in the presence and in the absence of a transmembrane Ca^{2+} concentration gradient. However, the apparent affinity to P_i differs in these two conditions, being in the range of 3–4 mM when a gradient is formed and in the range of 40–60 mM when measured in leaky vesicles (de Meis and Carvalho, 1974; de Meis et al., 1975). These experiments showed that the Ca^{2+}-ATPase uses the energy derived from the binding of Ca^{2+} to low-affinity Ca^{2+} sites to synthesize ATP. For the synthesis of ATP in sealed vesicles, the large difference of Ca^{2+} concentration on the two sides of the membrane is only needed to meet the differences of affinities in two different Ca^{2+} binding sites. The first is a high-affinity site ($K_{0.5}$ = 0.1-2 μM at pH 7.0) that faces the external surface of the vesicle. Binding Ca^{2+} to this site activates phosphorylation by ATP and in-

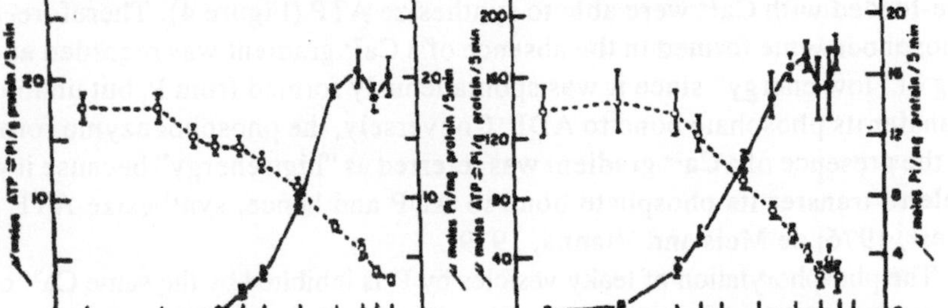

Figure 5. Ca^{2+} dependence of ATP ⇌ P_i exchange. The assay medium composition was 30 mM Tris maleate buffer (pH 6.8), 10 mM ATP, 6 mM $^{32}P_i$, 20 mM $MgCl_2$, and 0.3 mg/ml SRV protein. For the ATPase activity (○), γ^{32}ATP was used. For ATP ⇌ P_i exchange (●), $^{32}P_i$ was used. The reaction was performed at 37 °C (*left*) solubilized SRV; (*right*) leaky SRV. For details, see de Meis and Carvalho (1974).

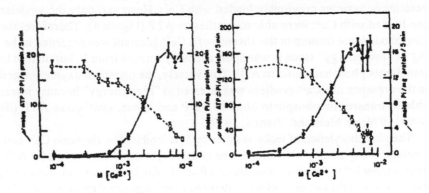

hibit the phosphorylation by P_i. The second is a binding site of low affinity that faces the vesicle lumen ($K_{0.5}$ =1-2 μM at pH 7.0), which permits the transfer of the phosphate from the phosphoenzyme formed by P_i to ADP leading to the synthesis of ATP.

VII. NET SYNTHESIS OF ATP IN THE ABSENCE OF A Ca^{2+} GRADIENT

Knowles and Racker (1975) reported that leaky vesicles prepared from purified Ca^{2+}-dependent ATPase catalyzes the synthesis of a small amount of ATP in the absence of a Ca^{2+} gradient. This was achieved by a two-step procedure where, initially, the enzyme was phosphorylated by P_i at pH 6.3 in the presence of $MgCl_2$ and EGTA. Subsequently, upon addition of ADP and 3.3–10 mM $CaCl_2$ (Ca^{2+} jump), it was found that about half of the phosphate of the phosphoenzyme was transferred to ADP, forming ATP. Using a similar methodology, Taniguchi and Post (1975) showed that the Na^+- K^+-ATPase catalyzes the synthesis of ATP and ATP ⇌ P_i exchange in the absence of an ionic gradient. These results confirmed the results obtained with ATP ⇌ P_i exchange in the absence of a Ca^{2+} gradient (de Meis and Carvalho), and indicated that each of the individual reactions involved in ATP synthesis can be catalyzed by the ATPase in the absence of a Ca^{2+} gradient.

The binding of Ca^{2+} to the high- and low-affinity sites of the ATPase can be altered by changing the pH of the medium (Meissner, 1973; Verjovski-Almeida and de Meis, 1977). Both sites exhibited an increased affinity for Ca^{2+} with high pH and a reduced affinity at lower pH values. Based on this finding, de Meis and Tume (1977) measured net synthesis of ATP by the pH jump procedure. With a constant Ca^{2+} concentration on both sides of the membrane, sudden alkalinization of the medium promotes the synthesis of ATP in one cycle. In this case, the concentration of Ca^{2+} in the medium is maintained constant but the affinity of the enzyme for Ca^{2+} is altered by varying the pH of the medium. Later, it was shown that the ATPase can also synthesize ATP after temperature and water-activity jumps (de Meis et al., 1980; de Meis and Inesi, 1982). These data indicate that the Ca^{2+}-ATPase might use the energy derived from pH, temperature or water-activity differences across the membrane to drive the synthesis of ATP.

In the beginning of the 1970s it was realized that after the binding of a ligand a protein can assume a multiplicity of conformations with structural rearrangements that can be transmitted and recogonized in different points of the protein structure (Weber, 1972). Thus, the energy derived from ligand-protein interaction can be used by an enzyme to perform work. In the case of the Ca^{2+}-ATPase, binding of Ca^{2+} to the low-affinity Ca^{2+} sites of the enzyme is required for the synthesis of ATP (de Meis and Carvalho; de Meis and Vianna). The standard free energy derived from the binding of two Ca^{2+} ions to the low-affinity sites is -8.2 Kcal/mol (de Meis, 1985). This is sufficient for the synthesis of one mol ATP for each mol of enzyme. Therefore, during the reversal of the Ca^{2+} pump, the binding energy of Ca^{2+} is used for the conversion of the phosphoenzyme from low into high energy and to the synthesis of ATP. From these observations it becomes apparent that the energy needed to drive the different intermediary reactions of the Ca^{2+}-ATPase catalytic cycle could be derived from the binding of ligands (Ca^{2+}, ATP, ADP, Mg^{2+}, and P_i) and not from the chemical energy derived from the cleavage of phosphate bonds as previouly envisaged (Hasselbach and Makinose, 1962; Makinose, 1973).

VIII. THE CATALYTIC CYCLE

The study of the reversal of the Ca^{2+} pump led to the proposal of a basic reaction sequence shown in Figure 6 (Carvalho et al., 1976; de Meis and Vianna, 1979; de-Meis, 1985). During catalysis the enzyme cycles through two distinct conformations, E_1 and E_2 (originally E and *E, Carvalho et al., 1976). The enzyme form E_1 binds calcium with high affinity on the outer surface of the vesicles and can be phosphorylated by ATP but not by P_i. The enzyme form E_2 binds calcium with low affinity in the inner surface of the vesicles and can be phosphorylated by P_i but not by ATP. The key feature of this cycle is the mechanism by which energy is transduced. The sequence of events proposed in earlier models of active transport assumed that energy was released at the moment of hydrolysis of the phosphate compound. The energy released would then be absorbed by the enzyme and used to perform work. In this view, the energy of hydrolysis of phosphate compounds would be the same regardless of whether the phosphate compound was in solution in the cytosol or bound to the enzyme surface (Lippman, 1941; Hasselbach and Makinose, 1962; Jardetzky, 1966; Makinose, 1973). The finding that the energy of hydrolysis of the phosphoenzyme formed from P_i varies during the reversal of the Ca^{2+}-ATPase indicates that the energy becomes available for the translocation of calcium through the membrane before the cleavage of the phosphoenzyme. Thus, during the catalytic cycle there is a large change of the equilibrium constant for the hydrolysis (K_{eq}) of the acyl phosphate residue of the phosphoenzyme and the work is linked with this transition of K_{eq} and not with the hydrolysis of the phosphate compound (Table 1). After the description of the catalytic cycle, it was found that

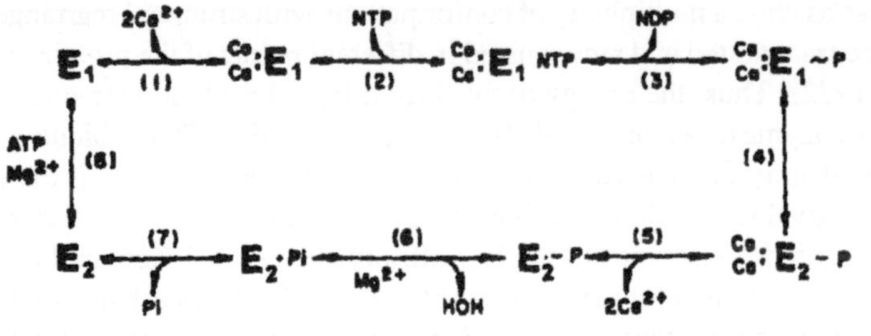

Figure 6. Reaction sequence (Carvalho et al., 1976; de Meis and Vianna, 1979).

Table 1. Variability of the Energy of Hydrolysis of Phosphate Compounds During the Catalytic Cycle of Energy Transducing Enzymes[1]

Enzyme	*Reaction*	*Solution or Enzyme Bound, Before Work*		*Enzyme Bound, After Work*	
		K_{eq} (M)	ΔG^0 (kcal/mol)	K_{eq}(M)	ΔG^0 (kcal/mol)
Ca^{2+}-ATPase, Na^+/K^+-ATPase	Aspartyl phosphate hydrolysis	10^6	-8.4	1.0	0
F_1-ATPase, Myosin	ATP hydrolysis	10^6	-8.4	1.0	0
Inorganic Pyrophosphatase	PP_i hydrolysis	10^4	-5.6	4.5	-0.9
Hexokinase	ATP + gluc→ gluc-P + ADP	2×10^3	-4.6	1.0	0

Notes: [1] For details, see de Meis (1989, 1993). The relationship between the standard free energy of hydrolysis (ΔG^0) and the equilibrium constant of the reaction (K_{eq}) is: $\Delta G^0 = -RT \ln K_{eq}$ where R is the gas constant (1.981) and T the absolute temperature.

the change of energy of the phosphoenzyme (Reaction 4) was related to a change in water activity in the microenvironment of the catalytic site (de Meis et al., 1980). The energy of hydrolysis of different phosphate compounds is determined by the differences in solvation energy between reactants and products (George et al., 1970; Hayes et al., 1978; de Meis et al., 1980, 1985; de Meis, 1984, 1989, 1993; Romero and de Meis, 1989) Thus, a change in water activity in the catalytic site of the enzyme leads to a change in solvation energy of the phosphoenzyme and to a change in its energy of hydrolysis. Experimental evidence supporting this possibility was obtained in various laboratories. These include measurements of the K_{eq} of the acyl phosphate formation in the media containing different concentrations of organic solvents (de Meis et al., 1980, 1982; Kosk-Kosicka et al., 1983; Dupont and Pougeois, 1983; Chiesi et al., 1984; Champeil et al., 1985; Moraes and de Meis, 1987), fluorescence measurements of probes (Kurtenbach and Verjovsky-Almeida, 1985; Highsmith, 1986; Wakabayashi et al., 1986), and an increase of hydrophobic labeling with photoactivable reagents when the enzyme was stabilized in different conformations (Andersen et al., 1986).

IX. THE PARTIAL REACTIONS OF THE CYCLE

These were previously described in detail in various reviews (de Meis and Vianna, 1979; de Meis, 1981; Tanford, 1984; Inesi, 1985). In this section we will discuss only some aspects of the catalytic cycle and recent findings.

A. Reaction 1

It was shown by equilibrium experiments that two Ca^{2+} ions bind with high affinity and cooperativity to each ATPase molecule in the absence of ATP (Inesi et al., 1980; Forge et al., 1993). The stoichiometry and positive cooperativity of Ca^{2+} binding were explained in terms of a sequential model whereby one Ca^{2+} binds first and increases the affinity of the enzyme for a second Ca^{2+} (Inesi et al., 1980). The affinity and cooperativity of Ca^{2+} binding are highly pH dependent, indicating that H^+ competes with Ca^{2+} for the binding sites (Meissner, 1973, Verjovski-Almeida and de Meis, 1977; Watanabe et al., 1981; Hill and Inesi, 1982). The sequential binding of Ca^{2+} from the cytosolic (outer with respect to the vesicles) medium in the absence of ATP was confirmed by rapid kinetic experiments and isotope exchange (Dupont, 1982; Inesi, 1987).

B. Reactions 2 and 3

ATP binds with high affinity to the catalytic site of the enzyme (Reaction 2). The true substrate of the Ca^{2+}-ATPase is the complex MgATP (Weber et al.; Yamamoto and Tonomura, 1967; Vianna, 1975). Transient kinetics experiments revealed that the concentration of MgATP needed for half-maximal phosphorylation of the enzyme falls in the range of 1 to 5×10^{-6} M (Kanazawa et al., 1971; Froelich and Taylor, 1975, Scofano *et al.*, 1979). The very high affinity of the enzyme for ATP permits the use of glucose 6-phosphate and hexokinase as an ATP regenerating system (Montero-Lomelli and de Meis, 1992). The phosphorylation of the ATPase is accomplished by transfer of phosphate from ATP to an aspartic residue located in the catalytic site of the enzyme (Reaction 3). Reversal of the reactions 2 and 3 accounts for the ATP-ADP exchange observed by Hasselbach and Makinose (1962) and Ebashi and Lipmann (1962).

C. Reactions 4 and 5

During the process of Ca^{2+} translocation through the membrane, the Ca^{2+} binding site of high affinity is converted into a site of low affinity. This was shown both during the hydrolysis of ATP (Ikemoto, 1975) and during the reversal of the Ca^{2+} pump (de Meis and Carvalho, 1974). Upon addition of ATP to purified ATPase, part of the bound Ca^{2+} is released and is bound back to the enzyme after ATP hydrolysis. Subsequent addition of ATP led to another cycle, which suggested that the enzyme cycles between states with high and low affinity to Ca^{2+} (Figure 7). The catalytic site and the region of the enzyme involved in the translocation of Ca^{2+} through the membrane are located in separate regions of the protein (Inesi and Kirtley, 1992). Phosphorylation of the enzyme by ATP is followed by a long-range conformational change of the protein after which the Ca^{2+} binding sites are reoriented, facing now the lumen of sarcoplasmic reticulum vesicles (Reaction 4). During this process, the Ca^{2+} ions remain transiently occluded within the enzyme (Takakuwa and Kanazawa, 1979; Dupont, 1980). Reorientation of the Ca^{2+} sites is associated with a 10^3 fold decrease in the enzyme affinity for Ca^{2+},

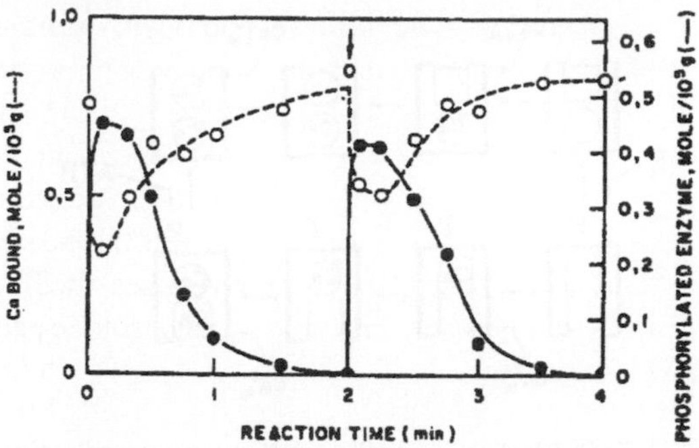

Figure 7. Relation between phosphorylation and Ca^{2+} binding of the purified ATPase. The reaction was initiated by adding 10 μM ATP to a solution containing 0.1 M KCl, 5 mM $MgCl_2$, 10 μM EGTA, 0.4 mg protein/ml, and 20 mM Tris-Maleate (pH 7.0). No calcium was added, but the reaction mixture contained 8 nmol Ca^{2+}/ml as determined by atomic absorption spectrophotometry. The free Ca^{2+} concentration of the solution, calculated as described previously from the amount of calcium present and with the appropriate CaEGTA and CaATP affinity constants, was 0.64 μM. It was assumed that all of the calcium present was available to the ligands of the solution. For the Ca^{2+} binding study, the radioactivity of ^{45}Ca in the solution was adjusted to 1.5×10^5 CPM/ml. The reaction was carried out at 4°C and, at the times indicated in the abscissa, 1-ml portions of the solution were filtered through Millipore filters. At 2 min, fresh ATP (10 μM) was added (arrow). Ca^{2+} binding at 0 time was measured by filtration before addition of ATP, and a blank was determined by filtration of the same solution devoid of protein (6.5 percent of the total ^{45}Ca was bound to the filter). The amount of phosphoenzyme present was determined under the same conditions as those used for Ca^{2+} binding except that no ^{45}Ca was added and 10 μM (γ-^{32}P)ATP was present. (●) E-P; (○) Ca^{2+} binding. From Ikemoto (1975).

favoring its dissociation into the vesicles lumen (Reaction 5). Dissociation of the Ca^{2+} into the vesicles lumen occurs before the phosphoenzyme hydrolysis. Thus, the work is performed before the cleavage of the acyl phosphate residue (Makinose, 1973; Kurzmack et al., 1977). The kinetics of Ca^{2+} dissociation from the phosphorylated ATPase into the lumen of the vesicles was shown to follow a sequential pattern, analogous to that of Ca^{2+} binding to the non-phosphorylated enzyme E_1 (Figure 8) (Inesi, 1987). Orlowski and Champeil (1991), and Hanel and Jencks (1991) were unable to observe the sequential pattern of release. This discrepancy was recently resolved by Forge et al. (1995) who demonstrated that inability to observe sequential Ca^{2+} release from the phosphorylated enzyme into the lumen of the vesicles is related to the use of unsuited experimental conditions. The phosphoenzymes E_2-P, 2Ca E_2-P and 2Ca E_1~P are distinguished by the sensitivity to ADP, where only 2Ca E_1~P is able to transfer its phosphate group to ADP. High intravesicular Ca^{2+} concentrations, as observed when a gradient is formed through the membrane, leads to the binding of Ca^{2+} to the low-affinity site

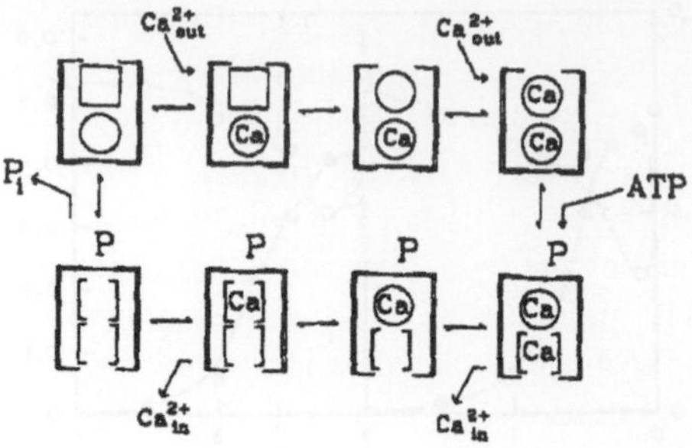

Figure 8. Sequential Ca^{2+} binding and dissociation upon utilization of ATP by the Ca^{2+}-ATPase. Squares and circles represent tight and very tight binding, respectively. Brackets represent weak binding and a change in orientation upon enzyme phosphorylation by ATP. For details see Inesi (1987).

and to the accumulation of the ADP-sensitive phosphoenzyme (Shigekawa and Dougherty, 1978).

D. Reactions 6 , 7, and 8

Hydrolysis of the acyl phosphate (Reactions 6 and 7) occurs after Ca^{2+} dissociation, generating the E_2 which is converted into E_1 in order to start a new catalytic cycle. The hydrolysis of the phosphoenzyme is much slower when an NTP other than ATP is used as substrate (de Meis and de Mello, 1973). When ITP is used as substrate the rate of hydrolysis and the phosphoenzyme level in leaky vesicles is lower than with ATP (Souza and de Meis, 1976). However, phosphorylation by P_i under this condition is possible in an amount much higher than that found using ATP as substrate. This can be explained by the fact that ATP accelerates the conversion of E_2 into E_1, which may be the rate-limiting step when substrates other than ATP are used (Scofano et al.). Therefore, if ITP or very low ATP concentrations are used, a larger amount of free E_2 forms is available, permitting the phosphorylation of the enzyme by P_i under steady state. Steady-state measurements of ATP hydrolysis revealed two different apparent Km values for ATP (Inesi et al., 1967; Yamamoto and Tonomura, 1967). The rate of hydrolysis increases to a plateau level as the ATP concentration is raised from 0.1 μM up to 50–100 μM. A second increase in the hydrolysis rate occurs as the ATP concentration approaches the millimolar range. This was related to the presence of two distinct classes of ATP binding sites. The first, a high-affinity catalytic site, and the second a low-affinity regulatory site. Controversy exists as to whether this is related to the presence of two ATP-binding sites per enzyme molecule (Dupont, 1977; Carvalho-Alves et al., 1985; Dupont et al., 1985; Coll and Murphy, 1985) or to the binding of ATP to one of the different enzyme forms E_2 in the region of the catalytic site available after dissociation of ADP (McIntosh and Boyer, 1983; Bishop et al., 1987; Seebregts and McIntosh, 1989). In the second case the enzyme would also be able to bind ATP but with a lower affinity than the form

E_1, and the rate of conversion of ATP.E_2 into ATP.E_1 would be faster than the rate of conversion of E_2 into E_1 (Scofano et al.; Carvalho-Alves and Scofano, 1983) . At present the mechanism of regulation by ATP is still controversial.

The NTP $\rightleftharpoons$ P_i exchange involves the reactions 2 to 7 flowing forward (ATP hydrolysis) and backward (ATP synthesis). The ratio between hydrolysis and synthesis of ATP is determined by the rate of conversion of E_2 into E_1 at the end of each catalytic cycle. It will be higher than one if some enzyme units E_2 are converted to $2CaE_1$ instead of being rephosphorylated by P_i and flowing backward. In the presence of high. ATP concentrations that accelerate Reaction 8, the rate of hydrolysis can be twenty to fifty times higher than the rate of synthesis., If, However, ITP is used as substrate or if the vesicles are treated with Ag^+ this ratio can be 2-4, due to a decrease in the rate of conversion of E_2 into E_1 in Reaction 8 (Carvalho et al.; de Meis and Sorenson, 1975).

E. Ca^{2+}/H^+ Countertransport

In 1978, Madeira reported that during Ca^{2+} uptake ejection of H^+ occurs, leading to a transient alkalinization of the lumen of the vesicles and formation of a proton gradient across the membrane. However, the H^+ ejection does not completely counterbalance the charge of the Ca^{2+} transported in the opposite direction. The electrogenicity of the calcium pump and the formation of a proton gradient during Ca^{2+} transport was later confirmed by several laboratories (Zimniak and Racker, 1978; Chiesi and Inesi, 1980; Ueno and Sekine, 1981a, 1981b; Meissner, 1981; Yamagushi and Kanazawa, 1984; Morimoto and Kasai, 1986). Using purified Ca^{2+} ATPase reconstituted into liposomes with a low H^+ permeability, it was demonstrated that for each Ca^{2+} transported, 1 H^+ (Chiesi and Inesi; Yu et al., 1994) or 1 to 1.5 H^+ (Levy et al., 1990) is ejected in the opposite direction . The membrane potential and H^+ gradient are progressively dissipated in the course of the Ca^{2+} uptake, due to the high permeability of the sarcoplasmic reticulum membrane to H^+, K^+ and Cl^- (Meissner, 1981). It has been proposed that Ca^{2+}/H^+ countertransport would be an intrinsic property of the Ca^{2+} sites of the ATPase (Chiesi and Inesi, 1980). After Ca^{2+} release in the lumen of the vesicles, H^+ would bind to the Ca^{2+} sites, and reorientation of the Ca^{2+} sites promotes H^+ dissociation in the assay medium (Chiesi and Inesi; Yu et al., 1994). The apparent pK of the residue(s) releasing H^+ into the medium was found to be 6.1, whereas the apparent pK of the residue(s) binding lumenal H^+ was approximately 7.7 (Yu et al., 1994). These findings led to the proposal that the residues involved in Ca^{2+}/H^+ countertransport would possess a lower affinity for H^+ in their outward orientation and a higher affinity for H^+ in the lumenal orientation.

X. UNCOUPLING OF THE Ca^{2+} PUMP

After the cloning and the determination of the amino acid sequence of the sarcoplasmic reticulum Ca^{2+} ATPase, structural studies and models indicate that the protein encompasses at least three distinct functional domains, the catalytic, the transport domain, and the stalk region, that links the catalytic site to the region of the enzyme that translocates Ca^{2+} (MacLennan et al., 1985; Clarke et al., 1989; Andersen and Vilsen, 1990; Inesi and Kirtley, 1992; Toyoshima et al., 1993; Vilsen, 1995). The portion of the protein that contains the catalytic site faces the cytoplasmic side of the membrane, while the transmembrane helices form a channel-like structure that allows Ca^{2+} translocation across the membrane (Clarke et al., 1989; Inesi and Kirtley, 1992; Toyoshima et al., 1993). The catalytic and the Ca^{2+} sites are distant from each other and the coupling of these regions is achieved by a long-range interaction traveling through the stalk region (Inesi and Kirtley, 1992). More recently, it has been shown that these interactions can be modified by different drugs and the ATPase can mediate a Ca^{2+} efflux as a Ca^{2+} channel (Gerdes et al., 1983; Gerdes and Moller, 1983; Gould et al., 1987a; Inesi and de Meis, 1989; de Meis et al., 1990; de Meis, 1991; Wolosker et al., 1992; de Meis and Inesi, 1992; Cardoso and de Meis, 1993; Wolosker and de Meis, 1994; de Meis and Suzano, 1994).

An efflux not coupled to the synthesis of ATP was first demonstrated by Hasselbach and coworkers (1972) with the use of arsenate, a phosphate analog, that induces Ca^{2+} efflux from sarcoplasmic reticulum vesicles, without synthesis of ATP. Arsenate inhibits competitively the phosphorylation of the enzyme form E_2 by P_i (Pick and Bassilian, 1982; Alves and de Meis, 1987). Further evidence supporting uncoupling of the Ca^{2+} pump emerged from measurements of exchange between the external and internal Ca^{2+} pools during steady state Ca^{2+} accumulation (Gerdes et al., 1983; Gerdes and Moller, 1983; Galina and de Meis, 1992). In these experiments it was possible to show that part of the Ca^{2+} accumulated by the vesicles leaks through the ATPase without the concomitant synthesis of ATP, thus characterizing an uncoupled route ("slippage of the pump"). The coupling between Ca^{2+} exchange and ATP synthesis was found to vary greatly depending on the enzyme phosphorylation level (Galina and de Meis, 1992). Conditions that favor phosphorylation of the enzyme by P_i, such as high P_i concentration, dimethyl sulfoxide 20 percent (v/v) or the use of ITP as substrate, decrease the Ca/ATP ratio from 11.4 to 2.0.

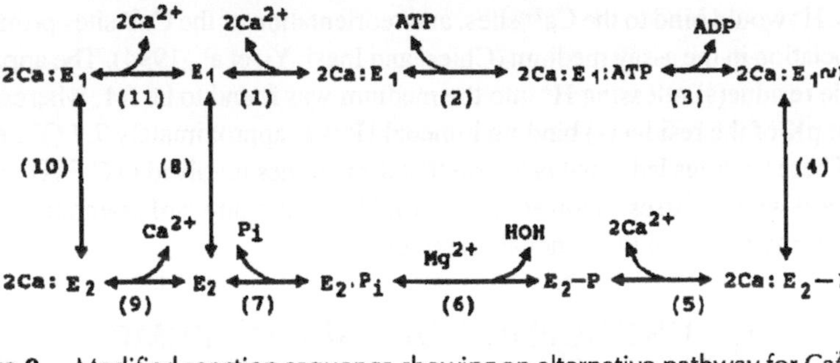

Figure 9. Modified reaction sequence showing an alternative pathway for Ca^{2+} efflux (Reactions 9, 10, 11). For details see Gerdes and Moller, 1983; Benech *et al.* (1991), Galina and de Meis (1992).

These findings led to the proposal of an alternative pathway by the Ca^{2+} efflux through the Ca^{2+}-ATPase (Figure 9). One of them is coupled to the synthesis of ATP and mediated by the reversal of reactions 7, 6, 5, 4, and 3. The second route is not coupled to synthesis of ATP and is mediated by reactions 9,10, and 11 (Gerdes et al., 1983; Gerdes and Moller, 1983; Galina and de Meis, 1992).

Recently, it was shown that uncoupling of the Ca^{2+} pump is also observed by a mutation $Tyr^{763} \rightarrow Gly$, located between the transmembrane and stalk regions of the enzyme (Andersen, 1995).

XI. UNIDIRECTIONAL Ca^{2+} EFFLUX

The release of Ca^{2+} through the ATPase can be measured during unidirectional Ca^{2+} efflux. In order to avoid the influence of "ryanodine-sensitive Ca^{2+} channels" (Fleischer and Inui, 1989; Palade et al., 1989), which are located in the terminal cistern of the sarcoplasmic reticulum, a preparation enriched with longitudinal tubules ("light" sarcoplasmic reticulum vesicles) was used (Eletr and Inesi, 1972). This preparation does not contain ryanodine-sensitive Ca^{2+} channels, nor does it exhibit the phenomenon of Ca^{2+}-induced Ca^{2+}-release (Inesi and de Meis, 1989; de Meis, 1991, Wolosker et al., 1992).

Light sarcoplasmic reticulum vesicles are preloaded with Ca^{2+}, sedimented by centrifugation, and further diluted in the medium containing excess EGTA. In the absence of substrates or ligands of the ATPase, Ca^{2+} is released from the vesicles without concomitant synthesis of ATP (Inesi and de Meis, 1989; de Meis et al., 1990; de Meis, 1991; Wolosker et al., 1992; de Meis and Inesi, 1992; Cardoso and de Meis, 1993; Wolosker and de Meis, 1994). Under this condition, Ca^{2+} leaks through the ATPase channel at a slow rate ("leaky state"). This efflux of Ca^{2+} seems to be an intrinsic property of the Ca^{2+} transport ATPase, since purification and reconstitution of the enzyme in phospholipid vesicles do not abolish the observed Ca^{2+} efflux (Inesi et al., 1983; Gould et al., 1987b). Addition of the natural ligands of the ATPase, such as cations, blocks the Ca^{2+} efflux through the pump. The concentration dependency of Ca^{2+} for inhibition of Ca^{2+} efflux is similar to that observed for binding of Ca^{2+} to the high-affinity sites of the enzyme. This suggests that the formation of an $Ca{:}E_1$ complex. and the energy derived from cation binding to the enzyme decreases Ca^{2+} efflux through the "uncoupled route" (Reactions 9, 10, 11; Figure 9), by promoting the closure of the ATPase channel. Other cations, such as Mg^{2+}, K^+ and Na^+, that are known to bind to the enzyme are also able to impair the efflux through the uncoupled route, but the concentrations needed are three to five orders of magnitude higher than Ca^{2+}. The order of potency of the cations in decreasing Ca^{2+} efflux is: $Ca^{2+} >> Mg^{2+} > K^+ = Na^+$ Li.

Phosphorylation of the ATPase by P_i decreases the Ca^{2+} efflux through reactions 9 to 11 (Figure 9) and increases the Ca^{2+} efflux coupled to ATP synthesis. (de Meis et al., 1990; Benech et al., 1991) because it favors the formation of $Ca{:}E_2$-P forms and decreases the amount of Ca:E_2 forms needed for the release of Ca^{2+} through the uncoupled route.

Ruthenium red, thought to be an inhibitor of Ca^{2+} channels, also decreases uncoupled Ca^{2+} efflux from light sarcoplasmic reticulum vesicles and competes with Ca^{2+}, Mg^{2+} and K^+ for their effects on the Ca^{2+}-ATPase (de Meis, 1991). Direct binding of ruthenium red to the Ca^{2+}-ATPase is displaced by the ligands Ca^{2+}, Mg^{2+} and K^+. The uncoupled Ca^{2+} efflux is also inhibited by the polyamines spermine, spermidine and putrescine, which are found in different animal tissues, including skeletal muscle (Tabor and Tabor, 1964). The concentration range in which polyamines block Ca^{2+} efflux through the ATPase is the same as that observed in junctional vesicles, that contain the ryanodine-sensitive Ca^{2+} channels (Fleischer and Inui, 1989; Palade et al., 1989). This provides evidence that binding of polyamines to the ATPase stabilizes the ATPase channel in a closed state, in a manner similar to that observed with the Ca^{2+} channel located in the terminal cistern of the sarcoplasmic reticulum.

The different inhibitors of Ca^{2+} efflux through the ATPase are listed in Table 2.

The ATPase channel seems to comprise three distinct functional states. The first is observed in the absence of ligands of the ATPase and represents the *leaky state*, by which Ca^{2+} leaks through the ATPase at a low rate. The second is a *closed state*, in which ligand-binding energy promotes the closure of the ATPase channel, thereby arresting the Ca^{2+} leakage. The third is an *open state*, which possesses very

Table 2. Inhibitors of Ca^{2+} Efflux Through the Ca^{2+} ATPase Channel

Agents	*Concentration*[1]	*References*[2]
Cations		(A)
Ca^{2+}	0.5-5 μM	
Mg^{2+}	0.5-5 mM	
K^+, Na^+	5-100 mM	
Li^+	20-200 mM	
Nucleotides (ATP, ADP)	5-100 μM	(B)
Polyamines		
Ruthenium red	3 μM	(C)
Spermine	20 μM	
Spermidine	200 μM	
Putrescine	1 mM	
Thapsigargin	1-2 μM	(D)
Dimethyl sulfoxide	20-40 percent v/v	(E)
Inorganic phosphate	1-4 mM	(E)

Notes: [1] The concentration of each agent that is effective at pH 7.0.
[2] References: (A). Inesi and de Meis (1989), de Meis et al. (1990), de Meis (1991), Wolosker et al. (1992), Wolosker and de Meis (1994). (B). de Meis et al. (1990), de Meis (1991), Wolosker et al. (1992). (C). de Meis (1991), Wolosker et al. (1992), Cardoso and de Meis (1993). (D). de Meis and Inesi (1992), Wolosker and de Meis (1994), Engelender et al (1995). (E). de Meis et al. (1990); Benech et al. (1991).

high permeability to Ca^{2+}. This state is stabilized by different hydrophobic drugs that bind to the catalytic site of the enzyme. Factors which alter the microequilibrium between the enzyme conformations E_1 and E_2, such as H^+ concentration, also modify the kinetics of channel activation and inactivation.

Since phosphorylation by P_i blocks the uncoupled Ca^{2+} efflux, it would be expected that drugs that alter the phosphorylation of the ATPase by P_i also modify the Ca^{2+} efflux through the pump. A large number of hydrophobic drugs have been shown to interact with the catalytic site of the ATPase (de Meis et al., 1988; Petretski et al., 1989; Wolosker et al., 1990; de Meis, 1991; Wolosker et al., 1992). Phenothiazines, local anesthetics and β-blockers inhibit competitively the phosphorylation of the ATPase by P_i (de Meis, 1991; Wolosker et al., 1992). Concomitantly, these drugs also promote a multi-fold increment in Ca^{2+} efflux (Figure 10). The effects of hydrophobic drugs on Ca^{2+} efflux are antagonized by ligands and substrates of the ATPase that impair the leakage of Ca^{2+} through the enzyme, such as cations, nucleotides, polyamines and ruthenium red. This suggests that the drugs tested do not promote unspecific leakage of the membrane, but rather promote the opening of the ATPase channel. The Ca^{2+} concentration range required to completely block drug-induced Ca^{2+} efflux is narrow (1-10 μM) and remains within the range for physiological Ca^{2+} oscillations, thus providing an interesting mechanism of blockage of Ca^{2+} efflux after a critical high cytoplasmic Ca^{2+} concentration is reached.

Activation of the Ca^{2+} efflux through the ATPase is also observed with fatty acids naturally found in muscle cells (Cardoso and de Meis, 1993). The effect of fatty acids on Ca^{2+} efflux is blocked by essentially the same agents that block the effects of phenothiazines and local anesthetics. Uncoupling of the Ca^{2+} pump is also ob-

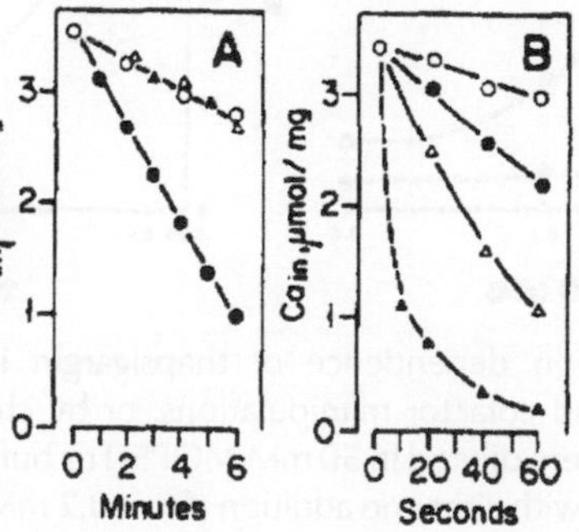

Figure 10. Inhibition of Ca^{2+} efflux by substrates and ligands (**A**) and enhancement by trifluoperazine (**B**). The reaction medium contained 50 mM MOPS-Tris pH 7.2, 2 mM EGTA and sarcoplasmic reticulum vesicles preloaded with ^{45}Ca phosphate. The figure shows the Ca^{2+} remaining in the vesicles after different incubation intervals at 35°0 C. A: ●, 0.1 mM P_i and 0.1 mM $MgCl_2$; , 4 mM P_i and 4 mM $MgCl_2$; -, 0.1 mM P_i, 0.1 mM $MgCl_2$ and 100 mM KCl; ▲, EGTA was replaced by 0.2 mM $CaCl_2$. B: without addition (○), or plus 20 TM (●), 40 μM (-) and 100 μM (▲) trifluoperazine. For details, see de Meis (1991).

Table 3. Activators of Ca^{2+} Efflux Through the Ca^{2+} ATPase Channel

Agents	*Concentration*[1]	*References*[2]
Arsenate	1-5 mM	(A)
Phenothiazines	20-100 µM	(B)
Imipramine	50-800 µM	(B)
Local Anesthetics	0.5-40 mM	(C)
Propranolol	4 mM	(B)
Arachidonic acid and Arachidic acid	8-32 µM	(D)
Heparin	1-10 µg/ml	(E)
Ag^{+}	3-10 µM	(F)

Notes: [1] The concentration of each agent that is effective at pH 7.0

[2] References: (A). Hasselbach et al. (1972), Pick and Bassilian (1982), Alves and de Meis (1987). (B). de Meis (1991), Wolosker and de Meis (1994), Engelender et al. (1995). (C). Wolosker et al. (1992). (D). Cardoso and de Meis (1993). (E). de Meis and Suzano (1994). (F). Gould et al. (1987b).

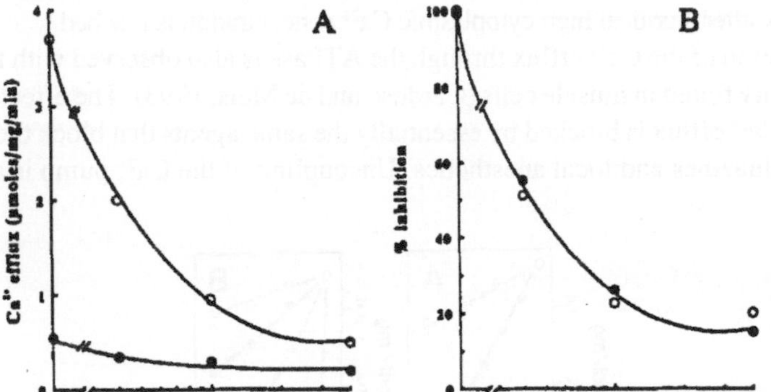

Figure 11. Concentration dependence of thapsigargin inhibition of Ca^{2+} efflux activated by substrate and cofactor manipulations, or by chlorpromazine. Preloaded sarcoplasmic reticulam were diluted in 50 mM MOPS-Tris buffer pH 7.0, 0.1 mM P_i, 0.1 mM $MgCl_2$, 2 mM EGTA, with either no addition (●) or 0.2 mM chlorpromazine (○). For details see de Meis and Inesi (1992).

served with sulphated polysaccharides, such as heparin (de Meis and Suzano, 1994). A list of activators of Ca^{2+} efflux is summarized in Table 3.

At acid pH, the ATPase channel is not sensitive to cations and remains sensitive to hydrophobic drugs (Wolosker and de Meis, 1994). This pH-dependent efflux pathway may play a role in conditions that lead to intracellular acidosis, as for instance in muscle ischemia.

Thapsigargin, a highly specific inhibitor of the Ca^{2+} transport ATPase from the sarcoplasmic reticulum (Thastrup, 1987; Sagara et al., 1992), inhibits both the efflux of Ca^{2+} coupled to ATP synthesis and the uncoupled Ca^{2+} efflux (de Meis and Inesi, 1992). Thapsigargin interacts with the E_2 form of the ATPase, blocking the conformational transition E_1 - E_2. This drug also prevents the efflux of Ca^{2+} promoted by phenothiazines (Figure 11) and fatty acids, thereby providing additional evidence that the uncoupled Ca^{2+} efflux is associated with the transition E_2 - E_1 of the ATPase. More recently, it has been shown that thapsigargin probably interacts with the transmembrane region (M3 portion) of the Ca^{2+} ATPase (Norregaard et al., 1994).

XII. Ca^{2+}-ATPase ISOFORM DIVERSITY

Different Ca^{2+}-ATPase isoforms are expressed in several tissues (MacLennan et al., 1985; Burk et al., 1989; Lytton et al., 1992). At present, the physiologic significance of isoform diversity is unclear. Recently, it has been shown that for the platelet Ca^{2+}-ATPase isoforms the stoichiometry between Ca^{2+} released and ATP synthesized during the reversal of the Ca^{2+} pump is one and not two as found in the muscle ATPase (Benech et al., 1995; Engelender et al., 1995). The Ca^{2+}-ATPase isoforms of platelets are also uncoupled by phenothiazines, but unlike the sarcoplasmic reticulum, they are not antagonized by thapsigargin or by the binding of Ca^{2+} to the high-affinity sites of the enzyme (Engelender et al., 1995). When extrapolated to the *in vivo* situation, these drugs may act by emptying intracellular Ca^{2+} compartments regardless of the raised Ca^{2+} concentration in the cytosol, while in muscle, the effect of the drugs is blocked after the cytosolic Ca^{2+} concentration exceeds 1×10^{-6} M.

XIII. SUMMARY

The first part of this chapter provides an overview of Ca^{2+} transport in skeletal muscle, with a focus on the discovery of the reactions involved in phosphoenzyme formation, reversal of the Ca^{2+} pump and ATP $\rightleftharpoons P_i$ exchange that led to the proposal of the E_1 –E_2 model. In the second part, the findings of uncoupling of the Ca_{2+} pump and the evidence for a functional transmembrane cannel within the Ca^{2+} ATPase are discussed.

REFERENCES

Andersen, J. P., Vilsen, B., Collins, J. H., & Jorgensen, P. L. (1986). Localization of E_1-E_2 conformational transitions of sarcoplasmic reticulum Ca^{2+} ATPase by tryptic cleavage and hydrophobic labeling. J. Membr. Biol. 93, 85-92.

Andersen, J.P. (1995). Functional consequences of alterations to amino acids at the M5S6 boundary of the Ca^{2+} ATPase of sarcoplasmic reticulum. Mutation Tyr763→Gly uncouples ATP hydrolysis from Ca^{2+} transport. J. Biol. Chem. 270, 908-914.

Andersen, J.P., & Vilsen, B. (1990). Primary ion pumps. Curr. Op. Cell. Biol. 2, 722-730.

Alves, E. W., & de Meis, L. (1987). Effects of arsenate on the Ca^{2+} ATPase of sarcoplasmic reticulum. Eur. J. Biochem. 166, 647-651.

Barlogie B., Hasselbach W., & Makinose M. (1971). Activation of calcium efflux by ADP and inorganic phosphate. FEBS Lett. 12, 267-268.

Bastide, F., Meissner, G., Fleischer, S., & Post, R. L. (1973). Similarity of active site of phosphorylation of the adenosine triphosphate for transport of sodium and potassium in kidney to that for transport of calcium ions in sarcoplasmic reticulum of muscle. J. Biol. Chem. 248, 8485-8491.

Bendall, J. R. (1952). Effect of the "Marsh factor" on the shortening of muscle fibers models in the presence of adenosine triphosphate. Nature 170, 1058-1060.

Bendall, J. R. (1953). Further observation on a factor (the 'Marsh factor') effecting relaxation of ATP-shortened muscle fiber models, and the effect of Ca and Mg ions upon it. J. Physiol. 131, 232-254.

Benech J.C., Galina A., & de Meis L . (1991). Correlation between Ca^{2+} uptake, Ca^{2+} efflux and phosphoenzyme level in sarcoplasmic reticulum vesicles. Biochem. J. 274, 427-432.

Benech, J.C., Wolosker, H., & de Meis, L. (1995). Reversal of the Ca^{2+} pump of blood platelets. Biochem. J. 306, 35-38.

Bishop, J.E., Al-Shawi, M.K., & Inesi, G. (1987). Relationship of the regulatory nucleotide site to the catalytic site of the sarcoplasmic reticulum Ca^{2+} ATPase. J. Biol. Chem. 262, 4658-4663.

Bozler, E. (1954). Relaxation in extracted muscle fibers. J. Gen. Physiol. 38, 149-159.

Burk, S.E., Lytton, J., MacLennan, D.H., & Shull, G.E. (1989). cDNA cloning, functional expression, and mRNA tissue distribution of a third organelar Ca^{2+} pump. J. Biol. Chem. 264, 18561-18568.

Cardoso, C.M., & de Meis, L. (1993). Modulation by fatty acids of Ca^{2+} fluxes in sarcoplasmic reticulum vesicles. Biochem. J. 296, 49-52.

Carvalho, M.G.C., Souza, D.O., & de Meis, L. (1976). On a possible mechanism of energy conservation in sarcoplasmic reticulum membrane. J. Biol. Chem. 251, 3629-3636.

Carvalho-Alves, P.C., & Scofano, H. (1983). Phosphorylation by ATP of sarcoplasmic reticulum ATPase in the absence of calcium. J. Biol. Chem. 258, 3134-3139.

Carvalho-Alves, P.C., Oliveira, C.R.G., & Verjovski-Almeida, S. (1985). Stoichiometric photolabeling of two distinct low and high affinity nucleotide sites in sarcoplasmic reticulum ATPase. J. Biol. Chem. 260, 4282-4287.

Champeil, P., Guillain, F., Venien, C., & Gingold, M. P. (1985). Interaction of magnesium and inorganic phosphate with the calcium-deprived sarcoplasmic reticulum adenosinetriphosphatase as reflected by organic solvent induced perturbation. Biochemistry 24, 69-81.

Chevalier, J., & Butow, R. A. (1971). Calcium binding to the sarcoplasmic reticulum of rabbit skeletal muscle. Biochemistry 10, 2733-2737.

Chiesi, M., & Inesi, G. (1980). Adenosine 5'-triphosphate dependent fluxes of manganese and hydrogen ion in sarcoplasmic reticulum vesicles. Biochemistry 19, 2912-2918.

Chiesi, M., Zurini, G., & Carafoli, E. (1984). ATP synthesis catalyzed by the purified erythrocyte calcium-ATPase in the absence of calcium gradients. Biochemistry, 23, 2595-2560.

Clarke, D.M., Loo, T.W., Inesi, G., & MacLennan, D.H. (1989). Location of high-affinity Ca^{2+}-binding sites within the predicted transmembrane domain of the sarcoplasmic reticulum Ca^{2+} ATPase. Nature 339, 476-478.

Coll, R. J., & Murphy, A.J. (1985). Sarcoplasmic reticulum Ca^{2+} ATPase: Product inhibition suggests an allosteric site for ATP activation. FEBS Lett. 187, 131-134.

de Meis, L., & de Mello, M. C. F. (1973). Substrate regulation of membrane phosphorylation and Ca^{2+} transport in the sarcoplasmic reticulum. J. Biol. Chem. 248, 3691-3701.

de Meis, L., & Carvalho, M. G. C. (1974). Role of the Ca^{2+} concentration gradient in the ATP⇌P_i exchange catalyzed by the sarcoplasmic reticulum. Biochemistry 13, 5032-5038.

de Meis, L., & Masuda, H. (1974). Phosphorylation of the sarcoplasmic reticulum membrane by orthophosphate through two different reactions. Biochemistry 13, 2057-2062.

de Meis, L., Carvalho, M. G. C., & Sorenson, M. M. (1975). In: *Concepts of Membranes in Regulation and Excitation,* (Silva, M. R. and Kurtz, G. S., Eds.). Raven Press, New York, pp. 7-19.

de Meis, L., & Sorenson, M. M. (1975). ATP ⇌ P_i exchange and membrane phosphorylation in sarcoplasmic reticulum vesicles: activation by silver in the absence of a Ca^{2+} concentration gradient. Biochemistry 14, 2739-2744.

de Meis, L. (1976). Regulation of steady state level of phosphoenzyme and ATP synthesis in sarcoplasmic reticulum vesicles during reversal of the calcium pump. J. Biol. Chem. 259, 6090-6097.

de Meis, L., & Tume, K. (1977). A new mechanism by which a H^+ concentration gradient drives the synthesis of adenosine triphosphate, pH jump, and adenosine triphosphate synthesis by the Ca^{2+}-dependent adenosine triphosphatase of sarcoplasmic reticulum. Biochemistry 16, 4455-4463.

de Meis, L., & Vianna, A.L. (1979). Energy interconvertion by the Ca^{2+} -transport ATPase of sarcoplasmic reticulum Ann. Rev. Biochem. 48, 275-292.

de Meis, L., Martins, O.B., & Alves, E.W. (1980). Role of water, hydrogen ions, and temperature on the synthesis of adenosine triphosphate by the sarcoplasmic reticulum Adenosinetriphosphatase in the absence of a calcium gradient. Biochemistry 19, 4252-4261.

de Meis, L. (1981). *The Sarcoplasmic Reticulum. Transport in the Life Sciences,* Vol. 2 (Bittar, E., Ed.). John Wiley & Sons, New York.

de Meis, L., & Inesi, G. (1982). ATP synthesis by the sarcoplasmic reticulum ATPase following Ca^{2+}, pH, temperature, and water activity jumps. J. Biol. Chem. 257, 1289-1294.

de Meis, L., Otero, A. S., Martins, O. B., Alves, E. W., Inesi, G., & Nakamoto, R. (1982). Phosphorylation of sarcoplasmic reticulum ATPase by orthophosphate in the absence of a Ca^{2+} gradient. Contribution of water activity to the enthalpy and entropy changes. J. Biol. Chem. 257, 4993-4998.

de Meis, L. (1984). Pyrophosphate of high and low energy. Contributions of pH, Ca^{2+}, Mg^{2+}, and water to free energy of hydrolysis. J. Biol. Chem. 259, 6090-6097.

de Meis, L., Behrens, M. I., Petretski, J. H., & Politi, M. J. (1985). Contribution of water to free energy of hydrolysis of pyrophosphate. Biochemistry 24, 7783-7789.

de Meis, L., Gomez-Puyou, M.T., & Gomez-Puyou, A. (1988). Inhibition of F1 ATPase and sarcoplasmic reticulum ATPase by hydrophobic molecules. Eur. J. Biochem. 171, 343-349

de Meis, L. (1989). Role of water in the energy hydrolysis of phosphate compounds. Energy transduction in biological membranes. Biochim. Biophys. Acta 973, 333-349.

de Meis, L., Suzano V.A., & Inesi G. (1990). Functional interactions of catalytic site and transmembrane channel in the sarcoplasmic reticulum ATPase. J. Biol. Chem. 265, 18848-18851.

de Meis L. (1991). Fast efflux of Ca^{2+} mediated by the sarcoplasmic reticulum Ca^{2+} ATPase. J. Biol. Chem. 266, 5736-5742.

de Meis, L., & Inesi, G. (1992). Functional evidence of a transmembrane channel within the Ca^{2+} transport ATPase of sarcoplasmic reticulum. FEBS Lett. 299, 33-35.

de Meis L. (1993). The concept of energy-rich phosphate compounds: water, transport ATPases, and entropic energy. Arch. Biochem. Biophys. 306, 287-296.

de Meis, L., & Suzano, V.A. (1994). Uncoupling of muscle and blood platelets Ca^{2+} transport ATPases by heparin: regulation by K^+. J. Biol. Chem. 269, 14525-14529.

Duggan, P. F., & Martonosi A. (1970). Sarcoplasmic reticulum. IX The permeability of sarcoplasmic reticulum membrane. J. Gen. Physiol. 56, 147-176.

Dupont, Y. (1977). Kinetics and regulation of sarcoplasmic reticulum ATPase. Eur. J. Biochem. 72, 185-190.

Dupont, Y. (1980). Occlusion of divalent cations in the phosphorylated calcium pump of sarcoplasmic reticulum. Eur. J. Biochem. 109, 231-238.

Dupont, Y. (1982). Low-temperature studies of the sarcoplasmic reticulum calcium pump. Mechanism of calcium binding. Biochim. Biophys. Acta 688, 75-87.

Dupont, Y., & Pougeois, R. (1983). Evaluation of H_2O activity in the free or phosphorylated catalytic site of the Ca^{2+} ATPase. FEBS Lett. 156, 93-98.

Dupont, Y., Pougeois, R., Ronjat, M., & Verjovski-Almeida, S. (1985). Two distinct classes of nucleotide binding sites in sarcoplasmic reticulum Ca^{2+} ATPase revealed by 2', 3'- O - (2,4,6-trinitrocyclohexadienylidene)-ATP. J. Biol. Chem. 260, 7241-7249.

Ebashi, S. (1958). A granule-bound relaxation factor in skeletal muscle. Arch. Biochem. Biophys. 76, 410-423.

Ebashi, S. (1960). Calcium binding and relaxation in the actomyosin system. J. Biochem. 48, 150-151.

Ebashi, S. (1961). Calcium binding activity of vesicular relaxing factor. J. Biochem. 50, 236-244.

Ebashi, S., & Lipmann, F. (1962). Adenosine triphosphate-linked concentration of calcium ions in a particulate fraction of rabbit muscle. J. Cell. Biol. 14, 389-400.

Ebashi, S., & Endo, M. (1968). Calcium ions and muscle contraction. Prog. Biophys. Mol. Biol., 18, 123-183.

Ebashi, S. (1985). *Structure and Function of Sarcoplasmic Reticulum.* Academic Press, Orlando.

Eletr S., & Inesi G. (1972). Phospholipid orientation in sarcoplasmic reticulum membranes: Spin label ESR and proton NMR studies. Biochim. Biophys. Acta 282, 174-179.

Engelender, S., Wolosker, H., & de Meis, L. (1995). The Ca^{2+} ATPase isoforms of platelets are located in distinct functional Ca^{2+} pools and are uncoupled by a mechanism different from that of skeletal muscle Ca^{2+} ATPase. J. Biol. Chem. 270, 21050-21055.

Fleischer, S., & Inui, M. (1989). Biochemistry and biophysics of excitation- contraction coupling. Ann. Rev. Biophys. Biophys. Chem. 18, 334-364.

Forge, V., Mintz, E., & Guillain, F. (1993). Ca^{2+} binding to the sarcoplasmic reticulum ATPase revisited. I. Mechanism of affinity and cooperativity modulation by H^+ and Mg^{2+}. J. Biol. Chem. 268, 10953-10960.

Forge, V., Mintz, E., Canet, D., & Guillain (1995). Lumenal Ca^{2+} dissociation from the phosphorylated Ca^{2+} ATPase of the sarcoplasmic reticulum is sequential. J. Biol. Chem. 270, 18271-18279.

Froelich, J.P., & Taylor, E.W. (1975). Transient state kinetics studies of sarcoplasmic reticulum adenosine triphosphatase. J. Biol. Chem. 250, 2013-2021.

Galina A., & de Meis L . (1992). Ca^{2+} translocation and catalytic activity of the sarcoplasmic reticulum ATPase. J. Biol. Chem. 266,17978-17982.

Garrahan, P. J., & Glynn, I. M. (1967). The incorporation of inorganic phosphate in fo adenosine triphosphate by reversal of the sodium pump. J. Physiol. 192, 237-256.

George, P., Witonsky, R. J., Trachtman, M., Wu, C., Dorwatr, W., Richman, L., Richman, W., Shurayh, F., & Lentz, B. (1970). "Squiggle-H_2O". An inquiry into the importance of solvation effects in phosphate ester anhydride reactions. Biochim. Biophys. Acta 223, 1-15.

Gerdes, U., Nakhala, A.M., & Moller, J.V. (1983). The Ca^{2+} permeability of sarcoplasmic reticulum vesicles. Ca^{2+} outflow in the non-energized state of the calcium pump. Biochim. Biophys. Acta 734,180-190.

Gerdes, U., & Moller, J.V. (1983). The Ca^{2+} permeability of sarcoplasmic reticulum vesicles. Ca^{2+} efflux in the energized state of the calcium pump. Biochim. Biophys. Acta 734,191-200.

Gould, G.W., McWhirter, J.M., & Lee, A.G. (1987a). A fast passive Ca^{2+} efflux mediated by the (Ca^{2+}+ Mg^{2+})-ATPase in reconstituted vesicles. Biochim. Biophys. Acta 904, 45-54.

Gould, G.W., Colyer, J., East, J.M., & Lee, A.G. (1987b). Silver ions trigger Ca^{2+} release by interaction with the Ca^{2+}- Mg^{2+}-ATPase in reconstituted systems. J. Biol. Chem. 262, 7676-7679.

Hanel, A.M., & Jencks, W.P. (1991). Dissociation of calcium from the phosphorylated adenosine triphosphatase of sarcoplasmic reticulum: Kinetic equivalence of the calcium ions bound to the phosphorylated enzyme. Biochemistry 30, 11320-11330.

Hasselbach, W., & Makinose, M. (1961). Die calciumpumpe der "Erschlaffungsggrana" des muskels und ihre abhangigkeit vor der ATP-spaltung. Biochem. Z. 333, 518-528.

Hasselbach, W., & Makinose, M. (1962). ATP and active transport. Biochem. Biophys. Res. Comm. 7, 132-136.

Hasselbach, W., & Makinose, M. (1963). Ueber der mechanismus des calciumtransportes durch die membranen des sarkoplasmatischen reticulums. Biochem. Z. 339, 94-111.

Hasselbach, W. (1964). Relaxing factor and relaxation of muscle. Progr. Biophys. Biophys. Chem. 14, 167-222.

Hasselbach W., Makinose M., & Migala A. (1972). The arsenate induced calcium release from sarcoplasmic vesicles. FEBS Lett. 20, 311-315.

Hayes, D.M., Kenyon, G.L., & Kollman, P.A. (1978). Theoretical calculations of the hydrolysis energies of some "high-energy" molecules. 2. A survey of some biologically important hydrolytic reactions. J. Am. Chem. Soc. 100, 4331-4340.

Highsmith, S. (1986). Solvent accessibility of the adenosine 5"- triphosphate catalytic site of the Ca^{2+} ATPase. Biochemistry 25, 1049-1054.

Hill, T.L., & Inesi, G. (1982). Equilibrium cooperative binding of calcium and protons by sarcoplasmic reticulum ATPase. Proc. Natl. Acad. Sci. USA 79, 3978-3982.

Ikemoto, N. (1975). Transport and inhibitory Ca^{2+} sites on the ATPase enzyme isolated from the sarcoplasmic reticulum. J. Biol. Chem. 250, 7219-7224.

Inesi, G., Goodman, J.J., & Watanabe, S. (1967). Effect of diethyl ether on the adenosine triphosphatase activity and the calcium uptake of fragmented sarcoplasmic reticulum of rabbit skeletal muscle. J. Biol. Chem. 242, 4637-4643.

Inesi, G., Maring, E., Murphy, A. J., & McFarland, B. H. (1970). A study of the phosphorylated intermediate of the sarcoplasmic reticulum calcium pump. Arch. Biochem. Biophys. 138, 285-294.

Inesi, G., Kurzmack, M., Coan, C., & Lewis, D. (1980). Cooperative calcium binding and ATPase activation in sarcoplasmic reticulum vesicles. J. Biol. Chem. 255, 3025-3031.

Inesi, G., Nakamoto, R., Hymel, L., & Fleischer, S. (1983). Functional characterization of reconstituted sarcoplasmic reticulum vesicles. J. Biol. Chem. 258, 14804-14809.

Inesi, G. (1985). Mechanism of Ca^{2+} transport. Ann. Rev. Physiol. 47, 573-601.

Inesi, G. (1987). Sequential mechanism of calcium binding and translocation in sarcoplasmic reticulum adenosine triphosphatase. J. Biol. Chem. 262, 16338-16342.

Inesi G., & de Meis L. (1989). Regulation of steady state filling in sarcoplasmic reticulum. J. Biol. Chem. 264, 5929-5936.

Inesi, G., & Kirtley, M.R. (1992). Structural features of cation transport ATPases. J. Bioenerg. Biomem. 24, 271-283

Jardetzky, O. (1966). Simple allosteric model for membrane pumps. Nature 211, 969-970.

Kanazawa, T., Yamada, S., Yamamoto, T., & Tonomura, Y. (1971). Reaction mechanism of the Ca^{2+}-dependent ATPase of sarcoplasmic reticulum from skeletal muscle. V. Vectorial requirements for calcium and magnesium ions of three partial reactions of the ATPase: formation and decomposition of phosphorylated intermediate and ATP-formation from ADP and the intermediate. J. Biochem.(Tokyo) 70, 95-123.

Kanazawa, T., & Boyer, P. (1973). Occurrence and characteristics of a rapid exchange of phosphate oxygens catalyzed by sarcoplasmic reticulum vesicles. J. Biol. Chem. 248, 3163-3172.

Kanazawa, T. (1975). Phosphorylation of solubilized sarcoplasmic reticulum by orthophosphate and its thermodynamic characteristics. J. Biol. Chem. 250, 113-119.

Kielley, W. W., & Meyerhof, O. (1948). A new magnesium-activated adenosinetriphosphatase from muscle. J. Biol. Chem. 174, 387-388.

Knowles, A. F., & Racker, E. (1975). Dependence of calcium permeability of sarcoplasmic reticulum vesicles on external and internal calcium concentration. J. Biol. Chem. 250, 113-119.

Kosk-Kosicka, D., Kurzmack, M., & Inesi, G. (1983). Kinetic characterization of detergent-solubilized sarcoplasmic reticulum adenosine triphosphatase. Biochemistry 22, 2559-2567.

Kumagai, H., Ebashi, S., & Takeda, F. (1955). Essential relaxing factor in muscle other than myokinase and creatinephosphokinase. Nature 176, 166-170.

Kurtenbach, E., & Verjovsky-Almeida, S. (1985). Labeling of a thiol residue in the sarcoplasmic reticulum ATPase by pyrene maleimide. J. Biol. Chem. 260, 9636-9641.

Kurzmack, M., Verjovsky-Almeida, S., & Inesi, G. (1977). Detection of an initial burst of Ca^{2+} translocation in sarcoplasmic reticulum. Biochem. Biophys. Res. Comm. 78, 772-776.

Levy, D., Seigneuret, M., Bluzat, A., & Rigaud, J.L. (1990). Evidence for proton countertransport by the sarcoplasmic reticulum Ca^{2+} ATPase during Ca^{2+} transport in reconstituted proteoliposomes with low ionic permeability. J. Biol. Chem. 265, 19524-19534.

Lipmann, F. (1941). Metabolic generation and utilization of phosphate bond energy. Adv. Enzymol. 1, 99-162.

Lytton, J., Zarain-Herzberg, A., Periasamy, M., & MacLennan, D.H. (1992). Functional comparison between isoforms of the sarcoplasmic or endoplasmic reticulum family of calcium pumps. J. Biol. Chem. 267, 14483-14489.

MacLennan D.H., Brandl C.J., Korczak B., & Green N.M. (1985). Amino-acid sequence of a Ca^{2+}+ Mg^{2+} -dependent ATPase from rabbit muscle sarcoplasmic reticulum, deduced from its complementary DNA sequence. Nature 316, 696-700.

Madeira, V.M.C. (1978). Proton gradient formation during transport of calcium by the sarcoplasmic reticulum. Arch. Biochem. Biophys. 185, 316-325.

Makinose, M., & Hasselbach, W. (1965). Der einfluss von oxalat auf den calcium-transport isolierter vesikel des sarkoplasmatischen reticulum. Biochem. Z. 343, 360-382.

Makinose, M. (1966). Second Internl. Biophys. Congr. Vienna, 276.

Makinose, M. (1969). The phosphorylation of the sarcoplasmic reticulum vesicles during active calcium transport. Eur. J. Biochem. 10, 74-82.

Makinose, M. (1971). Calcium efflux dependent formation of ATP from ADP and orthophosphate by the membranes of sarcoplasmic vesicles. FEBS Lett. 12, 269-270.

Makinose M., & Hasselbach, W. (1971). ATP synthesis by the reverse of the sarcoplasmic reticulum pump. FEBS Lett. 12, 271-272.

Makinose, M. (1972). Phosphoprotein formation during osmo-chemical energy conversion in the membrane of the sarcoplasmic reticulum. FEBS Lett. 25, 113-115.

Makinose, M. (1973). Possible functional states of the enzyme of the sarcoplasmic reticulum. FEBS Lett. 37, 140-143.

Marsh, B. B. (1952). The effects of adenosinetriphosphate on the fibre volume of a muscle homogenate. Biochim. Biophys. Acta 9, 247-260.

Martonosi, A. (1967). The role of phospholipids in the ATPase activity of skeletal muscle microsomes. Biochem. Biophys. Res. Comm. 29, 753-761.

Martonosi, A. (1969). Sarcoplasmic reticulum. VII. Properties of a phosphoprotein intermediate implicated in calcium transport. J. Biol. Chem. 244, 613-620.

Masuda, H., & de Meis, L. (1973). Phosphorylation of sarcoplasmic reticulum membrane by orthophosphate. Inhibition by calcium ions. Biochemistry 12, 4581-4585.

McIntosh, D., & Boyer, P. D. (1983). Adenosine 5'-triphosphate modulation of the catalytic intermediates of calcium ion adenosinetriphosphatase of sarcoplasmic reticulum subsequent to enzyme phosphorylation. Biochemistry 22, 2867-2875.

Meissner, G. (1973). ATP and Ca^{2+} binding by the Ca^{2+}-pump protein of sarcoplasmic reticulum. Biochim. Biophys. Acta 298, 906-926.

Meissner, G. (1981). Calcium transport and monovalent cation and proton fluxes in sarcoplasmic reticulum vesicles. J. Biol. Chem. 256, 636-643.

Mitchell, P. (1961). Coupling of phosphorylation to electron and hydrogen transfer by a chemi-osmotic type of mechanism. Nature 191, 144-148.

Mitchell, P. (1966). *Chemiosmotic Coupling in Oxidative and Photosynthetic Phosphorylation.* Glynn Research Laboratories, Bodmin, Cornwall.

Montero-Lomeli, M., & de Meis, L. (1992). Glucose 6-phosphate and hexokinase can be used as an ATP-regenerating system by the Ca^{2+} ATPase of sarcoplasmic reticulum. J. Biol. Chem. 267, 1829-1833.

Moraes, V. L. G., & de Meis, L. (1987). ATP synthesis by the (Na^+ - K^+)ATPase in the absence of an ionic gradient. Effects of organic solvent. FEBS Lett. 222, 163-166.

Morimoto, T., & Kasai, M. (1986). Reconstitution of sarcoplasmic reticulum Ca^{2+} ATPase vesicles lacking ion channels and demonstration of electrogenicity of the Ca^{2+} pump. J. Biochem. 99, 1071-1080.

Muscatello, U., Andersson-Cedergren, E., & Azzone, G. F. (1962). The mechanism of muscle-fiber relaxation adenosinetriphosphatase and relaxing activity of the sarcotubular system. Biochim. Biophys. Acta 63, 55-74.

Nagai, T., Makinose, M., & Hasselbach, W. (1960). De physiologische Erschlaffungsfaktor und die muskelgrana. Biochim. Biophys. Acta 43, 223-238.

Norregaard, A., Vilsen, B., & Andersen, J.P. (1994). Chimeric Ca^{2+} ATPase/Na^+, K^+ - ATPase molecules. Their phosphoenzyme intermediates and sensitivity to Ca^{2+} and thapsigargin. J. Biol. Chem. 269, 26598-26601.

Orlowski, S., & Champeil, P. (1991). The two calcium ions initially bound to the nonphosphorylated sarcoplasmic reticulum Ca^{2+} ATPase can no longer be kinetically distinguished when they dissociate from phosphorylated ATPase toward the lumen. Biochemistry 30, 11331-11342.

Palade ,P., Dettbarn, C., Brunder, D., Stein, P., & Hals, G. (1989). Pharmacology of calcium release from sarcoplasmic reticulum. J. Biochem. Biomembr. 21, 295-320.

Petretski, J. H., Wolosker, H., & de Meis, L . (1989). Activation of Ca^{2+} uptake and inhibition of reversal of the sarcoplasmic reticulum Ca^{2+} pump by aromatic compounds. J. Biol. Chem. 264, 20339-20343.

Pick, U., & Bassilian, S. (1982). The effects of ADP, phosphate and arsenate on Ca^{2+} efflux from sarcoplasmic reticulum vesicles. Eur. J. Biochem. 131, 393-399.

Plank, B., Hellmann, G., Punzengruber, C., & Suko, J. (1979). ATP P_i, ⇌ and ITP ⇌ P_i exchange by cardiac sarcoplasmic reticulum. Biochim. Biophys. Acta 550, 259-265.

Portzehl, H. (1957). Die biondung des Erschlaffungsfaktors von Marsh an die muskelgrana. Biochim. Biophys. Acta 26, 373-377.

Post, R. L., Sen, A. K., & Rosenthal, S. (1965). A phosphorylated intermediate in adenosine triphosphate-dependent sodium and potassium transport across kidney membranes. J. Biol. Chem. 240, 1437-1445.

Romero, P.J., & de Meis, L. (1989). Role of water in the energy of hydrolysis of phosphoanhydride and phosphoester bonds. J. Biol. Chem. 264, 7869-7873.

Sagara, Y., Fernandez-Belda, F., de Meis, L., & Inesi, G. (1992). Characterization of the inhibition of intracellular Ca^{2+} transport ATPases by thapsigargin. J. Biol. Chem. 267,12606-12613.

Scofano, H., Vieyra, A., & de Meis, L. (1979). Substrate regulation of the sarcoplasmic reticulum ATPase. transient kinetic studies. J. Biol. Chem. 254, 10227-10231.

Seebregts, C.J., & McIntosh, D. (1989). 2', 3'-O-(2,4,6 - trinitrophenyl)-8-azido-adenosine-mono-, di-, and triphosphates as photoaffinity probes of the Ca^{2+} ATPase of sarcoplasmic reticulum. J. Biol. Chem. 264, 2043-2052.

Shigekawa, M., & Dougherty, J.P. (1978). Reaction mechanism of Ca^{2+}-dependent ATP hydrolysis by skeletal muscle sarcoplasmic reticulum in the absence of added alkali metal salts. II. Kinetic properties of the phosphoenzyme formed at the steady state in high Mg in the absence of added alkali metal salts. II. Kinetic properties of the phosphoenzyme formed at the steady state in high Mg^{2+} and low Ca^{2+} concentrations. J. Biol. Chem. 253, 1451-1457.

Souza, D. G., & de Meis, L. (1976). Calcium and magnesium regulation of phosphorylation by ATP and ITP in sarcoplasmic reticulum vesicles. J. Biol. Chem. 251, 6355-6359.

Skou, J.C. (1957). The influence of some cations on an ATPase from peripheral nerves. Biochim. Biophys. Acta 23, 394-401.

Tabor, H., & Tabor, C.W. (1964). Spermidine, spermine and related amines. Pharmacol. Rev. 16, 245-300.
Takakuwa, Y., & Kanazawa, T. (1979). Reaction mechanism of the (Ca^{2+}-Mg^{2+})ATPase of sarcoplasmic reticulum. The role of Mg^{2+} that activates the hydrolysis of the phosphoenzyme. Biochem. Biophys. Res. Comm. 88, 1209-1216.
Tanford, C. (1984). Twenty questions concerning the reaction cycle of the sarcoplasmic reticulum calcium pump. CRC Crit. Revs. Biochem. 17, 123-151.
Taniguchi, K., & Post, R. L. (1975). Synthesis of adenosine triphosphate and exchange between inorganic phosphate and adenosine triphosphate in sodium and potassium ion transport adenosine triphosphatase. J. Biol. Chem. 250, 3010-3018.
Thastrup, O. (1987). The calcium mobilizing and tumor promoting agent, thapsigargin elevates the platelet cytoplasmic free calcium concentration to a higher steady state level. A possible mechanism of action for the tumor promotion. Biochem. Biophys. Res. Commun. 13, 654-660.
Toyoshima, C., Sasabe, H., & Stokes, D.L. (1993). Three dimensional cryo-electron microscopy of the calcium ion pump in the sarcoplasmic reticulum membrane. Nature (Lond.) 362, 469-471.
Ueno, T., & Sekine, T. (1981a). A role of H^+ flux in active Ca^{2+} transport into sarcoplasmic reticulum vesicles. I. Effect of an artificially imposed H^+ gradient on Ca^{2+} uptake. J. Biochem. 89, 1239-1246.
Ueno, T., & Sekine, T. (1981b). A role of H^+ flux in active Ca^{2+} transport into sarcoplasmic reticulum vesicles. II. H^+ ejection during Ca^{2+} uptake. J. Biochem. 89, 1247-1252.
Verjovski-Almeida, S., & de Meis, L. (1977). pH-induced changes in the reactions controlled by the low- and high-affinity Ca^{2+} binding sites in sarcoplasmic reticulum. Biochemistry 16, 329-334.
Vianna, A.L. (1975). Interaction of calcium and magnesium in activating and inhibiting the nucleotide triphosphatase of sarcoplasmic reticulum vesicles. Biochim. Biophys. Acta 410, 389-394.
Vilsen, B. (1995). Structure-function relationships in the Ca^{2+}-ATPase of sarcoplasmic reticulum studied by the use of the substrate analogue CrATP and site-directed mutagenesis. Comparison with the Na^+, K^+-ATPase. Acta Physiol. Scand. 624, (suppl.) 1-146.
Wakabayashi, S., Ogurus, T., & Shigekawa, M. (1986). Factors influencing calcium release by the ADP-sensitive phosphoenzyme intermediate of sarcoplasmic reticulum ATPase. J. Biol. Chem. 261, 9762-9769.
Watanabe, T, Lewis, D., Nakamoto, R., Kurzmack, M., Fronticelli, C., & Inesi ,G. (1981). Modulation of calcium binding in sarcoplasmic reticulum adenosinetriphosphatase. Biochemistry 20, 6617-6625.
Weber, A. (1959). On the role of calcium in the activity of adenosine 5'-triphosphate hydrolysis by actomyosin. J. Biol. Chem. 234, 2764-2769.
Weber, A., Herz, R., & Reiss, I. (1966). Study of the kinetics of calcium transport by the isolated fragments of sarcoplasmic reticulum. Biochem. Z. 345, 329-369.
Weber, A. (1971a). Regulatory mechanisms of the calcium transport system of fragmented rabbit sarcoplasmic reticulum. I. The effect of accumulated calcium on transport and adenosine triphosphate hydrolysis. J. Gen. Physiol. 57, 50-63.
Weber, A. (1971b). Regulatory mechanisms of the calcium transport system of fragmented sarcoplasmic reticulum. II Inhibition of outflux in calcium free media. J. Gen. Physiol. 57, 64-70.
Weber, G. (1972). Ligand binding and internal equilibria in proteins. Biochemistry 11, 864-878.
Wolosker, H., Petretski, J.H., & de Meis, L. (1990). Modification of ATP regulatory function in sarcoplasmic reticulum Ca^{2+} ATPase by hydrophobic molecules. Eur. J. Biochem. 19, 873-877.
Wolosker, H., Pacheco, A.G.F., & de Meis, L. (1992). Local anesthetics induce fast Ca^{2+} efflux through a nonenergized state of the sarcoplasmic reticulum Ca^{2+} ATPase J. Biol. Chem. 267, 5785-5789.
Wolosker, H., & de Meis, L. (1994). pH-dependent inhibitory effects of Ca^{2+}, Mg^{2+} and K^+ on the Ca^{2+} efflux mediated by the sarcoplasmic reticulum ATPase. Am. J Physiol. (Cell Physiol.) 35, C1376-C1381.

Yamada, S., & Tonomura, Y. (1972). Reaction mechanism of the Ca^{2+}-dependent ATPase of sarcoplasmic reticulum from skeletal muscle. VII Recognition and release of Ca^{2+} ions. J. Biochem. 72, 417-425.

Yamagushi, M., & Kanazawa, T. (1984). Protonation of the sarcoplasmic reticulum Ca^{2+} ATPase during ATP hydrolysis. J. Biol. Chem. 259, 9526-9531.

Yamamoto, T., & Tonomura, Y. (1967). Reaction mechanism of the Ca^{2+}-dependent ATPase of sarcoplasmic reticulum from skeletal muscle.. I. Kinetic studies. J. Biochem. 62, 558-575.

Yamamoto, T., & Tonomura, Y. (1968). Reaction mechanism of the Ca^{2+}-dependent ATPase of sarcoplasmic reticulum from skeletal muscle.. II. Intermediate formation of phosphoryl protein. J. Biochem. 64, 137-145.

Yu, X., Hao, L., & Inesi, G. (1994). A pK change of acidic residues contributes to cation countertransport by the Ca^{2+} ATPase of sarcoplasmic reticulum. J. Biol. Chem. 269, 16656-16661.

Zimniak, P., & Racker, E. (1978). Electrogenicity of Ca^{2+} transport catalyzed by the Ca^{2+} ATPase from sarcoplasmic reticulum. J. Biol. Chem: 253, 4631-4637.

THE ATP BINDING SITES OF P-TYPE ION TRANSPORT ATPases:

PROPERTIES, STRUCTURE, CONFORMATIONS, AND MECHANISM OF ENERGY COUPLING

David B. McIntosh

Advances in Molecular and Cell Biology
Volume 23A, pages 33-99.

ISBN: 0-7623-0287-9

I. INTRODUCTION

The P-type ion-transport ATPases comprise a family of ATP-dependent pumps which serve to keep the cationic cellular milieu in the correct balance against the prevailing concentration gradient. They share a conserved DKTGT sequence around the phosphorylated aspartyl residue, other conserved sequences, and a similar membrane topography, as deduced from hydropathy plots. They generally transport different cations in two directions and can be electrogenic. The transport of the first cation is activated by ATP binding and phosphorylation, and the other, in the second part of the cycle, is associated with dephosphorylation.

One type of polypeptide with a molecular size of about 100 kDa is sufficient for pumping activity of some pumps. The Ca pumps fall into this category. In the Na–K and mammalian gastric H–K pumps, which are closely related, the 100 kDa polypeptide (α-subunit) is intimately associated in a 1:1 ratio with a smaller protein (β-subunit), which appears to be necessary for processing of the α-subunit to the cell surface (Hiatt et al., 1984; Geering et al., 1987; Renaud et al., 1991), as well as playing a role in extracellular K^+ binding (Jaisser et al., 1992) and conferring cation-dependence and ouabain sensitivity of phosphoenzyme formation (Blanco et al., 1994). Cardiac sarcoplasmic reticulum (SR) Ca-ATPase is regulated by phospholamban, a 52-residue single-membrane-pass protein (Tada et al., 1974; Sasaki et al., 1992). Skeletal muscle SR Ca-ATPase (Wawrzynow et al., 1992), Na–K-ATPase (Collins and Leszyk, 1987), and plasma membrane H-ATPase of yeast (Navarre et al., 1992) are associated with similar proteolipids. In bacterial K pumps, notably the Kdp system of *E coli*, the polypeptide equivalent to the 100 kDa species of the Na–K, H–K, and Ca pumps is smaller, approximately 70 kDa (KdpB), and its association with two smaller polypeptides, KdpC (perhaps equivalent to the β-subunit of Na–K-ATPases) and KdpA may be necessary for activity (Epstein et al., 1990). However, the K pump of *Streptococcus faecalis*, which is closely homologous to the KdpB protein of *E coli*, consists of a single polypeptide of 63 kDa (Hugentobler et al., 1983; Solioz et al., 1987). There is no evidence to suggest that the smaller polypeptides play a direct role in ATP binding.

The larger pumps, such as the SR Ca-ATPase and Na–K-ATPase, probably span the membrane ten times (MacLennan et al., 1985; Stokes et al., 1994), whereas the smaller bacterial ones and Cu-ATPases appear to have 6 or 8 passes (Hesse et al., 1984; Odermatt et al., 1993). All ATPases contain two large extramembranous cytoplasmic loops between M2 and M3 (termed β-Strand domain), and M4 and M5 (large cytoplasmic loop) membrane traverses, which together constitute the catalytic and ATP binding sites.

The reaction cycle of P-type ATPases follows the minimal scheme:

$$\begin{array}{ccc} \text{M1(cis)} + \text{ATP} + \text{E}_1 & \leftrightarrow & \text{E}_1\text{P} + \text{ADP} + \text{M1(trans)} \\ \updownarrow & & \updownarrow \\ \text{M2(cis)} + \text{P}_i + \text{E}_2 & \leftrightarrow & \text{E}_2\text{P} + \text{M2(trans)} \end{array}$$

The distinction between the two enzyme forms, denoted as 1 and 2, varies. In this chapter, the E_1 forms represent the intermediates that bind and transport the first cation, M1, some of which are phosphorylated by ATP and not by P_i and react with ADP, and the E2 forms as those that bind and transport the counterion, M2, some of which are phosphorylated by P_i and not by ATP, and do not react with ADP.

Complications arise when analogues of ATP or P_i are used. For example, ρ-nitrophenyl phosphate is hydrolysed slowly by the SR Ca-ATPase but otherwise in a manner similar to ATP, with the formation of a phosphoenzyme and hydrolysis coupled to Ca transport (Inesi, 1971, Nakamura and Tonomura, 1978). In the case of the Na–K-ATPase and gastric H–K-ATPase, a high rate of hydrolysis occurs in the presence of K^+ (M2) and not Na^+ or H^+ (M1), without the formation of a phosphoprotein intermediate, suggesting it is acting more like a P_i analogue, but without actually phosphorylating the enzyme (Robinson, 1970; Post et al., 1972; Shaffer et al., 1978). Furylacryloylphosphate is recognized in a similar way (Inesi et al., 1980a; Odom et al., 1981), but dimethyl sulphoxide changes the mode of catalysis of SR Ca-ATPase into one more like that of the Na–K-ATPase, in which the coupling with Ca (M1) is lost (Inesi et al., 1980a). The ATP analogue 3'-O-(3-[N-(4-azido-2-nitrophenyl)amino]propionyl)-ATP (arylazido-ATP) phosphorylates the SR Ca-ATPase in the absence of Ca^{2+} and yet supports Ca^{2+} transport when Ca^{2+} is added to the medium (Oliveira et al., 1988). Similarly, the gastric H–K-ATPase, but not the Na–K-ATPase and SR Ca-ATPase, uses acetyl phosphate as a P_i analogue and is phosphorylated in a manner which is stimulated by K^+ without cation transport (Asano et al., 1992). Canine cardiac SR Ca-ATPase shows GTP hydrolysis activated by Mg^{2+}, not Ca^{2+}, without a detectable phosphoenzyme intermediate but Ca^{2+} is found to be transported when present (Tate et al., 1985). The Ca^{2+} component is increased by non-solubilizing concentrations of nonionic detergent (Tate et al., 1989).

Other pumps, such as plasma membrane Ca-ATPase of erythrocytes (Caride et al., 1982) and H-ATPase of yeast (Ferreira-Pereira et al., 1994), appear to be more like the Na–K- and H–K-ATPases in utilizing ρ-nitrophenyl phosphate as a P_i analogue. In the case of the former pump, hydrolysis of this substrate is activated by hydrolysis of ATP, suggesting that enzyme turnover is necessary to generate forms of the enzyme reactive to ρ-nitrophenyl phosphate (Caride et al., 1982). Evidently, the specificity of the E_1 and E_2 forms is not absolute and varies for each pump, each analogue of ATP and P_i, and the medium composition.

The specificity of the E_1 species for different analogues of ATP, which follow the normal cycle, is remarkably broad in many pumps. The SR Ca-ATPase and Na–K-ATPase show a similar specificity for CTP, GTP, ITP, and UTP and reasonably fast transport rates are observed for the Ca^{2+} pump with all substrates (Jensen and Nørby, 1971; Hegyvary and Post, 1971 de Meis and de Mello, 1973). The analogues $8N_3$-ATP and 8Br-ATP, which probably bind in the syn conformation about the glycosidic bond and not the anti of ATP, are hydrolyzed and transport Ca^{2+} as well as ATP with the SR Ca-ATPase (Briggs et al., 1980; Champeil et al., 1988), and $8N_3$-ATP is also a good substrate of the Na–K-ATPase (Scheiner-Bobis and Schoner, 1985). While the affinity of the pumps for ATP analogues can be much lower than for ATP, this need not be correlated with low transport rates and the turnover of the enzyme can be close to that of ATP. For example, both the Na–K-ATPase and SR Ca-ATPase exhibit a low affinity for acetyl phosphate, but high rates of phosphorylation and transport are observed (Friedman and Makinose, 1970; Rephaeli et al., 1986; Bodley and Jencks, 1987). Bulky additions to the ribose hydroxyls can increase the affinity, and for the SR Ca-ATPase, either inhibit phosphorylation completely as in the case of 2', 3'-O-(2,4,6-trinitrophenyl)-ATP (TNP-ATP), or markedly slow it down as for arylazido-ATP and TNP-$8N_3$-ATP (Watanabe and Inesi, 1982; Oliveira et al., 1988; McIntosh and Woolley, 1994). Removing either hydroxyl decreases the rate of phosphorylation (Coan et al., 1993). In the case of the Na–K-ATPase, addition of an arylazido-β-alanyl to the 3' position has little effect on the hydrolysis rate compared with ATP (Munson, 1981). Large substitutions in the purine ring, for example 6-[(3-carboxy-4-nitrophenyl)thiol]-ITP (Koepsell et al., 1982), and, most surprising, even the disulfide of 6-thioinosine triphosphate (Patzelt-Wenczler and Schoner, 1981), can also be accommodated as substrates. As will be seen below, analogues of ATP, modified in the position of the phosphates, can serve as high-affinity labels. The high degree of tolerance of these pumps for different substrates suggests that the ATP binding sites are rather open, with adjoining pockets in which bulky groups may be accommodated.

Other pumps can exhibit a higher specificity. A Ca-ATPase from rat hepatocyte plasma membranes appears to be absolutely specific for ATP and almost no activity is observed with GTP, UTP, or CTP (Kessler et al., 1990). A high affinity Ca-ATPase from *Leishmania donovani*, a protozoal parasite, exhibits no activity with UTP and CTP, low activity with GTP, and does not hydrolyze ρ-nitrophenyl phosphate (Ghosh et al., 1990). High rates of hydrolysis of ATP occur in the presence of Ca^{2+} and absence of Mg^{2+}.

Most higher animal P-type ATPases, for example Na–K-ATPase, plasma membrane Ca-ATPases (cardiac sarcolemma excepted) (Caroni and Carafoli, 1981; Dixon and Haynes, 1989), SR Ca-ATPases, and gastric H–K-ATPase are phosphorylated by low (micromolar) concentrations of MgATP, whereas the H-ATPases of the plasma membrane of plants, fungi, and yeast, the various bacterial pumps, the Ca-ATPase of *Leishmania donovani*, and possibly the mammalian Cu-ATPase, require higher (millimolar) concentrations of MgATP. Transport in the former group is activated by concentrations of ATP higher than that required for phosphorylation, resulting in a complex dependence of activity on MgATP concentration. Those pumps in the latter group that have been investigated do not appear to show the regulation by ATP. For example, ATP does not modulate the $P_i \leftrightarrow$ HOH exchange characteristics or dephosphorylation of the yeast pump (Amory et al., 1982), contrary to what is observed with SR Ca-ATPase (McIntosh and Boyer, 1983). However, some yeast and fungal H-ATPases display a sigmoidal dependence of activity on MgATP concentration, due either to a complex interaction of Mg^{2+} at an activating site and MgATP and ATP competing at the catalytic site, positive cooperativity between subunits, or parallel catalytic paths of two different conformations of the ATPase (Brooker and Slayman, 1983; Koland and Hammes, 1986).

In the more thoroughly studied pumps that are phosphorylated by low concentrations of ATP and accelerated by higher concentrations, the steps in the catalytic cycle responsive to secondary ATP binding have been largely elucidated. In the Na–K-ATPase, ATP binding to a low affinity site on E_2.K results in a large acceleration of K^+ (or Rb^+) dissociation or deocclusion and conversion to E_1 (Hegyvary and Post, 1971; Post et al., 1972; Karlish et al., 1978; Glynn and Richards, 1982; Forbush 1987). Deocclusion is the most significant rate-limiting step in the cycle of this pump. High ATP inhibits E_2P hydrolysis (Askari and Huang, 1984), P_i phosphorylation (Campos and Beaugé, 1994), and accelerates the rate of vanadate dissociation in the presence of Na^+ (Huang and Askari, 1984). Initial rate of phosphorylation by ATP is enhanced by high concentrations of this nucleotide (Peluffo et al., 1994). The inhibition of dephosphorylation through binding to a low affinity site is observed in the gastric H–K-ATPase (Helmich-de Jong et al., 1986a).

For the SR Ca-ATPase, most steps in the cycle are accelerated by ATP. These include the Ca-dependent E_2 to E_1 transition with ATP binding to a high affinity site (Scofano et al., 1979; Inesi et al., 1980b; Guillain et al., 1981; Champeil et al., 1983; Stahl and Jencks, 1984), ADP reaction with E_1P following inhibition of the pump with reactive-red 120 (Coll and Murphy, 1991), Ca^{2+} release to the lumen and the E_1P to E_2P transition (Wakabayashi et al., 1986; Champeil and Guillain, 1986; Bodley and Jencks), E_2P hydrolysis (de Meis and de Mello, 1973; Shigekawa et al., 1978; McIntosh and Boyer, 1983; Champeil et al., 1988) and P_i binding (Pickart and Jencks, 1984). In the red cell plasma membrane Ca-ATPase, ATP binding to low affinity sites accelerates the hydrolysis of E_2P and this effect is enhanced by calmodulin (Herscher et al., 1994).

The nature and locus of the secondary regulatory ATP site has been the subject of extensive investigation over many years, especially with the Na–K- and SR Ca-ATPases, and yet still remains unresolved. The issue revolves around the relationship of the catalytic site, often with broad specificity, to the regulatory site, generally of low affinity. The latter could be a separate allosteric site, a modified form of the catalytic site subsequent to phosphorylation and ADP dissociation, or a combination of both in which there is some overlap between sites (Post et al., 1972). There is also the possibility of a cooperative dimer in which interactions between polypeptides lock the pump subunits in alternative ATP binding and transport conformations, and ATP modulations are exerted via neighboring catalytic sites (see Scheiner-Bobis et al., 1987). Central to the problem is the number and nature of the ATP binding sites which will be considered in the next section.

II. NUMBER AND NATURE OF ATP BINDING SITES

Uncertainties regarding the amount of active ATPase in preparations impinge on any discussion of ATP site stoichiometry. The level of phosphoenzyme during steady state ATPase activity is generally less than half that expected for full phosphorylation of every ATPase monomer. Three factors influence this result, namely, the method of protein determination, the level of nonphosphorylated intermediates, and the amount of inactive ATPase and non-ATPase protein. Lowry determination of protein content overestimates the true value by 1.2-1.4 fold (Hardwicke and Green, 1974; Moczydlowski and Fortes, 1981a). The level of nonphosphorylated, active ATPase can be minimized by optimizing conditions. In the case of the SR Ca-ATPase, phosphorylation levels performed with P_i and dimethyl sulphoxide in ethylene glycol-bis(β-aminoethyl ether)N,N'-tetra-acetic acid (EGTA) with ATP at high Ca^{2+} and KCl concentrations, or with acetyl phosphate at high Ca^{+} give the same high value and this has been used as a measure of active ATPases (Barrabin et al., 1984; Teruel and Inesi, 1988; McIntosh et al., 1992). Dispersion of Ca-ATPases in detergent, under conditions which eliminate ATPase-ATPase interactions and stabilize the monomer, gives the same value and does not uncover previously catalytically-silent ATPases (Jørgensen et al., 1978; Kosk-Kosicka et al., 1983; Andersen et al., 1986; McIntosh and Ross, 1988). Procedures which eliminate inactive, aggregated ATPase give higher yields of phosphoenzyme, equivalent to phosphorylation of nearly every ATPase monomer, namely up to 7.3 nmol/ mg of Lowry protein (Barrabin et al., 1984; Gafni and Boyer, 1984; Andersen et al., 1986).

The situation is similar for the Na–K-ATPase. In this case an indicator of the amount of active ATPase can be the level of phosphoenzyme formed from ATP or P_i in the presence of ouabain (Glynn and Richards, 1982; Askari et al., 1983), (^{48}V)vanadate binding in the presence of ouabain and (γ-^{32}P)CrATP binding/phosphorylation (Esmann and Skou, 1984; Vilsen et al., 1987). (^{3}H)ouabain binding is often used as an indication of the amount of Na–K-ATPase (eg, Glynn and Richards, 1982; Askari et al., 1983; Forbush, 1987), and generally this figure correlates with phosphoenzyme and vanadate binding levels. Isoenzymes with a low affinity for this glycoside can be present in a single preparation (Sweadner, 1979; Sweadner and Farshi, 1987; Ahlemeyer et al., 1992). Possible interactions between monomers can be eliminated in nonionic detergent (Brotherus et al., 1981; Craig, 1982). Again, monomerizing in detergent does not reveal otherwise catalytically silent

ATPases (Esmann and Skou, 1984; Vilsen et al., 1987). However, K^+ induces apparent heterogeneity of ATP binding sites of pig kidney preparations, which reverts to a homogeneous population of sites by solubilization (Ottolenghi and Jensen, 1983; Jensen and Ottolenghi, 1983). Solubilized monomeric preparations exhibit low affinity ATP inhibition of K-activated ρ-nitrophenyl phosphate hydrolysis (Esmann and Skou, 1984) and the complex ATP activation of catalysis (Ward and Cavieres, 1993).

In both SR Ca-ATPase and Na–K-ATPase preparations, purified or not, the actual level of active enzyme can vary from 1 to 7 nmole/mg of Lowry protein. Preparations with high specific activity exhibit high maximum phosphoenzyme levels (Barrabin et al., 1984; Vilsen et al., 1987). The extent to which "inactive" ATPase is unable to function is probably variable. It may bind nucleotides and exhibit other partial reactions. Partial reactions of the detergent-solubilized Na–K-ATPase, for example vanadate binding, are more resistant to inactivation than the overall hydrolysis of ATP (Esmann and Skou, 1984). Extensively digested Na–K-ATPase preparations occlude Rb^+ (Capasso et al., 1992), and the large cytoplasmic loop of Ca-ATPase, expressed in bacteria independently of the other part of the pump by genetic manipulation, refolds after denaturation in urea to yield a protein that binds nucleotides (Moutin et al., 1994). However, as mentioned, the ratio of nucleotide binding, vanadate binding, ouabain binding, and phosphoenzyme levels are usually related mole per mole.

Purified preparations of gastric H–K-ATPase generally exhibit lower levels of phosphoenzyme, to the extent of about 1.5 nmol/mg of Lowry protein, equivalent to the level of fluorescein 5' isocyanate (FITC) modification that inhibits ATPase activity (Jackson et al., 1983), but half the level of high-affinity eosin binding (Helmich-de Jong et al., 1986b) and ATP-dependent (^{14}C)SCH 28080 binding (Keeling et al., 1989).

The nature of the catalytic site and the low-affinity activating site, and the complex relationship between them, is perhaps best exemplified by studies of the binding of analogues of ATP that can bind tightly to either or both sites. In this respect chromium and cobalt complexes of ATP and P_i and TNP nucleotides, and their interactions with SR Ca-ATPase, Na–K-ATPase and gastric H–K-ATPase, have been most instructive.

CrATP and $Co(NH_3)_4ATP$ compete with ATP at the catalytic site of both the SR Ca-ATPase and Na–K-ATPase (Gantzer et al., 1982; Klevickis and Grisham, 1982). The affinity of these pumps for CrATP is not high according to the competition studies. The Na–K-ATPase exhibits a high affinity for the Co complex and the Ca-ATPase a low one. The high affinity of the Na–K-ATPase for the Co complex has facilitated elucidation of the structure of the nucleotide at the active site and its interaction with a nearby Mn^+ using Nuclear Magnetic Resonance (NMR) techniques (Klevickis and Grisham, 1982; Stewart and Grisham, 1988; Stewart et al., 1989). The nucleotide adopts an anti glycoside/adenine conformation and overall U-shape with the phosphates lying parallel to the adenine ring. The Mn^+ cation interacts most closely with N6 and N7 of the adenine moiety, possibly via water molecules. No enzyme inactivation is observed under the conditions used for the NMR studies.

Prolonged incubation of the pumps with the ATP complexes leads to other changes that are only slowly reversible. CrATP can form a very tight complex with the SR Ca-ATPase and Na–K-ATPase that results in loss of the ability of both enzymes to be phosphorylated with ATP (Pauls et al., 1980; Serpersu et al., 1982). In the case of the Na–K-ATPase, CrATP is bound stoichiometrically with the level of (^{48}V)vanadate and (^{3}H)ouabain binding (Vilsen et al., 1987), and with the SR Ca-ATPase two sites/EP_{max} are evident, only one of which is associated with Ca^{2+} occlusion (Vilsen and Andersen, 1992). Early studies considered that the chromium nucleotide phosphorylated the ATPase, but more recent work with the SR Ca-ATPase suggests that the phosphoryl group is not transferred (Vilsen and Andersen, 1992), in line with similar inactivation produced by chromium complexes of nonhydrolyzing ATP analogues (Pauls et al., 1986). CrATP binding, but significantly not the binding of the CrAMPPCP analogue, inhibits activities relating to the E_2 forms, for example K-activated ρ-nitrophenylphosphate hydrolysis (Pauls et al., 1980; Hamer and Schoner, 1993). CrATP binding to SR Ca-ATPase, at the site associated with Ca^{2+} occlusion, is blocked by P_i and vanadate, and partially by FITC modification of K515 (Chen et al., 1991). $Cr(NH_3)_4ATP$ inactivates NaK-ATPase more slowly than CrATP probably by water or protein functionalities exchanging with the ammonias (Serpersu et al., 1990).

Inactivation of Na–K-ATPase is also observed with $Co(NH_3)_4ATP$ as well as $Co(NH_3)_4PO_4$, but requires high concentrations. The process is inhibited by Na^+ and high concentrations of ATP, and occurs at a site distinct from the high affinity ATP site to which CrATP and CrAMPPCP bind because Na-dependent phosphorylation with ATP is not inhibited, nor, in the case of $Co(NH_3)_4PO_4$, is subsequent complexation with CrATP (Scheiner-Bobis et al., 1987; Buxbaum and Schoner, 1990, 1991). $Co(NH_3)_4ATP$ induces dimeric 2-D crystalline arrays, which is compatible with it stabilizing the E_2 state, but the crystals are different from those induced by vanadate (Skriver et al., 1989).

Thus it has been possible to demonstrate the simultaneous existence of CrAMPPCP and $Co(NH_3)_4ATP$ sites, each associated with E_1 and E_2 catalytic characteristics respectively. Schoner's laboratory has interpreted these findings as due to interactions of a dimer, where E_1 and E_2 forms coexist in a linked manner. Accommodation of the results within a monomer model requires the existence of two sites both of which appear to be related to the catalytic site, because the Cr complexes prevent ATP dependent phosphorylation and the Co-nucleotide and -P_i complexes prevent the binding of P_i and inhibit ρ-nitrophenyl phosphatase activity (Buxbaum and Schoner, 1991). An important point is that both chromium and cobalt complexes inactivate ATPase activity and hence it is unlikely that "inactive" ATPase can be invoked to explain the results.

In the case of the SR Ca-ATPase, the two kinetically distinct CrATP binding sites are both blocked by FITC derivatization of Lys-515. The nucleotide-protein linkages are sufficiently stable in sodium dodecyl sulfate for both sites to be located on the B tryptic subfragment (Vilsen and Andersen, 1992). In this case interacting subunits are not involved in generating the two types of sites, as biphasic release kinetics is observed with the detergent-solubilized monomer.

TNP-nucleotides have also provided insight into ATP binding sites. TNP-ATP, -ADP, and -AMP bind tightly to the catalytic site of the Na–K-ATPase, gastric H–K-ATPase, and SR Ca-ATPase (Moczydlowski and Fortes, 1981a and 1981b; Faller, 1989 and 1990; Dupont et al., 1982; Watanabe and Inesi, 1982; Dupont and Pougeois, 1983). TNP-ATP is not hydrolyzed by any of the pumps. With the Na–K-ATPase, the binding is close to stoichiometric with ouabain binding sites showing a 1:1 ratio with monomer (Moczydlowski and Fortes, 1981a). TNP-ATP competitively inhibits high-affinity ATP-dependent phosphorylation as well as K-stimulated ρ-nitrophenyl phosphate hydrolysis and P_i phosphorylation (Moczydlowski and Fortes, 1981b). It also competitively inhibits ATP activation of enzyme turnover. Moczydlowski and Fortes concluded that the catalytic site changes conformation and function during turnover, from a high-affinity catalytic site to a low-affinity regulatory site.

The stoichiometry of tight TNP-ATP binding to gastric H–K-ATPase is double the amount of maximal phosphoenzyme, and ATP displacement of bound TNP-nucleotide in Mg^{2+} is suggestive of two classes of sites (Faller, 1989 and 1990). An identical situation is observed with SR Ca-ATPase (Watanabe and Inesi, 1982; Dupont et al., 1982; Dupont et al., 1985). Extra, apparently noncatalytic, binding is not observed with TNP-AMP (Suzuki et al., 1990). Although the two binding sites have not been demonstrated with the Na–K-ATPase, TNP-ADP (K_d=35 μM) inhibits dephosphorylation of the FITC-modified enzyme (Scheiner-Bobis et al., 1993).

TNP-nucleotides interact tightly with the regulatory ATP site of SR Ca-ATPase since low concentrations of TNP-ATP activate ATP hydrolysis (Dupont et al., 1985). Low concentrations of TNP-ATP and -ADP accelerate dephosphorylation in the same manner as higher levels of ATP (Champeil et al., 1988).

The interaction of TNP-nucleotides with SR Ca-ATPase has a unique feature in that activation of Ca-ATPase turnover by Ca^{2+} and ATP, or phosphorylation with P_i in the presence of low concentrations of TNP-ATP, -ADP, or -AMP, causes a large enhancement of fluorescence of bound TNP-nucleotide, now usually referred to as superfluorescence (Watanabe and Inesi, 1982; Nakamoto and Inesi, 1984; Bishop et al., 1984; Wakabayashi et al., 1986; Bishop et al., 1987). Superfluorescence in either direction of catalysis is associated with the formation of E_2P (Nakamoto and Inesi, 1984; Bishop et al., 1984; Wakabayashi et al., 1986; Davidson and Berman, 1987), although a detailed kinetic study could not distinguish between phosphoenzyme species formed from ATP (Bishop et al., 1986). Further kinetic studies (Bishop et al., 1987) have suggested that the superfluorescent nucleotide is regulatory and bound at the phosphorylated catalytic site.

Like the parent nucleotides, the 8-azido derivatives of TNP-ATP, -ADP, and -AMP bind tightly to both the catalytic site and the regulatory site. The diphospho and triphospho species activate ATP hydrolysis, and all three display superfluorescence on phosphorylation to E_2P (Seebregts and McIntosh, 1989; McIntosh et al., 1992). The stoichiometry of binding of TNP-8N_3-ATP is greater than that of the mono- and di-phospho species. In contrast to the parent compounds, however, TNP-8N_3-ATP is a slow but well-coupled substrate of the pump (McIntosh and Woolley, 1994). Exposure to light results in the specific covalent attachment of all three nucleotides on K492. The covalently bound nucleotide exhibits superfluorescence on phosphorylation of the enzyme to E_2P. The nucleotide is located at the catalytic site as addition of Ca^{2+} to the enzyme derivatized with TNP-8N_3-ATP induces the hydrolysis of the γ-phosphate and partial transport at low pH. These results definitely establish that nucleotides can remain bound at the catalytic site during the transport cycle, in much the same way as fluorescein can when attached to K515.

FITC derivatization of Na–K-ATPase blocks high-affinity ATP effects, but does not inhibit low-affinity nucleotide effects. For example, high concentrations of ATP still inhibit p-nitrophenyl phosphate hydrolysis of FITC Na–K-ATPase, a reaction which, as mentioned above, is carried out in the E_2 conformation in the presence of K^+ (Scheiner-Bobis et al., 1993). While low-affinity ATP inhibition of dephosphorylation is absent in the FITC enzyme, TNP-ADP inhibition of this reaction remains. With the SR Ca-ATPase, derivatization with FITC blocks ATP acceleration of dephosphorylation (Champeil et al., 1988), and yet a low-affinity site is observed by NMR (Clore et al., 1982). The structures of the ATP bound at the catalytic site and low-affinity site determined in the latter study are similar and show an anticonformation about the glycosidic bond, and a conformation of the ribose of the 3'-endo N type, similar to the structure of the Co-ATP complex with the Na–K-ATPase determined by Grisham and coworkers mentioned above. In cardiac SR Ca-ATPase, FITC blocks Ca-dependent ATP hydrolysis but not GTP hydrolysis, an activity not activated by Ca^{2+}, and AMPPNP but not GMPPNP protected against FITC derivatization (Tate et al., 1991). However, AMPPNP binding to a high-affinity site on the FITC enzyme inhibited GTP hydrolysis.

In the Na–K-ATPase, the simultaneous binding of ATP to two sites with a stoichiometry of 2/ouabain binding site is seen from the kinetics of ouabain release from phosphoenzyme formed from ATP or P_i (Askari et al., 1988). The affinities of the sites for ATP were much lower than that of the unphosphorylated catalytic site.

Peluffo et al (1992) have found that at high ATP concentrations (1 mM) the stoichiometry of initial phosphorylation is 2.4 mol/ mol ouabain bound, 2–4-fold higher than that observed during steady-state hydrolysis or at 10 μM ATP. This superphosphorylation is prevented by FITC labeling and has stability characteristics of an acyl phosphate and suggests that there may be two phosphorylatable sites per monomer, one at a low-affinity ATP-binding site. Altered kinetic characteristics of the first cycle compared with the steady state may explain discrepancies between the rate of phosphoenzyme hydrolysis measured initially and that calculated during steady-state hydrolysis of ATP (Rossi and Nørby, 1993). Superphosphorylation has also been observed with the SR Ca-ATPase using UTP as substrate (Ferreira and Verjovski-Almeida, 1988).

Incubation of SR Ca-ATPase with high concentrations of MgATP, in the presence of activating concentrations of Ca^{2+}, results in the binding of the nucleotide in a site which is only exchangeable with the medium under turnover conditions (Aderem et al., 1979).

Recently other tight-binding inhibitors of SR Ca-ATPase have been found. Fluoride binds tightly to the enzyme in the presence of Mg^{2+} and inactivates ATPase activity (Murphy and Coll, 1992; Coll and Murphy, 1992; Kubota et al., 1993). ATP only partially protected against inhibition (Murphy and Coll, 1992). Most recent results indicate that 2 mol Mg and 4 mol F bind per mol phosphoenzyme, ATP binding site, vanadate binding site, and 2 mol Ca^{2+}, to produce complete inactivation (Daiho et al., 1993). Ca^{2+} binding blocks the binding of Mg^{2+} and F^- into the tight sites. The results can be interpreted in terms of two $Mg(F)_2$ sites per active ATPase or one site on an active ATPase and another on a catalytically silent ATPase, both of which are in communication through the Ca^{2+} sites.

Thapsigargin is a very tight-binding inhibitor of the SR Ca-ATPase (Sagara and Inesi, 1991; Davidson and Varhol, 1995). As in the case of ouabain and the Na–K-ATPase, complete inhibition is obtained with 1 mol/mol phosphoenzyme, indicating that only active ATPase binds thapsigargin.

In summary, the stoichiometry of phosphorylation of purified, highly active preparations of both SR Ca-ATPase and Na–K-ATPase, and the stoichiometry of tight-binding inhibitors, coupled with the failure to increase the levels when the preparation is monomerized in detergent, suggest that the pumps operate as monomers of the catalytic polypeptide. A low level of phosphoenzyme in a pure preparation is probably the result of nonoptimal conditions for maximizing phosphoenzyme levels and the presence of inactive ATPase. Nucleotides can reside close to, or gain access to, the catalytic site following phosphorylation, and this binding is probably the principal source of regulation of the catalytic cycle. Under certain conditions two nucleotides appear to be able to be accommodated at the active site and this could fit in with the rather large, open binding site witnessed in the previous section, and possibly the need to engage different domains at the active site at different stages of the catalytic cycle. Interactions between ATPase polypeptides through specific contacts may also play a role, and the presence of an allosteric ATP binding site cannot be ruled out.

III. INTRON/EXON BOUNDARIES AND PROTEIN DOMAINS IN THE LARGE CYTOPLASMIC LOOP

The genes of higher eukaryotes are usually composed of protein-encoding regions (exons) interrupted by larger noncoding regions (introns). It has been proposed that exons encode functional units, protein domains, or stable structural elements, which could reflect their origins as separate structural/functional entities (Gilbert, 1978; Blake, 1978; Go, 1983; Rogers, 1985). Analysis of kinase genes has revealed distinct exon/intron patterns corresponding to catalytic and nucleotide/cofactor domains and secondary structural elements within these domains (see Michelson et al., 1985; Lee at al., 1987). For example, in the phosphoglycerate kinase gene the two mononucleotide subdomains of the ATP-binding region are separated by an intron and most of the β-strands of each subdomain are also separated by introns (Michelson et al., 1985). Since many bacterial and yeast enzymes with the same function exhibit the same folding pattern as their eukaryotic counterparts, the split gene organization could have been present in the original ancestral cells and the prokaryotes and lower eukaryotes lost the introns in the course of evolution (Doolittle, 1978). Recent analysis of 32 ancient proteins suggests modular structure (de Souza et al., 1996).

An alternative theory is that introns were gained later, beginning just prior to the evolution of eukaryotes, by migrating fragments of DNA or transposons which randomly inserted into the genome (Stoltzfus et al., 1994). In the latter case, correspondence between exons and structural/functional domains would be coincidental, except for the possibility that there was some evolutionary advantage to the survival of structural/functional domains as intact lengths of DNA.

Assigning exons to distinct protein domains is complicated by more recent random intron insertions, loss of the putative originals, or shifts in the position of the introns. In general, the genomes of higher eukaryotes show a greater intron insertional complexity when compared with lower eukaryotes, such as plants.

Sequence analysis of many P-type ATPases has shown that four phylogenetic groups can be distinguished: 1. the various distantly related bacterial pumps and including a Cu-ATPase of humans; 2. the Ca^{2+} pumps of protozoa, yeast, endoplasmic reticulum, and sarcoplasmic reticulum, as well as a bacterial Mg-ATPase; 3. the proton pumps of plants, yeast, and fungi; and 4. the Na–K and H–K pumps of higher eukaryotes (Fagan and Saier, 1994). During eukaryotic evolution, Ca pumps of intracellular vesicles and proton pumps of the plant plasmalemma may have evolved first and separately, from a Mg-type bacterial pump. The Ca plasmalemma pumps could have arisen from the intracellular forms, later the Na–K pumps (perhaps to regulate animal cell volume, Stein, 1995), and then the H–K pumps. In the introns-early theory, this would mean that intron positions common to both the proton pump of plants and the intracellular Ca^{2+} pump could indicate the boundaries of original ancestral structural and functional domains, and most changes from the ancestral gene organization would appear in the genome of Na–K and H–K pumps. In the introns-late theory, insertions in the early eukaryotic gene would have been random, except to the extent that splitting of structural/functional units in the gene was unfavorable, and the later evolving pumps would exhibit greater randomization. New insertions into the latter pumps should bear no relationship to patterns in the older genes.

The intron/exon boundaries of the genes of the large cytoplasmic loop of H-ATPase of the tobacco plant *Nicotiana plumbaginifolia* (Perez et al., 1992),

Ca-ATPases from *Artemia franciscana* (SERCA2) (Escalante and Sastre, 1994), and rabbit fast-twitch sarcoplasmic reticulum (SERCA1) (Korczak et al., 1988) and of χ-1 subunit of human Na–K-ATPase (Ovchinnikov et al., 1988) are compared in relationship to their primary, aligned structures in Figure 1. The protein sequences are drawn to scale and insertions or deletions of two or more amino acids are indicated.

All six of the intron positions of the plant gene can be traced with some confidence in one or more of the other genes (dotted lines). This strongly suggests that the intron positions of the plant were present in the pump gene ancestral to all eukaryotes. In the brine shrimp gene there are three additional introns, all of which can be traced to the rabbit gene and in two cases can be traced to the human Na-K-ATPase gene. The rabbit SR Ca-ATPase gene has an extra intron before the TICAL sequence that is traceable to the Na–K-ATPase gene. The latter has the same number of introns in the large cytoplasmic loop as the SR Ca-ATPase gene, but insertions and deletions in the Na–K-ATPase gene in places close to intron boundaries of the other genes suggest that introns have been lost and gained, as well as there being shifts in their position. One example is the intron immediately after the DKT sequence in the first three pumps, which has been lost in the Na–K-ATPase gene. The intron positions between the DRKS and FAR sequences of the Na–K-ATPase gene are not easy to trace.

The longer animal proteins necessitated introducing several gaps into the plant sequence to achieve the alignments. The gaps are close to intron positions, either in the plant gene or in the others. In fact, nearly every gap shown in the plant protein can be aligned with an intron in one of the animal genes. Evidently, intron/exon boundaries are favored points for expansion of the protein, if permitted. There are introns in the conserved membrane/stalk regions in which elaborations were disal-

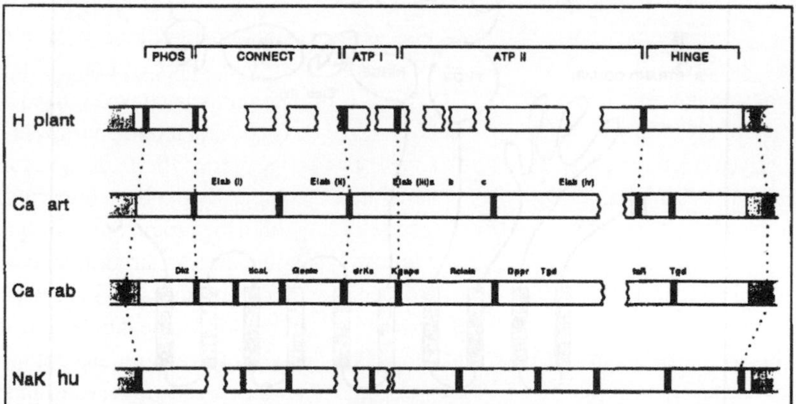

Figure 1. Intron positions in the large cytoplasmic loops. The sequences are aligned according to the position of the amino acids shown in capitals along the SR Ca-ATPase sequence.

lowed. The larger of the elaborations in the animal proteins have been labeled Elab (i), Elab (ii), and so forth, and can be expected to be outcrops on the protein surface and antigenic.

Overall, the evolution of the eukaryotic pumps from a common ancestor is clear and all the introns of plants were apparently gained before the split into H-ATPases and Ca-ATPases and therefore were in the ancestral eukaryotic gene. Whether they were present in the earlier primordial gene, prior to the development of archaebacteria and eubacteria, cannot be addressed. However there is some correspondence between predicted domain structure of the SR Ca^{2+}-ATPase based on sequence analysis and other criteria (MacLennan et al., 1985) and the intron positions in the plant gene, and therefore there may be some justification for categorizing these exons as domains. For this reason and for ease of referral, the exons of the plant gene have been given specific names and are labeled the Phosphorylation, Connect, ATP I, ATP II, and Hinge domains. The Connect, ATP I, and ATP II domains are split virtually in the middle by smaller gaps or elaborations, which, as noted above, can be traced to intron positions in the other genes. If the intron-early theory is correct, these could indicate separations into subdomain structure. The putative domain sequence that emerges is shown in the context of the two-dimensional structure of the entire sequence in Figure 2. How these sectors might be constituted and organized into an active site is examined in the light of information from a variety of sources in the next section.

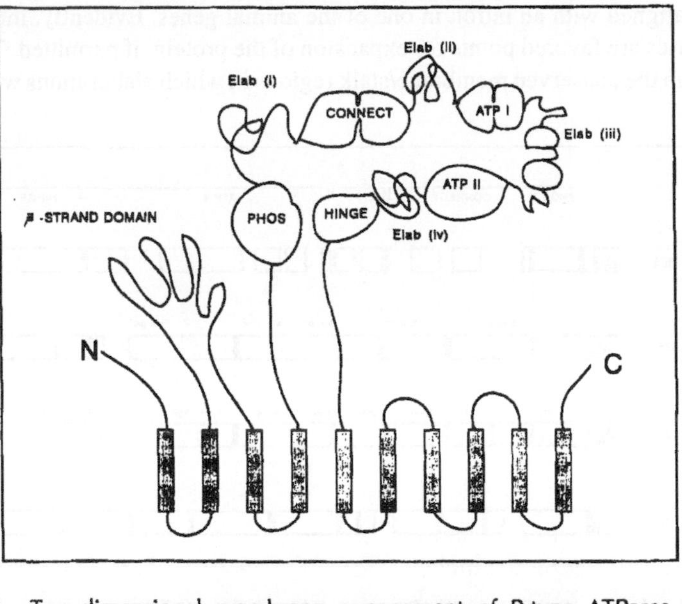

Figure 2. Two-dimensional membrane arrangement of P-type ATPases showing domain structure of the large cytoplasmic loop based on intron positions and elaborations seen in the larger pumps.

IV. STRUCTURAL ELEMENTS OF ATP BINDING AND CATALYSIS

In this section the affinity labeling, residue reactivity, antibody binding sites, and site-directed mutagenesis studies are examined with a view to trying to place the active site residues and structural components into a coherent three-dimensional structure.

The secondary structural predictions shown in the figures of this section are for the SR Ca-ATPase and have been obtained mainly from MacLennan et al (1985), Green (1989), and the more recent unpublished predictions of N.M. Green (personal communication).

First a word of caution. It is now apparent that affinity-labeling results can give inaccurate answers. For example, in the recA protein, Y264 is labeled by $2N_3$-ATP, and yet in the crystal structure of the protein with ADP, the residue is nearby but makes no specific contacts (Story and Steitz, 1992). The crystal structure of F1-ATPase of bovine heart mitochondria (Abrahams et al., 1994) has shown that Y311 labeled by $8N_3$-ATP is 5.5 Å away from the terminal phosphate at the opposite end to the pocket in which adenine binds. The labeling by $2N_3$-ATP appeared to be more faithful to the ADP-protein contacts. However, both $2N_3$-ATP and 5'-p-fluorosulfonylbenzoyladenosine (FSBA) label the same residue (Y368), which is in proximity to the 2-position of the adenine ring.

A. β-Strand Domain

The β-Strand domain is highly conserved and must play a fundamental role in the pumping mechanism of all pumps. The region is primarily implicated in the ATP binding site through affinity labeling of the SR Ca-ATPase (Ferreira and Verjovski-Almeida, 1988) and through the location of three conserved PGD, TGD, and TGE triplets in the region, motifs which have been linked to ATP binding sites and that recur twice later in the sequence in regions labeled by ATP analogues (see below) (Shull and Greeb, 1988). A role in P_i and vanadate binding is indicated through site-directed mutagenesis studies of SR Ca-ATPase and H-ATPase, and on vanadate resistance in the H-ATPases. The presence of a low-affinity ATP-modulatory site in this region is perhaps also suggested from the fact that many mutations in this area in the SR Ca-ATPase and yeast H-ATPase inhibit the E_1P to E_2P transition, a partial reaction of the cycle, which is regulated by ATP in the SR Ca-ATPase.

The region is predicted to contain eight β-strands (Figure 3). The most conserved region is located in and around strands 3, 4, and 5, particularly the five acidic residues, which occur in virtually every pump sequenced to date.

The region of SR Ca-ATPase derivatized by photoaffinity labeling with UTP is on the N-terminal side of the T2 tryptic cleavage site (R198-A199) and most of the extramembranous portion is in the β-Strand domain (Ferreira and Verjovski-Almeida,

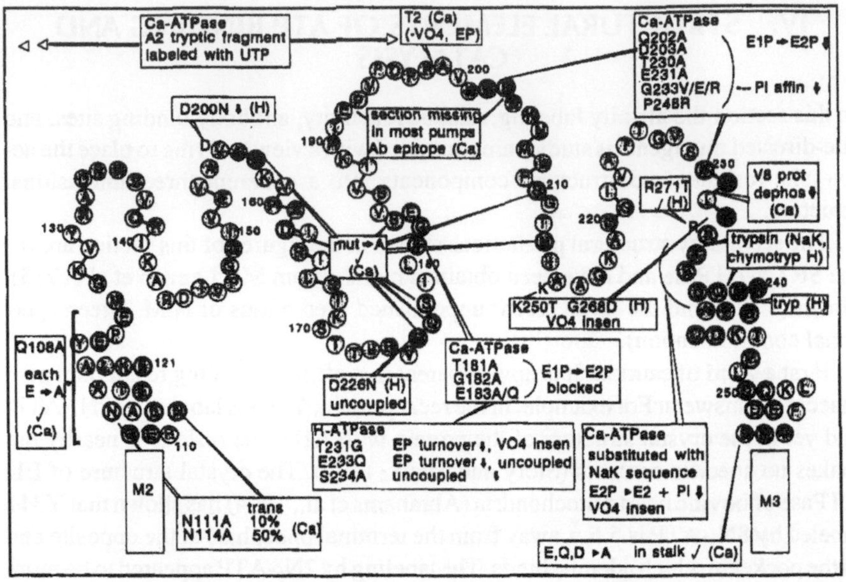

Figure 3. Analysis of β-strand domain. The sequence and numbering in this and other figures is of fast twitch SR Ca-ATPase (Brandl et al., 1986). Residues in a β-strand are connected in a zig-zag manner. The following abbreviations are used in this and other figures: Ab, antibody; +, no effect; insen, insensitive; ↓, inhibited; Ca, SR Ca-ATPase; NaK, Na–K-ATPase, H, yeast or fungal H-ATPase; HK, gastric H–K-ATPase; trans, transport; affin, affinity; EP, phosphoenzyme;prot, protease; dephos, dephosphorylation; lab, label;XL, cross-link; other abbreviations are in the text.

1988). UTP phosphorylates the SR Ca-ATPase to levels twice that of ATP at high Ca concentrations. Fluorescein 5'-isothiocyanate (FITC) derivatization of the ATPase blocks phosphorylation with both substrates, although phosphorylation with UTP is less sensitive at intermediate levels of derivatization. Irradiation of the ATPase with 3 mM UTP and high divalent cation concentration together with 5 mM dithiothreitol for 2 h results in covalent derivatization of approximately 1/10 of the phosphorylatable ATPase. At shorter irradiation times (five minutes), conditions which label approximately 1/100 of phosphorylatable ATPase, the labeling is competitively inhibited by 100 μM ATP. TNP-ATP (50 μM) is a more effective inhibitor but prior derivatization with FITC is ineffective. The label is located on the A2 tryptic fragment following irradiation for forth-five minutes.

There are several concerns. The low level of photolabeling means that only a small proportion of ATPases participated in the reaction. The concentration of UTP was in the millimolar range, which is high if one wants to obtain specific labeling, although the dithiothreitol may help to prevent nonspecific labeling. The localization of label to a particular fragment was done with long irradiation times, conditions under which the protective effect of nucleotide was not demonstrated.

Nevertheless, labeling did favor the A2 fragment and the lack of protection by FITC could indicate a regulatory nucleotide site which is phosphorylated under certain conditions.

There are two segments in this region of the SR Ca-ATPase which, when mutated, decrease enzyme activity due to inhibition of the E_1P to E_2P transition. The first occurs in and around the central TGE triplet. Changes T181A, G182A, E183A, and E183Q produce ATPases that display normal phosphorylation by ATP and P_i but with greatly reduced decay of the ADP-sensitive E_1P (Clarke et al., 1990a). Mutation S184A and changes to the surrounding conserved aspartates (D157N, D162N, D176N) are without effect, suggesting a non essential role in catalysis. In the yeast H-ATPase, mutation of the equivalent glutamate, E233Q, has similar effects, although some uncoupling of ATPase activity and H^+ transport is observed (Portillo and Serrano, 1988; conclusions revised in Serrano et al., 1992). A change to the next serine, S234A, results in greatly stimulated ATPase activity and uncoupling (Portillo and Serrano, 1989). Uncoupling is also observed with the D226N change. Change T231G results in 50% inhibition of ATPase activity, probably through inhibition of the E_1P to E_2P transition since the phosphoenzyme levels are raised, and there is a loss in vanadate sensitivity (Portillo and Serrano, 1989). D200N causes 50% inhibition of activity (Portillo and Serrano, 1988).

The second region centers on SR Ca-ATPase G233, which when mutated to valine, glycine, or arginine results in zero transport activity and defective processing of the ADP-sensitive E_1P (Andersen et al., 1989). Although phosphorylation levels from ATP are normal, those in the backward direction of catalysis with P_i are reduced due to greatly diminished affinity of the enzyme for P_i. In this case the changes may be the result of either an inhibition of both E_1P to E_2P and the E_1 to E_2 transitions or destabilization of E_2P. Changes to surrounding residues Q202A (40% inhibition), D203A, T230A, E231A, and P248R (in the stalk section) inhibited partially and S210A and D213A were without effect.

In support of a role for this region in phosphoenzyme processing is the finding that cleavage of the preceding bond between E231 and I232 of the sequence TEIG with V8 protease permits phosphorylation with ATP but blocks dephosphorylation (le Maire et al., 1990).

Substitution of the next few amino acids, KIRDQMA, with the equivalent residues in the Na–K-ATPase results in 30% inhibition of transport due to an inhibition of E_2P dephosphorylation (Andersen and Vilsen, 1993). The enzyme is insensitive to vanadate but affinity for P_i is normal, suggesting that the transition state is destabilized.

Interaction of the β-Strand domain with vanadate and P_i is also seen in the protection these anions afford to cleavage of SR Ca-ATPase at the T2 cleavage site (between R198 and A199) (Dux et al., 1985). Vanadate insensitivity, as well as a 3-fold lowering of K_m with decreased ATPase activity, is associated with a K250T or G268D change in yeast H-ATPase (Ulaszewski et al., 1987; Ghislain et al., 1987). The authors speculated that this could be due to the mutation inhibiting the E_1 to E_2

transition. The stimulatory effect of P_i binding on ATP hydrolysis is differentially modified by the mutations (Goffeau and de Meis, 1990).

A superficial position of some amino acid residues in this segment is indicated by the location of tryptic and chymotryptic cleavage sites in yeast H-ATPase and Na-K-ATPase just after the critical glycine (Davis and Hammes, 1989; Jørgensen and Petersen, 1982). Cleavage of the E_1-dependent chymotryptic site of Na–K-ATPase inhibits phosphoenzyme processing (Jørgensen and Andersen, 1988).

Both stalk regions contain a surfeit of acidic residues which individually do not appear to play a major role in pump function as single mutations to alanine have little effect (Clarke et al., 1989). Two asparagines in the first stalk, N111A and N114A, were more important with the mutants exhibiting 10 and 50% inhibition of wild type activity respectively.

In conclusion, the β-Strand domain is a highly conserved region with acidic residues in loops and strands which could ligate divalent cations. A direct role of these residues in phosphoryl transfer has been ruled out but there is a strong indication that this domain is an integral part of the phosphorylation domain. It is possible that the β-strands form a contiguous sheet with the β-strand just prior to the DKT sequence, and act as a brace for the latter (see below). Many residues in the β-Strand domain are critical for phosphoenzyme processing, and probably play a secondary role in P_i and vanadate binding. The former partial reactions are accelerated by high concentrations of ATP in SR Ca-ATPase and the location of a low-affinity ATP binding site in this domain is a possibility reinforced by the labeling of this region with UTP.

B. Phosphorylation Domain

The sequence DKTGTLTXN/G is almost invariant among P-type ATPases. Homology with other ATP binding proteins, in particular the P-loop motif of GXXXXGKT/S which binds the di- or triphosphate moiety of many proteins (Walker et al., 1982), is not obvious. A resemblance to the ATP binding site of *E coli* F_0F_1 H^+-ATPase (ITSTKTGSIT) has been explored, but without success (Noumi et al., 1988).

The crystal structures of human protein tyrosine phosphatase 1B (Barford et al., 1994) and *Yersinia* protein tyrosine phosphatase (Stuckey et al., 1994) have been elucidated. Protein tyrosine phosphatases all contain an 11-residue sequence motif (I/V)HCXAGXXR(S/T)G where the cysteine is phosphorylated during catalysis and the arginyl residue plays an essential role in the mechanism. The sequences of some phosphatases in this region are compared with the DKT region of P-type ATPases in Figure 4. The residues phosphorylated in the P-type ATPases and phosphatases are aligned. The position of the β-strand preceding the phosphorylated cysteine and the α-helix just after are indicated in the phosphatases whose structures are known. The crystal structure of this region of the *Yersinia* enzyme is shown in Figure 5A.

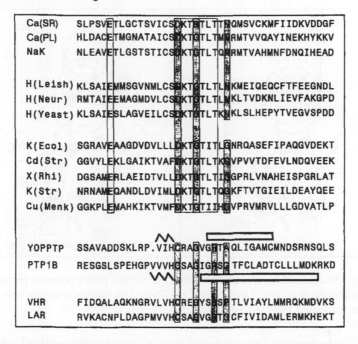

```
Ca(SR)      SLPSVETLGCTSVICSDKTGTLTTNQMSVCKMFIIDKVDDGF
Ca(PL)      HLDACETMGNATAICSDKTGTLTMNRMTVVQAYINEKHYKKV
NaK         NLEAVETLGSTSTICSDKTGTLTQNRMTVAHMNFDNQIHEAD

H(Leish)    KLSAIEMMSGVNMLCSDKTGTLTLNKMEIQEQCFTFEEGNDL
H(Neur)     RMTAIEEMAGMDVLCSDKTGTLTLNKLTVDKNLIEVFAKGPD
H(Yeast)    KLSAIESLAGVEILCSDKTGTLTKNKLSLHEPYTVEGVSPDD

K(Ecol)     SGRAVEAAGDVDVLLLDKTGTITLGNRQASEFIPAQGVDEKT
Cd(Str)     GGVYLEKLGAIKTVAFDKTGTLTKGVPVVTDFEVLNDQVEEK
X(Rhi)      DGSAMERLAEIDTVLLDKTGTLTIGGPRLVNAHEISPGRLAT
K(Str)      NRNAMEQANDLDVIMLDKTGTLTQGKFTVTGIEILDEAYQEE
Cu(Menk)    GGKPLEMAHKIKTVMFDKTGTIIHGVPRVMRVLLLGDVATLP

YOPPTP      SSAVADDSKLRP.VIHCRAGVGRTAQLIGAMCMNDSRNSQLS
PTP1B       RESGSLSPEHGPVVVHCSAGIGRSGTFCLADTCLLLMDKRKD

VHR         FIDQALAQKNGRVLVHCREGYSRSPTLVIAYLMMRQKMDVKS
LAR         RVKACNPLDAGPMVVHCSAGVGRTGCFIVIDAMLERMKHEKT
```

Figure 4. Sequence alignments of P-type ATPases and protein tyrosine phosphatases. Sequences of P-type ATPases have been taken from Green (1989), excepting for Cu-ATPase which comes from Vulpe et al. (1993). The *Yersinia* protein tyrosine phosphatase (YOPPTP) sequence is taken from Zhang et al (1994) and the secondary structure from Stuckey et al (1994), that of the Human protein tyrosine phosphatase (PTP1B) with secondary structure from Barford et al. (1994), the human dual-specific phosphatase, vaccinia H1-related (VHR), from Denu et al. (1995), and the leukocyte antigen-related protein (LAR) is also from Zhang et al (1994). A thick zig-zag line indicates β-strand and a reactangle indicates α-helix.

In the structure of the phosphatases there is a critical glycine residue three residues after the cysteine which allows a kink in the phosphate binding loop. In the P-type ATPases this position is also occupied by an invariant glycine. The arginine residue which follows the cysteine in the *Yersinia* enzyme plays a role in binding the phosphate, and the invariant lysine in the cation pumps could function similarly. The critical catalytic arginyl residue is substituted by threonine in the pumps. These phosphatases do not use Mg^{2+} for catalysis and it appears that the usual role of the divalent cation has been obviated by the positively charged guanidinium moiety. In the pumps the latter grouping could be replaced by Mg^{2+} and the shorter threonine would permit this replacement. A glycine often appears two residues after the arginine of the phosphatases, and in the bacterial pumps the threonine is almost invariably followed by glycine, otherwise in the other pumps it is a highly conserved asparagine.

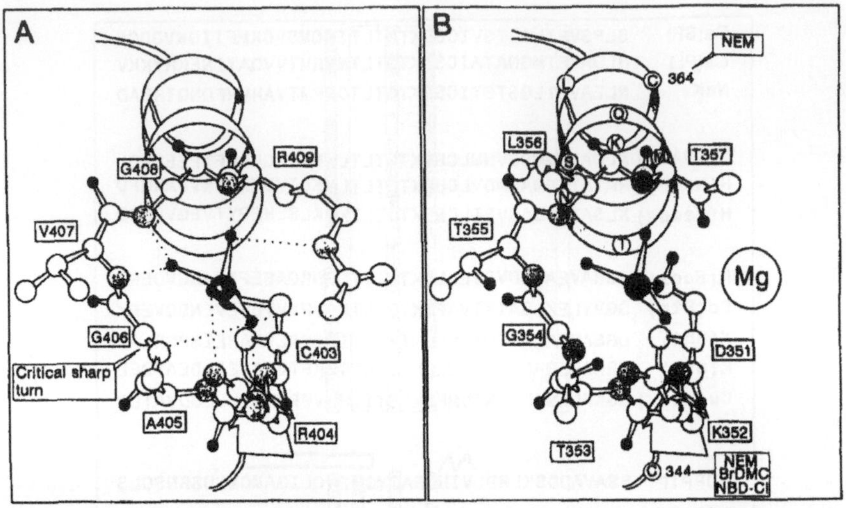

Figure 5. Structure of the phosphorylation site of *Yersinia* protein tyrosine phosphatase and hypothetical structure of that of SR Ca-ATPase. The *Yersinia* structure with bound tungstate is adapted from Stuckey et al (1994).

It should be noted that this conserved catalytic thiol loop has evolved independently in several different protein families including the dual-specific phosphatases, low molecular weight tyrosine-specific phosphatases, receptor type tyrosine-specific phosphatases and high molecular weight soluble tyrosine-specific phosphatases (PTPases), and rhodanases (Stuckey et al.; Su et al., 1994). The loop has homologies with the glycine rich P-loop of many ATP binding enzymes which may also have evolved independently in different families.

A possible structure of this region of the SR Ca-ATPase modeled on the *Yersinia* fold is shown in Figure 5B.

There are two cysteines (C344 and C364) in close proximity to the phosphorylation site of SR Ca-ATPase which are selectively reactive to maleimides (Saito-Nakatsuka et al., 1987; Kawakita and Yamashita, 1987) and their possible positions in this model are indicated. N-ethylmaleimide (NEM) modification of either one of the -SH groups (C344 or C364) has no effect on activity whereas modification of both is inhibitory by slowing the E_1P to E_2P transition (Kawakita et al., 1980; Kawakita and Yamashita, 1987; Davidson and Berman, 1987). C344 is selectively labeled with 7-chloro-4-nitrobenzo-2-oxa-1,3-diazole (NBD-Cl) (Wakabayashi et al., 1990), derivatization blocks the same transition, and the inhibition is reversed by solubilizing in $C_{12}E_8$, in this way similar to the NEM modification. Several fluorescent maleimides and spin-labeled maleimides probably also react with C344 and C364 (Saito-Nakatsuka et al.; Horvath et al., 1990; Bigelow and Inesi, 1991). Energy-transfer and spin-quenching characteristics suggest that the two cysteines may be up to 36Å apart. 4-Bromomethyl-6,7-dimethoxycoumarin (Br-DMC) specifically modifies C344 with 25% inhibition of ATPase activity but without affecting ρ-nitrophenyl phosphate hydrolysis (Stefanova et al., 1992), consistent with the inhibition being due to slowing of the E_1P to E_2P transition.

Thus the catalytic changes brought about by derivatization of these cysteines, namely inhibition of the E_1P to E_2P transition, are similar to the effect of mutations in the β-Strand domain, seen in the previous section. This is consistent with the β-Strand domain being an integral part of the phosphorylation site. Derivatization of the cysteines does not appear to affect ATP binding, consequently they must be located outside of the nucleotide binding site.

The highly conserved ET341 doublet (Figure 4), just before the NEM-reactive cysteine, is important in the complexation of vanadate as mutation S368F in yeast H-ATPase is vanadate-resistant and second-site revertants involved the neighboring glutamate (E367V), as well as V289F in the β-strand domain (Harris et al., 1991).

On the amino side is a buried tryptic cleavage site (R334) which is only exposed in the SR Ca-ATPase after a structural change induced by repetitive homogenization in EGTA, possibly associated with the loss of an occluded Ca^{2+} (Lesniak et al., 1994). This same cleavage site is uncovered in the Na–K-ATPase (K354) following labeling of the glycine in the conserved KGAPE sequence with $2N_3$-ATP (Tran et al., 1994). However, at least portions of the surrounding residues are normally medium exposed as the sequence V333-S338 of SR Ca-ATPase includes a monoclonal antibody epitope (Colyer et al., 1989; Tunwell et al., 1991). Binding of the antibody has no effect on activity.

N-(1-pyrene)maleimide modifies two ouabain-sensitive cysteine residues of lamb kidney Na–K-ATPase, one of which is C367 (the other is C656, see later), two residues on the amino side of the D369 which is phosphorylated (Kirley and Peng, 1991). The modification was performed in 5 mM P_i, suggesting the residue is pointing away from the phosphorylated aspartate, possibly in the same hydrophobic pocket to which NEM binds in the SR Ca-ATPase. Mutation of the same residue (C369) to serine in rat Na–K-ATPase has no effect on activity, apparent ATP-binding affinity, nor ouabain-binding affinity (Lane et al., 1993).

Just beyond the C364 of SR Ca-ATPase is K371 which reacts with 7-amino-4-methylcoumarin-3-acetic acid succinimidyl ester (AMCA) (Stefanova et al., 1993a) and C377 which cross-links intermolecularly with the same residue of a neighboring ATPase (Yamasaki et al., 1990).

C. Connect Region

This region has been condensed in the bacterial pumps and all Cu pumps to what appears to be three loops and two α-helices. The first is a hydrophobic helix which aligns with the VELATICAL sequence of SR Ca-ATPase and the second an amphipathic helix aligning with GEATETALTTLVE of the same pump. In the loop prior to the first helix of SR Ca-ATPase is the sequence KNDKPI, binding site of phospholamban in the cardiac pump (Toyofuka et al., 1994). The two putative helices are divided by an intron in the Artemia and rabbit SR Ca-ATPases, and the Na–K-ATPase (Figure 1).

The importance of this region in ATPase function and nucleotide binding is indicated from the location of a K-dependent T1 tryptic cleavage site just after the first helix of the Na–K-ATPase which inhibits enzyme turnover (Jørgensen, 1975). This cleavage of the K^+-bound ATPase is blocked by ATP binding to a low-affinity site. The cleavage has no effect on high-affinity ATP binding nor phosphorylation of the ATPase, and FITC still blocks these reactions (Jørgensen, 1977; Zolotarjova et al., 1995). However, it disrupts the low-affinity site as seen in the shift in $K_{0.5}$ (0.05 to 2 mM) for ATP protection of the FITC reaction (Zolotarjova et al., 1995). This disruption is carried through to all functions involving the E_2 form of the enzyme as well as certain partial reactions involving the E_1 species.

The glutamate at the beginning of the first helix is selectively derivatized with 5-(bromomethyl)fluorescein in a manner that is insensitive to ATP binding and which results in partial inhibition of activity (probably due to inhibition of the E_1P to E_2P transition) (Stefanova et al., 1993b). The probe does not report on the binding of any ligand including nucleotides nor on phosphorylation.

The end of the second putative helix in the Na–K-ATPase contains three cysteine residues, the first two of which may be linked together (Gevondyan et al., 1993). However, mutating the three individually to alanines has no effect on activity nor $K_{0.5}$(ATP) (Lane, 1993). The third cysteine is labeled by 5-iodoacetamidofluorescein (5-IAF) and reports on the E_1P to E_2P conformational transition without inhibiting activity (Tyson et al., 1989).

In summary, the connect region is a highly variable part of the protein except for what appears to be two helical stretches. The two putative helices are joined by a loop with a cleavage site sensitive to ATP binding at a low-affinity site. In this regard it may not be a coincidence that phospholamban exerts at least part of its modulatory effects on enzyme turnover of cardiac SR Ca-ATPase in this region by altering the E_2 to E_1 transition (Lu et al., 1993), a step, as we have seen, that is activated by ATP binding in SR Ca-ATPase. The partial inhibition by the fluorescein label on E439 may link this section conformationally to the β-strand and phosphorylation domains.

D. ATP I Domain

This domain is demarcated in the beginning by an intron found in the plant H-ATPase and *Artemia* and rabbit SR Ca-ATPases. The intron position is the site of elaborations in the Ca pumps (Elab (ii)), relative to the plant H pumps and higher animal Na-K pumps (Figure 1). Huge inserts exist in the malarial Ca and Na-K pumps (Krishna et al., 1993; Kimura et al., 1993). The predicted secondary structure and analysis of the amino acids in this region is shown in Figure 6.

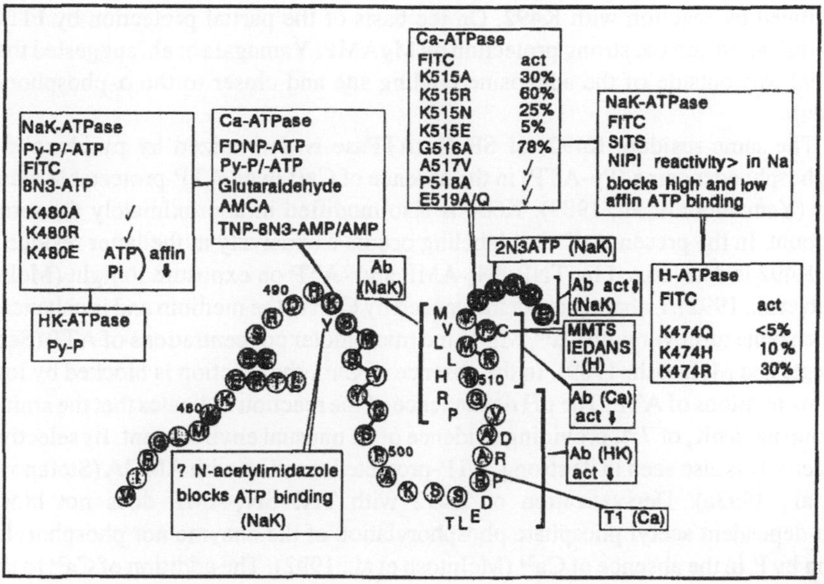

Figure 6. Analysis of ATP I domain. Note the demarcations for domains are taken from the intron positions of plant H-ATPase shown in Figure 1.

At the amino terminal end of this section, in Elab (ii), is a cysteine residue in the SR Ca-ATPase which becomes more reactive to iodoacetate following derivatization of another SH group (presumably within the ATP binding site) with the disulfide of 3'(2')-O-biotinyl 6-S-inosine triphosphate, suggesting a nucleotide-dependent structural change in this region (Kison et al., 1989). In keeping with this possibility, there is a tryptic cleavage site of the Na–K-ATPase located here (PK/IV) which is protected from cleavage by derivatization of the glycine in the KGAPE sequence with $2N_3$-ATP (Tran et al., 1994).

The beginning of formal secondary structure is not clear and is shown as a short helix, consisting of EFTLEF in the SR Ca-ATPase. The phenylalanine, F487, is highly conserved and can be traced through to bacterial pumps. Five residues further on is K492, also conserved, which is the point of attachment of a number of affinity labels in various pumps. The number of amino acids between the conserved F and K is constant in most pumps. The sequence of four charged residues in the SR Ca-ATPase, RDRK, suggests that this section is a loop projecting into the aqueous medium.

SR Ca-ATPase is inhibited on reaction of pyridoxal 5'-phosphate (Py-P) with K492 (Murphy, 1977, Yamagata et al., 1993). The reaction is inhibited by nucleotides, including MgAMP, but only partially by FITC labeling of K515. In fact, more extensive reaction results in the labeling of K515 as well as K492. Ca-dependent acetyl phosphate hydrolysis and P_i phosphorylation in the absence of Ca^{2+} are not inhibited by reaction with K492. On the basis of the partial protection by FITC modification and the strong protection by MgAMP, Yamagata et al., suggested that K492 was outside of the adenosine binding site and closer to the α-phosphoryl group.

The same residue, K492, of SR Ca-ATPase is derivatized by pyridoxal 5'-diphosphoadenosine (Py-ATP) in the absence of Ca^{2+} in an ATP-protectable manner (Yamamoto et al., 1989). K684 is also modified in approximately the same amount. In the presence of Ca^{2+} labeling occurs exclusively at the latter residue.

K492 is derivatized by TNP-8N_3-AMP and -ATP on exposure to light (McIntosh et al., 1992). Labeling is not influenced by Ca^{2+} in the medium and is enhanced by enzyme turnover with Ca^{2+}, Mg^{2+}, and micromolar concentrations of ATP (Seebregts and McIntosh, 1989). In the absence of Ca^{2+}, the reaction is blocked by low concentrations of ATP. The pH dependence of the reaction indicates that the amino group has a pK_a of 7.4, providing evidence of an unusual environment. Its selective reactivity is also seen in its strong, ATP-protected reaction with AMCA (Stefanova et al., 1993a). Derivatization of K492 with TNP-8N_3-AMP does not block Ca-dependent acetyl phosphate phosphorylation of the enzyme nor phosphorylation by P_i in the absence of Ca^{2+} (McIntosh et al., 1992). The addition of Ca^{2+} to the enzyme derivatized with TNP-8N_3-ATP results in the accelerated hydrolysis of the γ-phosphoryl group, indicating that the phosphoryl groups are correctly aligned in the active site with respect to D351, which is phosphorylated, and that phosphoryl transfer is tolerant of covalent attachment of the adenyl moiety to the lysyl residue (McIntosh and Woolley, 1994). However, the transport function of the enzyme is compromised and only a small amount of Ca^{2+} is released to the lumen at acidic pH. Derivatization of K492 with AMCA has no effect on ATPase activity at high ATP concentrations, verifying the nonessential nature of this lysine and possibly indicating that the bulk of the probe is not within the nucleotide binding site.

Another affinity label, 3'-O-(5-fluoro-2,4-dinitrophenyl)-ATP (FDNP-ATP), also specifically derivatizes Lys-492 of SR Ca-ATPase in the presence of Mg^{2+} with protection afforded by MgATP and Mg- acetyl phosphate and not by MgP_i (Yamasaki et al., 1994). The modification results in inhibition of acetyl phosphatase activity. Molecular modeling showed that the reactive fluoro atom could approach the α-phosphate in a stable nucleotide conformation, suggesting that K492 may be aligned with these oxygens.

K492 of SR Ca-ATPase is cross-linked to K678 by glutaraldehyde in a reaction which is blocked by nucleotide binding, including AMP, to the catalytic site (Ross and McIntosh, 1987a; 1987b; Ross et al., 1991; McIntosh, 1992). The probable mechanism of the reaction suggests that it is a zero distance cross-link, which places these two residues alongside each other in the tertiary structure at the active site. This supports the findings of Yamamoto et al (1989), mentioned above, which places K492 and K684 close together in the absence of Ca^{2+}. Cross-linkage is not affected by Ca^{2+} binding, in keeping with the Py-P (Yamagata et al., 1993) and TNP-8N_3-AMP reactions with K492 (Seebregts and McIntosh, 1989), and the lack of a

Ca^{2+} effect on energy transfer between AMCA (K492) and FITC (K515) (Stefanova et al., 1993a). This shows that the Ca-dependent changes in labeling obtained by Yamamoto et al., are due to changes in nucleotide position caused by alterations elsewhere in the active site, probably associated with K684. P_i and monovanadate have little effect on cross-linkage. Cross-linkage is strongly, but not completely, inhibited by FITC on K515, and it is possible to obtain the cross-linked species with attached FITC, again in agreement with lack of effect of the latter reaction on Py-P labeling of K492 seen above. The cross-link lowers the affinity of the enzyme for ATP approximately 1000-fold, and yet high levels of E_1P are measured due to the complete inhibition of the following step. The cross-link does not affect the binding of acetyl phosphate or p-nitrophenyl phosphate.

The equivalent residue of pig gastric H–K-ATPase, K497, is modified by Py-P in an ATP-and FITC-protectable manner (Maeda et al., 1988, Tamura et al., 1989). Phosphate binding has no effect. The same residue in the lamb kidney Na–K-ATPase, K480, is modified by Py-P and Py-ATP (Hinz and Kirley, 1990; Kaya et al., 1994). ATP protects against derivatization by both compounds. The modifications, carried out in the presence of NaCl, inactivate Na–K-stimulated ATPase activity and ATP-dependent phosphorylation but not phosphorylation from acetyl phosphate (Kaya et al., 1994). Surprisingly, ATP (200 μM) binding to the ATPase derivatized with Py-P or Py-ATP could be detected through fluorescence changes of the probes.

$8N_3$-ATP is a high-affinity substrate of the dog kidney Na-K-ATPase and in the presence of light attaches to K480 (Scheiner-Bobis and Schoner, 1985; Tran et al., 1994b). There is a requirement for Mg^{2+} for photoinactivation and protection is afforded by K^+ (Scheiner-Bobis and Schoner, 1985). The Mg^{2+} effect may be due to the cation changing the structure of the nucleotide directly or it could be due to its interaction with the protein. Mg^{2+} enhances the fluorescence and increases the high-affinity binding of eosin-Y at ATP sites (Skou and Esmann, 1983).

Mutation studies of K480 have been carried out in kidney Na–K-ATPase, and mutation K480E lowers the affinity for both ATP and P_i, but changes to alanine or arginine had no effect (Wang and Farley, 1992). ATPase activity was not affected by any of the mutations. Similarly, mutation of the same residue in the rat α-1 Na–K-ATPase to alanine (K482A) has no effect on ATPase activity, phosphoenzyme formation, or apparent affinity for ATP (Lane et al., 1993).

Overall, the results suggest that K492 is located at the catalytic nucleotide binding site in an unusual environment outside of, but close to, the phosphorylation site and possibly ligating a portion of the ATP molecule. If there is direct interaction with the nucleotide, ligation is not necessary for pump function. In most cases Ca^{2+} binding has no effect on the labeling of this residue. Even fairly large labels attached to this residue still permit ATP binding, showing the large size of the ATP binding pocket. While the residue does not play an important role in phosphoryl transfer, its movement in relation to R687, and probably not to K515, may be important for coupling with the transport sites later in the cycle.

K492 of SR Ca-ATPase is followed by eight residues in an alternating hydroxy/hydrophobic pattern. In most of the other pumps the amino acids are often hydrophobic and very seldom charged. A β-strand is predicted. The serine immediately after K492 is replaced by a tyrosine in Na–K-ATPases and it may be the one modified by N-acetylimidazole, leading to inhibition of high affinity ATP binding but not P_i phosphorylation or ρ-nitrophenyl phosphate hydrolysis in a medium containing K^+ (Argüello and Kaplan, 1990). Following this section of SR Ca-ATPase is the sensitive T1 tryptic cleavage site. This cleavage has no effect on catalytic activity. The superficial position of this region in the SR Ca-, gastric H–K-, and Na–K-ATPases is indicated by the binding of antibodies (epitopes NKMFVK, DPRDP, and PRHLL respectively) (Van Uem et al., 1990, 1991; Tunwell et al., 1991; Malik et al, 1993). The SR Ca-ATPase-directed antibody partially inhibits ATPase activity and its binding is competitive with another antibody binding to epitope P662-LAEQ in Elab (iv). The antibody in the case of the H–K pump has no effect on ATP binding and vice versa, but partially inhibits ATPase activity and phosphorylation by prefentially binding to an E_1K conformational state and presumably slowing a conformational change out of this species.

The conserved KGAPE bend is usually preceded by four hydrophobic residues, predicted to form a β-strand. C472 in the sequence ITCVKGAP of *Neurospora* H-ATPase is modified by 5-(2-((iodoacetyl)amino)ethyl)-aminonaphtheylene-1--sulfonic acid (IAEDANS) and methylmethanethiosulfonate (MMTS), and in the case of the latter reagent without loss of activity (Pardo and Slayman, 1989). Polyclonal antibodies directed against a GAPER peptide inhibit FITC reaction and ATPase activity of Na–K-ATPase (Xu and Kyte, 1989).

Reaction of FITC with the lysine of KGAPE is the best characterized chemical modification of the P-type ATPases. Covalent modification of K515 of SR Ca-ATPase (Mitchinson et al., 1982) occurs with concomitant inactivation of ATPase activity and the reaction is inhibited by high-affinity ATP binding (Pick and Karlish, 1980; Pick and Bassilian, 1981; Murphy, 1988; Champeil et al., 1988). The specificity of the reaction is due to the reagent acting as an affinity label rather than an unusual reactivity of K515 (Murphy, 1988), supporting the idea that the fluorescein moeity mimics the adenine base. FITC reacts with a 1:1 stoichiometry to active ATPases (Andersen et al., 1982, Murphy, 1988, Champeil et al., 1988) and partial modification does not influence the biphasic kinetics of ATPase activity as a function of ATP concentration (Andersen et al., 1982), a finding also observed for the red cell Ca-ATPase (Muallem and Karlish, 1983).

ATP binding to high-affinity sites and low-affinity effects, like ATP activation of dephosphorylation, are blocked by FITC modification (Andersen et al., 1982; Champeil et al., 1988). Nevertheless, at very high concentrations of ATP there is some depression of the fluorescence of attached FITC and a small acceleration of dephosphorylation (Champeil et al., 1988). Thus the steric hindrance to ATP binding caused by FITC is probably only partial, again attesting to the large size of the ATP binding pocket.

FITC modification of SR Ca-ATPase has little effect on acetyl phosphate-dependent Ca^{2+} transport, P_i-dependent phosphorylation, and vanadate binding in EGTA (Pick and Bassilian, 1981; Pick, 1981a; Pick, 1982; Champeil et al., 1988), showing the unessential nature of K515 in the transport mechanism in both directions of catalysis, a result somewhat at variance with site-directed mutagenesis results, as we shall see below.

Rather similar results involving FITC are obtained with several other pumps. FITC reaction with Na–K-ATPase occurs at the equivalent residue of canine kidney, K501 (Farley et al., 1984), and is blocked by ATP binding, probably through the direct ligation of K501 with the nucleotide since the inhibition by ATP is much more pronounced than that afforded by major conformational changes (Xu and Kyte, 1989). Na-K-ATPases from the electric ray and brine shrimp react similarly (Ohta et al., 1985). The reaction with the canine pump is not entirely specific and significant labeling also occurs at K480, equivalent to K492 of SR Ca-ATPase, and to a lesser extent at K766 at the beginning of membrane segment M5 (Xu, 1989). All three reactions are protected by ATP. Also, a significant portion of the fluorescein attached to Na–K-ATPase is quenched with a fluorescein-directed antibody (Abbott et al., 1991, Lin and Faller, 1993). Only the non-quenched species, presumably that attached to K501, and possibly K487 and K766, is responsive to ligand binding. The reaction with Na–K-ATPase results in inactivation of Na^+-dependent ATPase activity but not of Na^+-dependent acetyl phosphate hydrolysis (Taniguchi et al., 1988) nor K-stimulated P_i phosphorylation and p-nitrophenyl phosphate hydrolysis (Karlish, 1980). While high-affinity ATP binding is blocked by FITC, high concentrations of ATP are required to inhibit the FITC reaction (Zolotarjova et al., 1995), which suggests that the high-affinity site is deeper than the FITC site, and the low-affinity site perhaps overlaps more with the FITC region. However, this would contradict the fact that low-affinity ATP effects are still observed with the FITC enzyme. For example, high concentrations of ATP inhibit p-nitrophenyl phosphate hydrolysis, and TNP-ADP, which mimics ATP binding to a low-affinity site, inhibits dephosphorylation (Scheiner-Bobis et al., 1993). Derivatization at pH 6.5 or 9.0 leads to different extents of inactivation of Na^+,K^+-stimulated ATP hydrolysis and K^+-stimulated p-nitrophenyl phosphate hydrolysis, the reaction at lower pH inactivating both activities equally whereas at higher pH the former is more sensitive (Sen et al., 1981). Prolonged reaction with FITC also results in further inactivation of p-nitrophenyl phosphate hydrolysis (Karlish, 1980), possibly by reaction with K480.

FITC has been shown to react with equivalent residues of the gastric H–K-ATPase (Farley and Faller, 1985), erythrocyte Ca-ATPase (Filoteo et al., 1987), and *Neurospora* H-ATPase (Pardo and Slayman, 1988).

Other aromatic isothiocyanates, 4-acetamido-4'-isothiocyanatostilbene-2,2'-disulfonic acid (SITS) and N-(2-nitro-4-isothiocyanophenyl)-imidazole (NIPI) react with the same residue of Na–K-ATPase with concomitant inactivation of ATPase activity and in a manner which is protected by adenine nucleotides

(Pedemonte et al., 1992, Ellis-Davies and Kaplan, 1993). ATPase derivatized with the latter reagent neither binds ATP with high affinity nor activates deocclusion of Rb^+ by a low-affinity ATP site, providing evidence of an overlap of the high- and low-affinity sites.

A combination of site-specific (KGAPER) immunoadsorption and competitive labeling using acetic anhydride, FITC, and ATP has been used to determine the reactivity of K501 of canine kidney Na–K-ATPase (Xu and Kyte, 1989; Xu, 1989). The pK_a of the amino group is a normal 10.4, but the reaction with acetic anhydride is slowed 4- 5-fold compared with a model exposed amino group, suggesting it resides in a pocket that forms the active site on the surface of the protein (Xu, 1989).

The neighboring G502 of canine kidney Na–K-ATPase is labeled by $2N_3$-ATP with inactivation of ATPase activity, in a manner protected by ATP binding to a high-affinity site (Tran et al., 1994a). The azido nucleotide is a substrate of the enzyme and presumably G502 is positioned close to the 2-position of the adenine of the natural substrate ATP. The situation is different for H-ATPase of yeast plasma membrane, where $2N_3$-AMP and -ATP preferentially label D560 to K566 (Davis et al., 1990), part of the highly conserved Hinge TGD region, and for the SR Ca-ATPase, where $2N_3$-ADP interacts with T532 and T533 (Lacapère et al., 1993), just after the KGAPE region.

Site-directed mutagenesis of the KGAPE region in SR Ca-ATPase and H-ATPase of *Neurospora* has indicated the importance of some of these amino acids in ATPase function. Changes K515R, K515A, K515Q, and K515E in SR Ca-ATPase resulted in 65, 30, 25, and 5% wild-type transport activity respectively (Maruyama and MacLennan, 1988, Maruyama et al., 1989). All exhibited normal EP levels from ATP and P_i, suggesting that K515 may be more important for ATP-induced or other conformational changes required for energy coupling than for phosphoryl transfer *per se*, a result in line with the inhibition afforded by the GAPER-specific antibody studies with the Na–K-ATPase noted above. A complex role for K515 is also suggested from the finding that a K515A change results in the inhibition of transport with both ATP or acetyl phosphate. A G516A mutation gave 78% Ca^{2+} transport and A517V, P518A, and E519A/Q had no effect. Again, all had normal EP levels in both directions of catalysis.

The importance of the positive charge at the lysyl residue is also seen in the H-ATPase of *Neurospora* (Serrano and Portillo 1990). Changes K474R, K474H, and K474Q exhibited 30, 10, and <5% wild-type ATPase activity respectively. It is evident that the yeast enzyme is more sensitive to changes here than the SR Ca-ATPase. Random *in vitro* mutagenesis of the K474R allele has yielded second site suppressor mutations, one of which (V396I) is located 18 residues away from D378 which is phosphorylated (Maldonado and Portillo, 1995). This shows the importance of the lysine residue/phosphorylation domain interaction during pumping. Other second site mutations were located near K474, the end of M2, and P536L, a fairly conserved proline or glycine in the β-strand just prior to the DPPR segment.

E. ATP II Domain

The KGAPE sequence leads into a less-known and rather variable portion probably constituted as an amphipathic helix (SR Ca-ATPase: VIDRCNYVRVG), a bend (GTTRVPMTGP), amphipathic helix (VKEKILSVIKEW), a bend (GTGRDTLR), and a conserved hydrophobic helix (LALAT) (Figure 7). The bends are substantially smaller in H-pumps of yeast and plants and the bacterial pumps. The first elaboration in SR Ca-ATPase is designated Elab (iii)a, the second Elab (iii)b, and the third Elab (iii)c. The elaborations are antigenic (see below).

The threonine doublet, T532 and T533, in the SR Ca-ATPase, is labeled with $8N_3$-ADP in a Ca^{2+} medium (Lacapère et al., 1993). The authors considered that the segment between KGAPE and GTT is folded as a helix-bend-β-strand. This seems unlikely as there is no evidence of a bend (that is, no prolines or glycines) in this region in any of the pumps sequenced thus far. The two threonines are not conserved residues.

However, mutation C513A of Na–K-ATPase, predicted to be in the middle of the first helix just before the two threonines labeled with $2N_3$-ATP in SR Ca-ATPase, lowers the affinity for ATP 2-fold (Lane, 1993). This cysteine, or possibly C551 in the conserved β-strand, could be a NEM-reactive residue protected by ATP (Le, 1986). Identifying C513 rather than C551 as the targeted residue would fit with C551S mutation of Na–K-ATPase as having no effect on $K_{0.5}$ (ATP) (Lane, 1993). C513 and C551 have been reported to be cross-linked (Gevondyan et al., 1993), but this seems unlikely in view of the labeling and mutation results. There is evidence

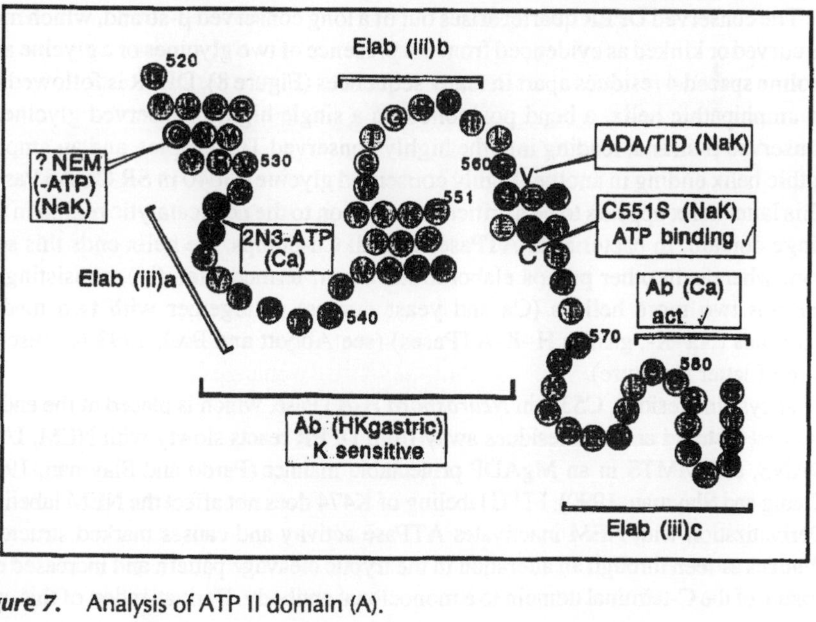

Figure 7. Analysis of ATP II domain (A).

from affinity labeling with the disulfide of thioinosine triphosphate for cysteines to be in the nucleotide binding site (Patzelt-Wenczler and Schoner, 1981)

Part of the following region consisting of a bend-helix-bend forms the epitope of an antibody in the gastric H–K-ATPase (Bayle at al., 1992). Antibody binding lowers ATPase activity but not the K_m(ATP), and does not affect the FITC reaction. K^+ inhibits antibody binding, and the latter reduces the K^+-induced quench of FITC fluorescence.

The next stretch begins with a highly conserved arginine which leads into a predicted α-helix. The residue after the arginine in the Na–K-ATPase is a valine (V547), which is labeled by the hydrophobic reagents 1-tritiospiro(adamantane-4,3'-diazirine) (ADA) and trifluoromethyliodophenyldiazarine (TID) (Nicholas, 1984; Modyanov et al., 1991). This site is also compatible with the location of one of the labeling sites of adamantylidene (Farley et al., 1980). The overall labeling was diminished in the presence of K^+, and to a lesser extent, ATP. As mentioned above, mutation C551S of Na–K-ATPase, which is close by, had no effect on ATP binding (Lane, 1993).

This stretch evidently forms a hydrophobic pocket and, if in proximity to the active site, could serve to bind hydrophobic extentions of ATP analogues, such as TNP.

Elab (iii)c in the SR Ca-ATPase is the binding site of a noninhibitory antibody which is not competitive with antibody binding in the KGAPE and 662-666 regions (Tunwell et al., 1991). An intron is inserted here in the gene of Artemia and rabbit SR Ca-ATPases (Figure 1) and may indicate a division of the ATP II region into two subdomains.

The conserved DPPR quartet arises out of a long conserved β-strand, which may be curved or kinked as evidenced from the presence of two glycines or a glycine and proline spaced 4 residues apart in many sequences (Figure 8). DPPR is followed by an amphipathic helix, a bend positioned on a single highly conserved glycine, a conserved β-strand, leading into the highly conserved TGD triplet, and an amphipathic helix ending in another highly conserved glycine (G640 in SR Ca-ATPase). This latter helix appears to be a critical connection to the next catalytic region in the hinge domain. In bacterial K-ATPases and all Cu-pumps, the helix ends this section, whereas in other pumps elaborations occur, termed Elab (iv), consisting of perhaps two more helices (Ca and yeast pumps) or together with two medial β-strands (Na–K-, gastric H–K-ATPases) (see Abbott and Ball, 1993 for discussion of latter structure).

A cysteine residue, C532, in *Neurospora* H-ATPase, which is placed at the end of the first β-strand and two residues away from DPPR reacts slowly with NEM, IAE-DANS, and MMTS in an MgADP protectable manner (Pardo and Slayman, 1989, Chang and Slayman, 1990). FITC labeling of K474 does not affect the NEM labeling. Derivatization with NEM inactivates ATPase activity and causes marked structural changes as seen through an alteration in the tryptic cleavage pattern and increased exposure of the C-terminal domain to a monoclonal antibody. Derivatization of this cys-

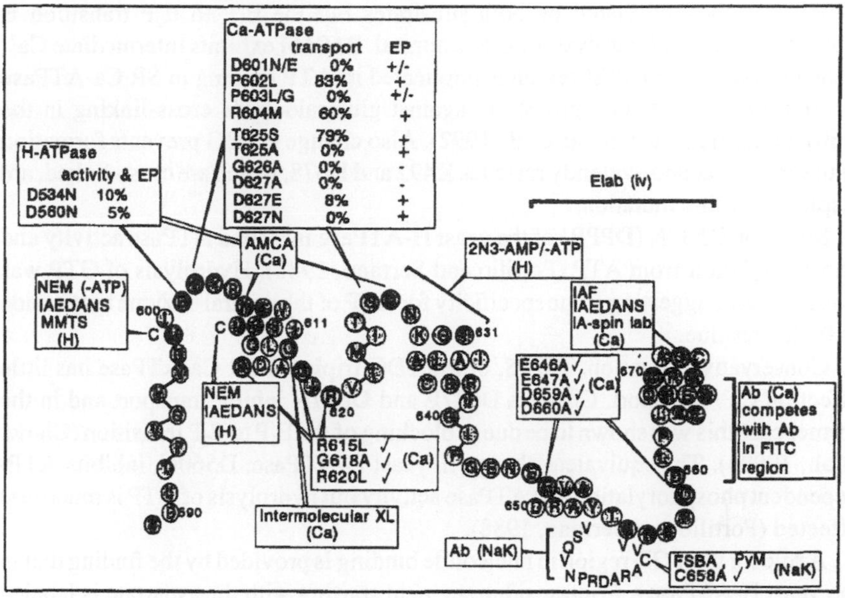

Figure 8. Analysis of ATP II domain (B).

teine and four others (C376, C409, C472, and C545) with the smaller reagent MMTS has little effect on activity indicating the sulfhydryls are not essential to catalysis.

The labeling of C532 in the fungal pump with NEM and IAEDANS suggests that the residue resides in a hydrophobic pocket. Its proximity to the RCLALA helix, which, as we have seen, is also labeled by hydrophobic reagents, may be significant. Perhaps this is the hydrophobic pocket which is part of the ATP binding site and into which the TNP moiety of TNP-ATP fits to produce the superfluorescence seen in the SR Ca-ATPase on phosphorylation. The proximity of the tryptophan of the segment KEW552, which is almost unique to SR Ca-ATPase (interestingly, the sequence REW appears in the mammalian Cu-ATPases in this region, Vulpe et al., 1993) may be significant in this regard.

The sequence, DPPR, of SR Ca-ATPase is followed by K605 which reacts mildly with AMCA, an indication of its superficial position (Stefanova et al., 1993a). The fourth residue after the DPPR sequence of *Neurospora* H-ATPase is a cysteine residue. It reacts strongly with IAEDANS and NEM but without any MgATP protection or loss in ATPase activity (Pardo and Slayman, 1989). A few residues further on in SR Ca-ATPase and toward the end of the helix is C614 which is cross-linked intermolecularly to C377 with a bifunctional maleimide in a manner protected by AMPPNP (Yamasaki et al., 1990).

In SR Ca-ATPase, changes D601E and P603G produce ATPases unable to be phosphorylated by either ATP or P_i (Clarke et al., 1990b). However, D601N and

P603L are phosphorylated by both substrates but the E_1P to E_2P transition is slowed. The rate of hydrolysis of E_2P is normal. R604M exhibits intermediate Ca^{2+} transport rates. The DPPR region is implicated in ATP binding in SR Ca-ATPase through the lack of ATP protection against glutaraldehyde cross-linking in the D601E enzyme (MacLennan et al., 1992). Also change P603G prevents formation of the cross-link and evidently residues K492 and R678, which are cross-linked, are displaced by this mutation.

Mutation D534N (**D**PPR) of the yeast H-ATPase inhibited ATPase activity and phosphorylation from ATP (Portillo and Serrano, 1988). Hydrolysis of GTP was less affected, suggesting some specificity for ATP of the natural enzyme may reside with this residue.

Conservative mutation, T625S, of the TDG triplet of SR Ca-ATPase has little effect on Ca^{2+} transport. Changes D627E and D627N inhibit transport and in the former case this was shown to be due to blocking of the E_1P to E_2P transition (Clarke et al., 1990b). The equivalent change in yeast H-ATPase, D560N, inhibits ATP-dependent phosphorylation and ATPase activity but hydrolysis of GTP is much less affected (Portillo and Serrano, 1988).

A role for the TGD region in nucleotide binding is provided by the finding that in the yeast H-ATPase a seven-residue segment starting with the aspartate is labeled by $2N_3$-AMP and -ATP in a manner protected by ATP (no Mg^{2+} in the medium) (Davis et al., 1990). The exact amino acid derivatized could not be identified. This result apparently conflicts with the $2N_3$-ATP labeling of the glycine of KGAPER in Na–K-ATPase (Tran et al., 1994), or else places the two residues close together. Nucleotide binding in proximity to this triplet is also indicated in the SR Ca-ATPase through the partial protection which the change G626P afforded to ATP inhibition of the glutaraldehyde cross-link (MacLennan et al., 1992).

The surface location of the following meander (Elab iv) in Na–K-ATPase is seen in its acting as an epitope (SQVNPRD, S646-D652) for a monoclonal antibody (Abbott and Ball, 1993). The antibody partially inhibits ATPase activity by stabilizing the E_2 conformation in competition with ATP (Ball, 1984, Abbott and Ball, 1992). C656, four residues further on in the Na–K pump, is derivatized with N-(1-pyrene)maleimide (PyM) along with C367, and the reactions are inhibited by ouabagenin binding on the opposite side of the membrane and vice versa (Kirley and Peng, 1991). The inhibition of ouabagenin binding on derivatization of both residues is enhanced if the labeling is performed under ligand conditions designed to induce the E_2K state, which includes Mg^{2+}, KCl, and ATP, suggesting the residue is not part of the ATP binding site. This contradicts the labeling of this residue with FSBA in an ATP protectable manner (Ohta et al., 1986). The reaction requires Mg^{2+}. Mutation of the residue to alanine (C658A) has no effect on activity (Lane et al., 1993). Mutation of surrounding acidic residues to alanine in SR Ca-ATPase is similarly without effect (Vilsen et al., 1991).

The end of the last helix of this region in SR Ca-ATPase contains three cysteines (C670, C674, and C675), the first two of which are labeled with 5'-(iodoacetimido)-fluorescein (5'-IAF) and IAEDANS respectively, without affecting ATPase activity but sensitive to ligand binding, including ATP (Suzuki et al., 1987, Bishop et al., 1988; Suzuki et al., 1994). The surface location of the beginning of this helix is indicated by the binding of a monoclonal antibody (epitope: P662-Q666) (Tunwell et al., 1991). Binding strongly inhibits ATPase activity and is competitive with the antibody binding in the KGAPE region (Colyer et al., 1989), placing the segments close together in the native structure.

F. Hinge Domain

The hinge domain is one of the most conserved sections of the sequence. Its conservation suggests that it is intimately involved in catalysis and essential conformational changes. It is composed of a bend, two β-strands, the absolutely conserved TGDGVND bend, followed by three α-helices, the last of which is a stalk section leading into M5 (Figure 9).

The first arginine, R678, of SR Ca-ATPase is cross-linked intramolecularly to K492 with glutaraldehyde, a reaction blocked by ATP (McIntosh, 1992). K684 is derivatized with Py-ATP in the presence of Ca^{2+} (Yamamoto et al., 1988). The proximity of K684 to K492 and R678 is also indicated by the finding that mutation of K684 to a bulky histidine blocks the glutaraldehyde cross-link (Vilsen et al., 1991). The role of K684 in the binding of nucleotide is not clear as changes K684R and K684A had no effect on TNP-ATP, ADP, or polyvanadate inhibition of the glutaraldehyde cross-link (Vilsen et al., 1991).

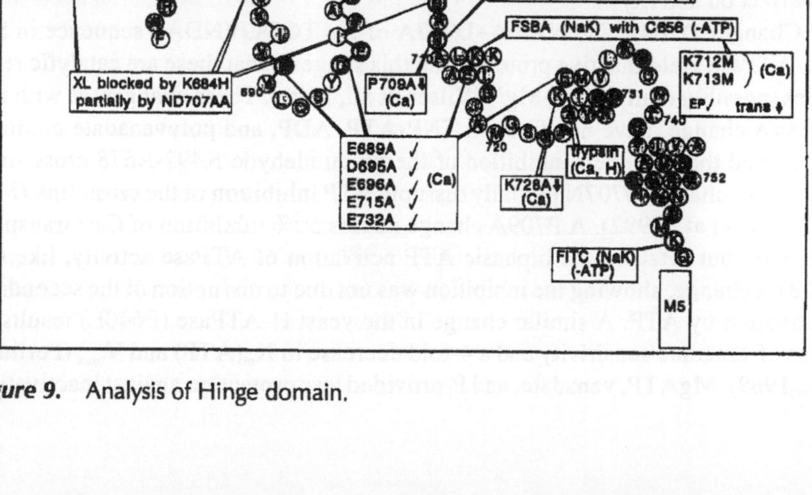

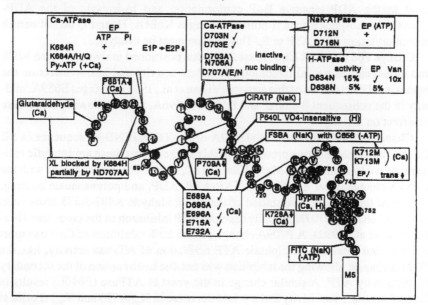

Figure 9. Analysis of Hinge domain.

Either R678 or K684 is likely to be involved in one of the zero distance intramolecular cross-links that form with either ATP-imidazolidate (Gutowski-Eckel et al., 1993) and an adduct of ATP and 1-ethyl-3-(3-(dimethylamino)-propyl)carbodiimide (Murphy, 1990). The mechanisms of both are such that the nucleotide is not part of the adduct formed. Both cross-links are dependent on Ca^{2+} binding to high-affinity sites and result in the inability of the pump to be phosphorylated by ATP or P_i. In the first case, the linkage results in an apparent molecular weight change from 110 to 130 KDa and in the latter to 175 KDa on electrophoresis in sodium dodecyl sulfate. The former change is similar to that found with the glutaraldehyde cross-link (110 to 125 KDa). Gutowski-Eckel et al. (1993) located the imidazolidate cross-link between the the phosphorylation site and a site within I624-M700. It is possible that there is a Ca^{2+}-dependent movement of K684 towards D351 which permits cross-linking of the two residues. This would fit with the labeling of K684 by Py-ATP in the presence of Ca^{2+}.

The first aspartate (D710) in the TGDGVND sequence of the pig kidney Na–K-ATPase is derivatized with gamma-(4-N-2-chloroethyl-N-methylamino)benzylamide-ATP (ClR-ATP) in the presence of Na^+ in an ATP protectable manner, and the second may be derivatized at high reagent concentrations in the presence of K^+, based on the failure of pepsin to cleave the preceding bond of the labeled fragment (Dzhandzhugazyan et al., 1988).

The functional importance of K684 of SR Ca-ATPase is seen in the consequences of introducing mutations (Vilsen et al., 1991). A K684R change permits phosphorylation by ATP but not P_i. The phosphoenzyme formed from ATP is stabilized in the ADP-sensitive E_1-P conformation and hydrolysis of the ADP-insensitive phosphoenzyme is normal. Changes K684A, H, and Q do not permit phosphorylation from ATP or P_i. This residue must be catalytic.

Change P681A of SR Ca-ATPase, a residue positioned midway between R678 and K684, causes partial inhibition of ATPase activity and has no effect on the biphasic ATP activation of this activity (Vilsen et al., 1991). Changes E689A, midway in the subsequent β-strand, and D695A+E696A, in the following bend, have no effect on activity.

Changes D704A and N706A+D707A of the TGDGVNDAP sequence in SR Ca-ATPase yield inactive proteins, and this suggests that these are catalytic residues, possibly complexing Mg^{2+} (Vilsen et al., 1991). These changes, as with the K684A change, have no effect on TNP-ATP, ADP, and polyvanadate binding, measured through their inhibition of the glutaradehyde K492-R678 cross-link. However change D707N partially disrupts ATP inhibition of the cross-link (MacLennan et al., 1992). A P709A change causes 50% inhibition of Ca^{2+} transport activity, but retained the biphasic ATP activation of ATPase activity, like the P681A change, showing the inhibition was not due to disruption of the secondary activation by ATP. A similar change in the yeast H-ATPase (P640L) results in loss of vanadate sensitivity and a 4-fold decrease in K_m(ATP) and V_{max} (Perlin et al., 1989). MgATP, vanadate, and P_i provided less protection against inactivation by trypsin, suggesting that the residue mediates protein stabilization induced by these ligands.

Just after the critical proline are two conserved lysines (however, mostly missing in the bacterial pumps), the first of which, along with C656, is labeled with FSBA in the dog kidney Na–K-ATPase in a manner protected by ATP and resulting in inactivation of ATPase activity (Ohta et al., 1986). Inactivation required Mg^{2+} and was enhanced by Na^+ and K^+. ρ-Nitrophenyl phosphate hydrolysis was not impaired. Three residues further on is a conserved acidic residue, which however in the SR Ca-ATPase does not appear to play an important role as change E715A has no effect (Vilsen et al., 1991)

In the next putative helix is a tryptic cleavage site in the SR Ca-ATPase (Lesniak et al., 1994) and H-ATPase (Davis and Hammes, 1989). The targeted lysine appears to be quite important as change K728A in SR Ca-ATPase results in partial inhibition (Vilsen et al., 1991). Mutation E732A a little further on has no effect.

Just prior to the large cytoplasmic loop penetrating the membrane at M5 is a fairly conserved lysine which is labeled with FITC in Na–K-ATPase (Xu, 1989). It is protected from labeling by ATP, presumably by an allosteric effect.

G. A Model of the ATP Binding Site

As mentioned previously, Grisham and coworkers have elucidated the three-dimensional structure of substitution-inert $Co(NH_3)_4ATP$ in the active site of Na–K-ATPase (Stewart and Grisham, 1988; Stewart et al., 1989). The nucleotide adopts an anticonformation about the glycosidic bond and the phosphates lie nearly parallel to the adenine moiety forming a U-shape. Besides the Cr atom, Mn^{2+} is located above and in the plane of the adenine ring, coordinated in a second sphere manner, probably via water molecules, to N6 and N7. The structure of the nucleotide obtained by Grisham et al is shown in Figure 10 together with a speculative arrangement of the amino acid residues of SR Ca-ATPase identified above as being in the active site.

A brief justification for the arrangement is as follows. K352 and T357 are placed alongside D351 in accordance with the model of the phosphorylation domain in Figure 5B. D703, N706, and D707 have been identified by mutational studies as being the most likely to complex the catalytic Mg^{2+}. K712 seems more peripheral from mutational studies in spite of being affinity labeled. K684 is catalytic and could serve to neutralize the developing negative charge of the terminal oxygen of ADP. A Ca-dependent shift in the position of K684 is suggested from the Ca^{2+} sensitivity of its labeling with Py-ATP. This movement could be towards D351 as deduced from the Ca-dependent intramolecular cross-linking obtained by either Murphy (1990) or Gutowski-Eckel et al., (1993). K492 and R678 are placed alongside through the glutaraldehyde cross-link. K684 is in proximity to them because mutation to histidine blocks the cross-link. With the same logic, D601 is close since change P603G blocks the cross-link. Conservative mutations to D601 and D627

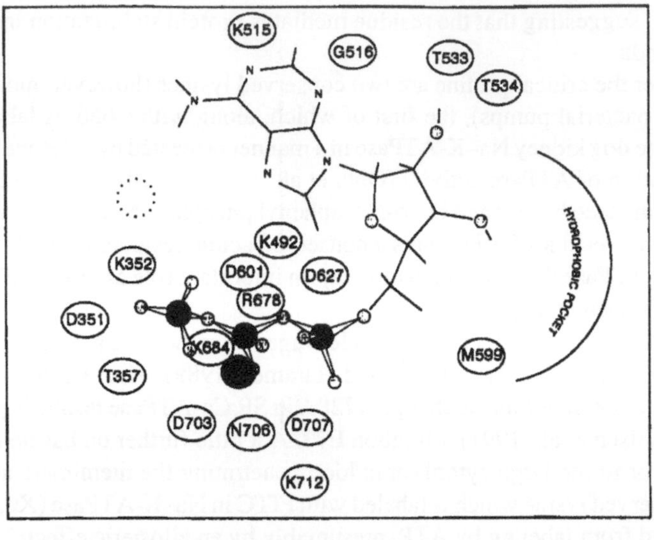

Figure 10. Residues implicated at the active site of SR Ca-ATPase. The structure of the nucleotide is adapted from Stewart et al. (1989). The filled in circle shows the position of the Cr ion and the dotted circle that of the Mn ion.

permit phosphorylation but block the E_1P to E_2P transition, and nonconservative changes prevent phosphorylation. These residues may play an important part in complexing the Mg^{2+} following ADP departure from the active site and when ATP reenters in a modulatory role. The Mn^{2+} ion, seen in the structure of Grisham and coworkers, may be in relation to D601 and D627.

The residues K492, K515, and G516 must be in relationship with the adenine ring. The binding site of the TNP moiety of TNP-ATP is envisaged to be a hydrophobic cleft near to the ribose hydroxyls. This binding could contribute significantly to the binding energy of the nucleotide and cause the adenine ring to be displaced from its most favored position. This may account for TNP-8N_3-ATP and Py-ATP labeling the same residue. FITC could occupy some of the hydrophobic locale. The hydrophobic pocket could account for the wide diversity of substrates tolerated by the pumps regulated by ATP, and could form part of the secondary activation site. The RCLALAT helix labeled by ADA and TID in the Na–K-ATPase may be in proximity to this site. T533 and T534 are poorly conserved and are probably located peripheral to the binding site in spite of being affinity labeled.

An attempt to arrange the large cytoplasmic loop into an active site is made in Figure 11. This should be seen in association with the speculative positioning of domains and elaborations in the three-dimensional structure obtained from image reconstructional analysis of vanadate crystals of the SR Ca-ATPase (Toyoshima et al., 1993) shown in Figure 12. In this model, the phosphorylation domain is envisaged to be braced by the β-strand domain, and the single β-strand just prior to the phosphoryla-

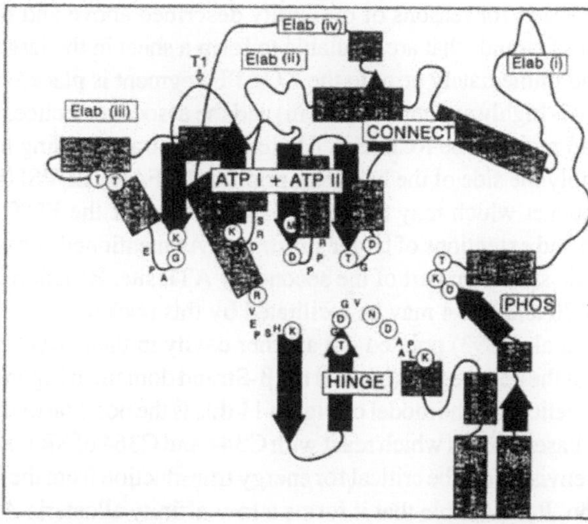

Figure 11. Speculative tertiary structure of the large cytoplasmic loop of SR Ca-ATPase.

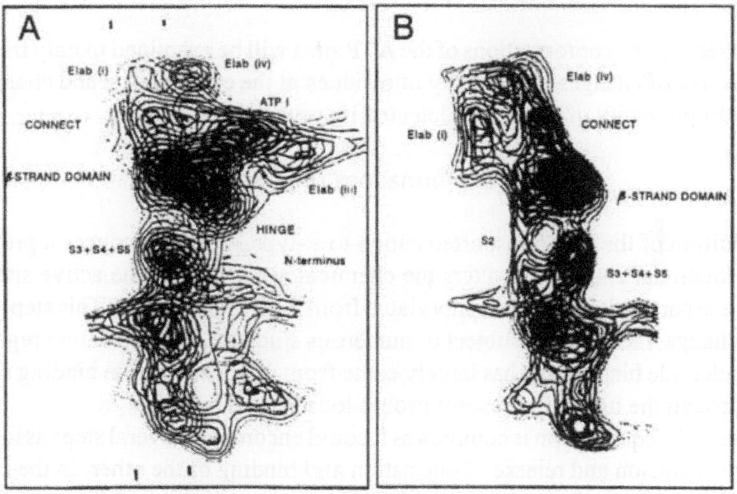

Figure 12. Speculative asignment of domains and elaborations in the SR Ca-ATPase. The structures, and not the asignments, are taken from Toyoshima et al., 1993 (reprinted with permission). The structure in B is a section through that in A between the vertical spacers.

tion site shown in Figure 11 could be contiguous with the sheet. Elab (i) and Elab (iv) are envisaged to make up the outcrops on the side and top respectively of the structure, seen in Figure 12. ATP I and ATP II domains have been linked together as a single four strand sheet, for reasons of proximity described above and because of the limited number of strands that are available to form a sheet in the large cytoplasmic loop. The strand immediately prior to the KGAPE segment is placed on the edge of the sheet since it is highly antigenic. Elab (iii) and the associated helices constitute the nose of the head region. The RCLALAT helix, the β-Strand leading into the DPPR loop, and possibly the side of the last helix prior to the FARVEPVSHK loop form a hydrophobic pocket which may accommodate portions of the FITC molecule attached to K515 and extentions of larger substrates. As mentioned above, it may also, with the catalytic site, form part of the secondary ATP site. Reaction of fluorescent probes with C670 and C674 may be facilitated by this pocket.

Toyoshima et al. (1993) pointed out another cavity in their structure, just below the main mass of the head region (labeled the β-Strand domain in Figure 12B) and between the stalk helices. In the model of Figure 11 this is the position of the binding site for maleimide-based probes which react with C344 and C364 of SR Ca-ATPase. The locus could be envisaged to be critical for energy transduction from the head region to the stalk section. It is possible that it forms a low-affinity allosteric ATP site.

V. CONFORMATIONS OF ATP BINDING SITES

In this section the conformations of the ATP sites will be examined mainly from the perspective of changes in reactivity of residues at the catalytic site and changes in the hydrophobicity of the site as detected by probes located there.

A. Conformations of E_1 and E_2

Addition of the first transported cation to P-type ATPases induces a profound conformational change that alters the chemical specificity of the active site such that the aspartyl residue is phosphorylated from ATP rather than P_i. This step, the E_2 to E_1 change, has been the subject of numerous studies, and information regarding the nucleotide binding site has largely come from direct nucleotide binding studies and through the use of fluorescent probes located at the site.

The E_1-E_2 equilibrium is complex as it could encompass several steps associated with deocclusion and release of one cation and binding of the other. In the case of the Na–K-ATPase, these steps could describe the deocclusion of 2 K^+ (or Rb^+) ions, the release of the two cations, and the binding of 3 Na^+ ions, with the possibility of all seven steps being associated with conformational rearrangements (Pratap and Robinson, 1993). With the SR Ca-ATPase the equilibrium is pH dependent (Pick and Karlish, 1982; Froud and Lee, 1986; Wakabayashi et al., 1990; Forge et al., 1993), as protons are the transported counterions (Yamaguchi and Kanazawa, 1984 and 1985; Levy et al., 1990).

One apparent difference between SR Ca-ATPase and Na–K-ATPase is the affinity of the E_2 form(s) for ATP. SR Ca-ATPase exhibits a high, micromolar affinity for ATP in EGTA which is independent of pH in the absence of Mg^{2+} and only mildly pH dependent in its presence (5- to 7-fold change from pH 6 to 8) (Meissner, 1973; Lacapère et al., 1990). Kinetic and equilibrium studies have indicated that there is little or no change in affinity for MgATP or CaATP on Ca^{2+} binding at pH 7.0 (Lacapère and Guillain, 1990 and 1993). A small (approximately 5-fold) lowering of affinity is observed for MgAMPPCP and MgADP binding at pH 8.0 (Ross and McIntosh, 1987b).

As mentioned earlier ATP accelerates Ca^{2+} binding by binding to a high-affinity site, which permits a relatively rapid phosphorylation of the Ca^{2+}-deprived enzyme in the presence of Mg^{2+} on addition of Ca^{2+} and ATP compared with the rate of Ca^{2+} binding in the absence of ATP. This acceleration can only come about by ATP binding to the E_2 form. ATP binding in EGTA results in stabilization of an E_1-like conformation as indicated by a fluorescent probe located on C344 (Wakabayashi et al., 1990).

The affinity for ATP in the Na-K-ATPase compared with the SR Ca-ATPase is governed to a much greater extent by the distribution of unphosphorylated E_1 and E_2 forms, which in turn depends on the concentrations of K^+ and Na^+ (or co-geners or non-specific ligands) as well as of ATP. The E_1 species binds 3 Na^+ and binds ATP with high affinity ($K_d \approx 0.2$ μM) with ensuing phosphorylation. In the absence of Na^+, the pump binds 2 mol K^+/mol EP_{max} tightly ($K_d \approx 10$ μM) in an occluded form that exchanges only slowly with the medium. Binding is blocked by high concentrations of Na^+ or ATP and these ligands also accelerate the deocclusion of K^+ (or Rb^+) (Post et al., 1972; Glynn and Richards, 1982; Forbush, 1987; Rossi and Nørby, 1993). Non-hydrolyzable analogues of ATP, AMPPNP, and AMPPCP, as well as ADP and formycin nucleotides, also accelerate deocclusion (Jørgensen, 1975; Karlish et al., 1978; Beaugé and Glynn, 1980; Forbush, 1987). The change from a low-affinity K^+ form to a high-affinity Na^+ form can be monitored at intermediate concentrations of formycin diphosphate through the increase in fluorescence on induction of the high-affinity form and nucleotide binding (Karlish et al., 1978). The induction of a high-affinity form of the nucleotide binding site can also be usefully monitored by the increase of fluorescence of two eosin molecules binding to this site (Esmann and Skou, 1983; Skou and Esmann, 1983; Esmann, 1994). The high fluorescent state is associated with enzyme species binding two or three Na^+ and no Rb^+ (Esmann, 1994). In the case of both the formycin nucleotides and eosin, binding to the K^+ form is observed at high concentrations but the fluorescence yield is low, so there is a definite change in hydrophobicity of the site in the E_1 to E_2 transition. Not all analogues of ATP exhibit a low affinity for the E_2 species. The affinity of the enzyme for TNP-ATP is changed 4-fold by K^+ binding, from 0.09 to 0.38 μM (Moczydlowski and Fortes, 1981a). Like the Ca-ATPase, ATP induces an E_1-like state in the Na–K-ATPase as seen by conformation-specific trypsin cleavage (Jørgensen, 1975).

One possibility for the difference between the two pumps is that the SR Ca-ATPase does not reach a conformation that could be likened to the occluded Rb^+

state of the Na–K-ATPase even at low pH. The specific Ca-ATPase inhibitor thapsigargin binds to the E_2 form of the Ca-ATPase and it is possible that the E_2.thapsigargin complex may be closer to the occluded form of the Na–K pump. In the presence of thapsigargin the Ca pump exhibits a 100-fold lower affinity for ATP, although the affinity for TNP-ATP is not affected (DeJesus et al., 1993). Ca binding to SR Ca-ATPase, as monitored by NBD-Cl attached to C344, is accelerated by ATP binding to a site with an affinity 30-fold lower than the affinity displayed by the E_1 state (Wakabayashi and Shigekawa, 1990). It is possible that the label altered the equilibrium toward an occluded E_2H form.

The difference between the E_1-E_2 changes in the Na–K- and SR Ca-ATPases is highlighted by the opposite flùorescence changes of FITC induced by the two forms. The change of the catalytic nucleotide binding site of Na–K-ATPase to a tight-binding and more hydrophobic form, as witnessed by eosin fluorescence changes, is also seen in the increase in fluorescence of FITC with Na^+ (or the reverse with K^+) (Karlish, 1980). FITC derivatization itself, like ATP binding, pulls the enzyme into an E_1-like state as seen in the lower apparent affinity for K^+ (Karlish, 1980; Jørgensen and Petersen, 1982). With the Ca-ATPase, Ca^{2+} addition quenches the fluorescence and removal of Ca^{2+} reverses this (Pick and Karlish, 1980). The changes are greatest at low pH (up to 8%), compatible with H^+ stabilizing the E_2 species by binding to the transport sites, similar to K^+ in the Na–K-ATPase (Pick and Karlish, 1982). The effect of iodide quenching on FITC. Ca-ATPase also indicates that the nucleotide binding site becomes more exposed on Ca^{2+} binding (Highsmith, 1986).

The fluorescence of TNPATP is enhanced approximately 6-fold on binding to SR Ca-ATPase and Ca^{2+} binding has no effect (Nakamoto and Inesi, 1984).

In the gastric H–K-ATPase, the changes in nucleotide binding site associated with the E_1-E_2 steps appear to be similar to the Na–K-ATPase. An occluded E_2K (or Rb^+) species has not been demonstrated due to an unfavorable equilibrium toward an E_1K form (Faller et al., 1991). However, Rb^+ occlusion occurs in the presence of vanadate (Rabon et al., 1993). As with the Na–K-ATPase, millimolar ATP and ADP inhibits occlusion and K^+ and Rb^+ binding quenches FITC fluorescence (Jackson et al., 1983; Rabon et al., 1993). Also, Mg^{2+} and Ca^{2+} binding, apparently to the monovalent cation transport sites, induce an E_1-like state with high FITC or eosin fluorescence, which is quenched by K^+ binding (Jackson et al., 1983; Helmich-de Jong et al., 1986b; Mendlein et al., 1990). Similarly, TNPATP binds very tightly (K_d<25 nM) to the enzyme with a 5-fold increase in fluorescence (Faller, 1990). K^+ binding (K_d=3 mM) causes a rapid quenching of the fluorescence which can be transiently reversed by Mg^{2+}.

Thus there is a consistent picture of the ATP binding site of the Na–K-ATPase and H–K-ATPase, whether the probe is FITC, eosin, or TNPATP, that the tighter binding in the E_1 form is accompanied by a change to a more hydrophobic, shielded pocket. The same site of the SR Ca-ATPase undergoes much less change in affinity and the change is to a more open conformation.

Not only does ATP binding accelerate the conversion to the E_1 species and stabilize this conformation in the absence of bound cations, it also appears to promote a further conformational change of the E_1.M1 species that may optimally position the terminal phosphoryl group for tranfer to the aspartyl residue. In fact, it is possible that it is this conformational change, rather than phosphorylation itself, which results in occlusion of the transported ions.

Since phosphorylation is usually an extremely rapid reaction with the natural ligands and substrate, the most convincing evidence of a nucleotide-induced conformational change prior to phosphorylation has come from the use of unnatural substrates and cations.

In the SR Ca-ATPase, Ca^{2+} binding causes 16 exposed SH groups to react with 5,5'-dithiobis(2-nitrobenzoate) as a single, moderately reactive class (Murphy, 1978). AMPPNP binding lowers the reactivity of all these groups by 50%, suggesting a conformational change extending over the entire exposed surface of the protein. An iodoacetamide spin label reacts in a hyperbolic manner with the Ca-bound ATPase to the extent of 2 mol/mol EP_{max} (Coan and Keating, 1982; Wawrzynow et al., 1993). The presence of AMPPNP or AMPPCP causes half of reactive SH groups to become much more reactive, and C674 has been identified as the reactive residue. In the slower reacting group, C674 may also be the predominant labeled cysteine suggesting heterogeneity of the ATPase population. Electron spin resonance spectroscopy of ATPase labeled in the absence of nucleotide has shown that the probe has a single motional state in Ca^{2+} and that addition of low concentrations of ATP, ADP, CrATP, or non-hydrolyzable analogues induces a second, motionally restricted spectrum in approximately 50% of the labeled molecules (Coan and Inesi, 1977; Coan and Keating, 1982; Chen et al., 1991; Lewis and Thomas, 1992; Wawrzynow et al., 1993; Mahaney et al., 1995). Only in the case of CrATP is Ca^{2+} not required for immobilization (Chen et al., 1991). The fraction of ATPases with a restricted spectrum is increased by AMPPCP binding to a second, low-affinity class of sites (Mahaney et al., 1995). The restricted component is relatively inaccessible to quenching by Ni ions, indicating the probe is locked in a pocket by nucleotide binding (Wawrzynow et al., 1993).

There is kinetic evidence that ATP binding to E_1.2Ca causes a conformational change which permits rapid phosphorylation (Petithory and Jencks, 1986). It is based partly on a 3-fold change in k_{off} for ATP and partly on a lower observed rate constant for phosphorylation in the presence of ADP than expected for a simpler mechanism (Pickart and Jencks, 1982; Petithory and Jencks, 1986). The kinetics of phosphorylation using LaATP, which phosphorylates the ATPase very slowly, is also indicative of a rate limiting conformational change prior to phosphorylation (Hanel and Jencks, 1990). However, the kinetics of phosphorylation using MgATP or CaATP at 5° C, a temperature at which such a conformational change might be expected to be more obvious, yielded no evidence of a slow step prior to phosphorylation (Lacapère and Guillain, 1990 and 1993). A significant amount of E_1.ATP can be measured during steady state ATPase turnover with CaATP as substrate due to the slow phosphorylation and dissociation of CaATP (Shigekawa et al., 1983). The structural changes that lead to the formation of the transition state and permit rapid phosphoryl transfer depend on the order of addition of ligands ATP, Mg, and Ca^{2+} (Reinstein and Jencks, 1993). In a physiological milieu, with Mg^{2+} and ATP always present, the changes may occur together as a single concerted movement.

It has been found that complexation of the SR Ca-ATPase with Ca^{2+} and nucleotide (AMPPCP is sufficient) exposes a disulfide bond which can be reduced with dithiothreitol (Daiho and Kanazawa, 1994). The reduction completely inhibits the pump by blocking the E_1P to E_2P transition.

As alluded to above, the divalent cation complexed with ATP has a profound effect on phosphorylation kinetics. CaATP phosphorylates the SR Ca-ATPase approximately 10-fold slower than MgATP (Yamada and Ikemoto, 1980; Shigekawa et al., 1983; Lacapère and Guillain, 1990). The rate with LaATP is 3-fold slower still or slightly faster depending on conditions (Hanel and Jencks, 1990). CrATP binds very tightly over several hours and without phosphorylating the enzyme and yet Ca^{2+} is occluded at the transport sites (Serpersu et al., 1982; Vilsen and Andersen, 1986, 1992; Chen et al., 1991). The decrease in k_{obs} for Ca^{2+} occlusion and/or phosphorylation in the order MgATP>CaATP>LaATP>CrATP is paralleled by slower nucleotide off rates and lower K_m values. Thus the tighter the binding the slower the conformational change leading to occlusion and/or phosphorylation.

Details of the structural changes at the active site that occur following ligand binding in this early part of the catalytic cycle are gradually coming to light. With the Na–K-ATPase, the reactivity of the predominant lysine reacting with FITC (K501) is not markedly altered by K^+ or Na^+ binding. In contrast, the reactivity of this residue to NIPI decreases 10-fold on binding K^+ compared with in the presence of Na^+ (Ellis-Davies and Kaplan, 1993). Its reactivity with SITS is virtually nil in the presence of K^+ (Pedemonte et al., 1992). Ca^{2+} binding to SR Ca-ATPase has little effect on FITC reaction with K515 (Pick, 1981a). It has no effect on glutaraldehyde cross-linking of K492 and R678 (Ross and McIntosh, 1987a). In contrast, cross-linking of D351 to a residue close to K684 with ATP-imidazolate is dependent on Ca^{2+} binding (Gutowski-Eckel et al., 1993). Evidently, the ATP I domain and the Hinge domain are moving as a unit with respect to the phosphorylation site. Ca^{2+} binding changes the specificity of reaction of Py-ATP from reacting with both K492 and K684 to reacting exclusively with K684 (Yamamoto et al., 1988, 1989).

B. Conformation of E_1P

Addition of acetyl phosphate to FITC-modified Na–K-ATPase, under conditions designed to accumulate E_1P, results in a small increase (Rephaeli et al., 1986) or biphasic decrease in fluorescence, attributable to two E_1P conformations with different FITC fluorescence (Taniguchi et al., 1988). A decrease in fluorescence of

Py-P covalently attached to K480 of the same enzyme is also observed on phosphorylation to E_1P (Kaya et al., 1994).

It was mentioned above that Ca^{2+} addition to SR Ca-ATPase derivatized with FITC quenches the fluorescence up to 8%. Addition of acetyl phosphate results in further slow quenching to approximately 16%, in parallel with Ca transport (Pick, 1981b). Remarkably, addition of EGTA causes a further 45% quench, presumably associated with the formation of E_1P with bound Ca^{2+}. P_i enhances the gradient-dependent quenching by increasing the amount of phosphoenzyme. Pick estimated that the E_1P species with bound Ca^{2+} exhibits 70% less fluorescence compared with the E_2 form. This suggests that the fluorescein moiety is in a relatively hydrophobic pocket in the E_2 and E_1 conformational states and becomes much more solvent exposed on phosphorylation as long as Ca^{2+} remains bound (either occluded or bound to luminal sites).

Phosphorylation of SR Ca-ATPase with acetyl phosphate causes a large increase in the reactivity of residues K492 and R678 to glutaraldehyde cross-linking (Ross et al., 1991), indicative of a structural change involving these residues. In agreement with the FITC results, it also shows that these residues remain medium-exposed on phosphorylation. The enhanced cross-linking is blocked by millimolar ATP (Ross and McIntosh, 1987b) suggesting that these residues may be involved in low-affinity ATP binding to the phosphorylated active site. Prior cross-linking accelerates acetyl phosphate phosphorylation of the ATPase (Ross et al., 1991), which is compatible with a reorientation of the two residues during phosphoryl transfer to a position which permits accelerated cross-linkage.

C. Conformation of E_2P

The transition from an ADP-sensitive phosphoenzyme (E_1P) to one which is insensitive to ADP (E_2P) must be accompanied by further conformational changes at the active site. The rate of the change with the SR Ca-ATPase is greatly influenced by the divalent cation at the active site, similar to what is observed for phosphorylation (Wakabayashi and Shigekawa, 1987; Hanel and Jencks, 1990). Mg^{2+} accelerates the E_1P to E_2P transition in the red cell plasma membrane Ca-ATPase (Herscher et al., 1994).

A good example of the dramatic change in the nucleotide binding site is seen in the SR Ca-ATPase where bound TNP-nucleotides undergo a large increase in fluorescence (superfluorescence) either during ATP-mediated turnover or following phosphorylation by P_i in the absence of Ca^{2+} (Watanabe and Inesi, 1982; Nakamoto and Inesi, 1984; Davidson and Berman, 1987). Vanadate complexation with the pump also increases TNP-nucleotide fluorescence, but to a lesser extent. The locus of the superfluorescent nucleotide was shown to be the same as the catalytic site through the use of TNP-8N_3-nucleotides to covalently attach the nucleotide to the site and thereafter elicit superfluorescence (Seebregts and McIntosh, 1989). Evidently, phosphorylation to E_2P creates a hydrophobic pocket into which the TNP moiety fits. The site has been estimated to release 18 water molecules on formation of E_2P from the relationship between the observed association constant for P_i phosphorylation and the water activity in dimethyl sulphoxide/water mixtures and from the spectral changes of TNP-ATP bound to the enzyme and in organic solvents (Dupont and Pougeois, 1983). There have been no reports of superfluorescence in any of the other pumps.

A change in the conformation of the site is also seen in the complete inhibition of glutaraldehyde cross-linking of K492 and R678 (Ross and McIntosh, 1987b). This inhibition is in line with the inability of the cross-linked pump to reach the E_2P state in either direction of catalysis (Ross et al., 1991).

There are also large changes in the fluorescence of FITC in the SR Ca-ATPase in changing from E_1P with bound Ca^{2+} to E_2P. The large quench in the fluorescence in the former species (70% quench relative to E_2) is mostly reversed by the transition (16% quench relative to E_2) (Pick, 1981b), indicating a change to a more hydrophobic environment. Restricted accessibility of the fluorescein moiety is also seen in the greater shielding of this reporter which phosphorylation with P_i provides to iodide (Highsmith, 1986).

The change in FITC fluorescence is the opposite for the Na–K-ATPase and H–K-ATPase. As noted above, Taniguchi and co-workers observe a small decrease in fluorescence on phosphorylation with acetyl phosphate at high Na^+ concentrations, which favors the accumulation of E_1 (Taniguchi et al., 1988). Addition of ouabain and formation of E_2P causes a further 30% decrease in fluorescence (-35% compared with E_1Na). Karlish and co-workers also observe a large fluorescence quench on formation of E_2P compared with E_1Na and E_1P (approximately 18%) (Rephaeli et al., 1986). The quench is greater than that observed with the addition of K^+ to the E_1 enzyme. In the case of the H–K-ATPase, the change produced by phosphorylation in the presence of eosin is intermediate between the E_1 and E_2 states (Mendlein et al., 1990).

In the H-ATPase of *Neurospora crassa*, Mg-vanadate blocks derivatization of the enzyme by periodate-oxidized AMP (Bidwai et al., 1989).

The relative fluorescence changes observed with FITC in the SR Ca-ATPase, Na–K-ATPase, and H–K-ATPase, those with TNP-ATP and SR Ca-ATPase, and those occurring with Py-P or Py-ATP at K480 of Na–K-ATPase are shown in Figure 13. It is difficult to derive general conclusions regarding the hydrophobicity of the various catalytic intermediates in the different pumps. This may be indicating that, at least in some cases, the different fluorescence levels reflect local changes in the vicinity of the probe that are unique for each pump and probe, and major changes that are fundamental to the pumping mechanism, such as opening or closing of the active site, or increased hydrophobicity of the entire active site, are not being reported. While it is tempting to relate the large fluorescence increase observed with TNP-ATP on formation of E_2P to a fundamental hydrophobicity change at the active site, there is little evidence for this. The nucleotide site is not occluded in E_2P since dephosphorylation is modulated by nucleotide binding to this

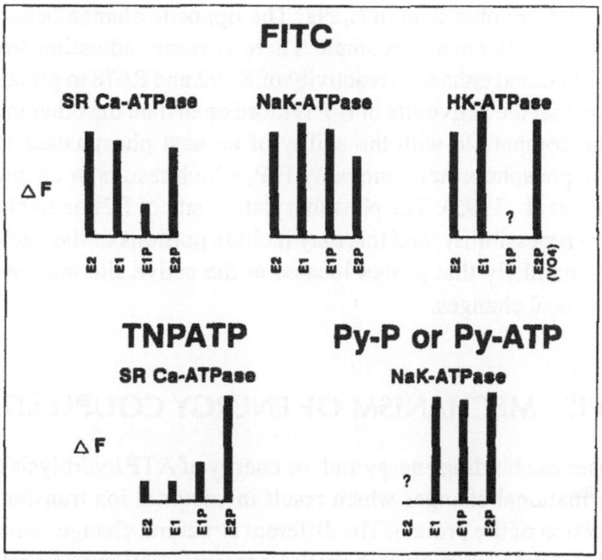

Figure 13. Conformational changes at the nucleotide binding site as reported by fluorescent probes. The FITC changes in SR Ca-ATPase are adapted from Pick (1981), in Na–K-ATPase from Taniguchi and Mardh (1993), and in H–K-ATPase from Jackson et al. (1983). The TNP-ATP information is adapted from Davidson and Berman (1987) and that for Py-P and Py-ATP from Kaya et al. (1994).

intermediate. Fluorescence energy transfer between spectroscopic probes covalently bound to various domains of SR Ca-ATPase, including FITC, failed to reveal structural changes that could be construed as fundamental to the pumping mechanism (Bigelow et al., 1992).

High levels of E_2P can be formed from P_i with the SR Ca-ATPase and Na-K-ATPase under favorable conditions. Acyl phosphate formation is highly unfavorable in solution and therefore the active site of the enzyme must constitute a special environment for this reaction to occur so readily. In the enzymatic reaction, there is a very large preference for water over methanol as phosphate acceptor (Chipman and Jencks, 1988), again suggesting a tightly controlled catalytic reaction. Thus the phosphorylation site, if not the nucleotide binding site, must be well-shielded from the medium in E_2P in these pumps. This might not apply to all P-type ATPases. For example, in the yeast H-ATPase the P_i phosphorylation reaction is highly unfavorable (Amory et al., 1982).

In conclusion, from a survey of the pumps, there is no coherent picture of the major conformational changes that occur at the active site during the cycle, changes that could be linked to hinge bending movements such as active site opening and closing and which could be considered to be fundamental to the pumping mechanism. The low-affinity nucleotide binding site of Na–K- and H–K-ATPases in E_2.K is mildly less hydrophobic than in E_1.Na. The opposite change occurs in the Ca-pumps and the affinity change is small. There is some indication from the huge quenching of FITC and enhanced reactivity of K492 and R678 to glutaraldehyde of SR Ca-ATPase that the active site of E_1P is more open than the other intermediates. This would be compatible with the ability of an acyl phosphatase to act on the Na–K-ATPase phosphoprotein, probably E_1P, which results in uncoupling of the pump (Nediani et al., 1995). The phosphorylation site of E2P is likely to have reduced medium accessibility, and this may include portions of the nucleotide binding site. It seems likely that probes located at the active site may in a large part monitor small local changes.

VI. MECHANISM OF ENERGY COUPLING

P-type ATPases use binding energy and the energy of ATP hydrolysis to drive a series of conformational changes which result in vectorial ion transfer through the membrane portion of the protein. The different structural changes emanating from the nucleotide binding site with each catalytic intermediate must be relayed by mechanical and electrostatic means to residues complexing and gating the transported ions. P-type ATPases have evolved to transport two types of cations in different directions in a single cycle, probably using the same channel, and a decrease in affinity for the one cation has been linked with an increase in affinity for the counterion. Therefore, changes at the catalytic site need to be connected with alternating access and a change in specificity at the transport sites. Following binding of the first transported cation (M1) and MgATP, the overall protein structure undergoes a series of relaxations with internal rearrangements promoting cross-protein shifts of specific regions in an ordered manner. As Jencks has emphasized, coupling requires that certain partial chemical and vectorial reactions occur together in one state and other reactions are linked at other specific stages of the cycle (Jencks, 1980).

It has been proposed that the water activity at the active site is crucial to energy transduction (de Meis et al., 1980; Dupont and Pougeois, 1983; de Meis, 1985). In this view, the pumps can be in either of two main conformations, namely E and E*, which differ in transport site orientation and affinity/specificity and in hydrophobicity or solvation of the active site. This would allow facile phosphorylation by P_i in the hydrophobic E* state and permit transfer of the phosphoryl group from the enzyme to ADP in the E state. In the forward direction of pumping the principal and essential change at the active site following phosphorylation would be desolvation and then resolvation after dephosphorylation and reorientation of transport sites. Whether water activity changes are a primary mechanism of energy transduction in P-type ATPases or are secondary to other essential changes at the active site may only be decided when the details of the changes at the active site are elucidated for several pumps.

The means by which chemical energy produces the forces and secondary structural displacements for long-distance structural and pK_a changes which result in vectorial movement of the tranported ions, will ultimately require linking details of the reaction kinetics with high-resolution structural changes. A variety of conformational probes and techniques have been used toward this goal and, in deference to the complexity of the problem, the results are difficult at this stage to tie together into a coherent picture.

In terms of the low-resolution shape of the pump obtained from image reconstructional analysis (Toyoshima et al., 1993) (Figure 12), it can be imagined that structural changes in the head region are communicated through the stalk section to the membrane helices and vice versa. Analysis of the mass distribution of the membrane helices and other considerations has resulted in a hypothetical arrangement of the helices (Stokes et al., 1994) and their approximate relationship to one another and the stalk segments is shown in Figure 14. We have seen that mutations to the many acidic residues of SR Ca-ATPase in the S2 and S3 stalks have little effect on transport. This leaves the much more conserved stalk helices S4 and S5 which lead out of and into the critical M4 and M5 cation binding helices in the membrane which are involved in cation binding.

Site-directed mutagenesis has identified the M4S4 segment as critical for communication of the E_1P to E_2P transition (Andersen and Vilsen, 1992). Some changes in the stalk helix disturb its amphipathic nature and the latter may be an important element of the required movement. We have seen how mutations in the β-strand domain, ATP II domain (specifically the DPPR region), and Hinge domain identify conserved residues in the catalytic site that are linked to this conformational change. The mutations concern mainly acidic residues or those nearby and are probably positioned as a phalanx around the Mg^{2+} at the catalytic site. We have seen

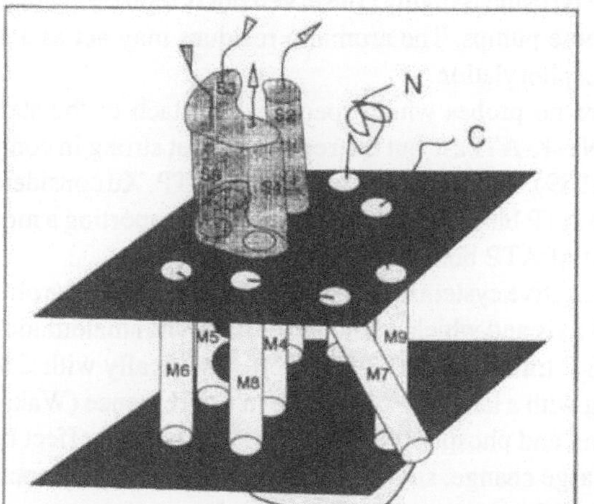

Figure 14. Speculative arrangement of membrane helices and stalk helices. The arrangements are based on an analysis by Stokes et al. (1994). The filled in circles represent Ca ions.

that the E_1P to E_2P transition is greatly influenced by the identity of the divalent cation. Thus an axis of linkage can be traced from the divalent cation/phosphoryl group/acidic residues down the S4 helix to the Ca^{2+} binding sites.

The change at the catalytic site not only concerns the residues directly complexing Mg^{2+} and the phosphoryl group but incorporates K492 and R678 which are cross-linked with glutaraldehyde, and probably others. The cross-link blocks the E_1P to E_2P transition completely and the Ca^{2+} at the transport site is forced to leave the channel from the side it entered on hydrolysis of the phosphoenzyme (McIntosh et al., 1991). Covalently attaching TNP-8N_3-ATP to K492 partially uncouples the pump probably by affecting the same step (McIntosh and Woolley, 1994) which means that K492 may have to separate from the nucleotide for the proper conformational transition to occur. It may be the reason for the rapid departure of ADP from the site. ATP can then rebind in a slightly different way to accelerate the transition.

The importance of nucleotide/protein interactions and binding site specificity in the coupling process is well illustrated by the uncoupling observed with some substrates. We have seen how ρ-nitrophenyl phosphate and furylacryloylphosphate are utilized by the Na–K-ATPase in a K-dependent reaction, how arylazido-ATP phosphorylates the SR Ca^{2+}-ATPase in the absence of Ca^{2+} and canine cardiac SR Ca-ATPase hydrolysis of GTP is Mg- and not Ca-dependent (see Introduction). Also, uncoupling produced by partial thermal denaturation carried out in the absence of Ca^{2+} (Berman et al., 1977; McIntosh and Berman, 1978) alters the fluorescence properties of bound TNP-ATP (Berman, 1986).

A tyrosine residue located at the M4-S4 joint of SR Ca-ATPase may be important for the gating process at the cation transport site. Mutation Y763G results in uncoupling of the pump where ATP hydrolysis is unimpaired but there is no transport (Andersen, 1995). The tyrosine is highly conserved but is replaced with tryptophan or phenylalanine in some pumps. The aromatic residues may act as a gate to the channel following phosphorylation.

At present there are no probes which specifically attach to the stalk helices. FITC labels K766 of Na–K-ATPase but the reaction is not strong in comparison to K501 and K480 (Xu, 1989). The labeling is blocked by ATP. Xu considered that the residue was part of the ATP binding pocket, but it may be reporting a movement of the S5 helix as a result of ATP binding.

There are a pair of reactive cysteines in SR Ca-ATPase close to the phosphorylation site and on the S4 axis and which when derivatized with maleimides or NBD-Cl inhibit the E_1P to E_2P transition. NBD-Cl reacts specifically with C344 and responds to Ca^{2+} binding with a large, 50% increase in fluorescence (Wakabayashi et al., 1990). ATP binding and phosphorylation to E_1P has a small effect but conversion to E_2P evokes a large change, similar to that seen with Ca^{2+} dissociation. The probe therefore distinguishes between the E_1 and E_2 forms of the pump as defined earlier. Ca^{2+} binding and the fluorescence change occurred simultaneously whereas the fluorescence decrease associated with Ca^{2+} dissociation was slower than the latter process indicating the conformational change occurs before Ca^{2+} binding and after Ca^{2+} dissociation (Wakabayashi and Shigekawa, 1990). Thus there are similarities in the conformational transitions of E_1 to E_2 and E_1P to E_2P although the mechanism to effect them must be different. For example, the glutaraldehyde cross-link has no effect on the E_1 to E_2 transition in contrast to the complete block of the E_1P to E_2P step.

The interpretation of fluorescence changes in terms of alterations of protein domain interactions is often difficult. Br-DMC also specifically labels C344 but seems to report on different conformational changes from NBD-Cl (Stefanova et al., 1992). The probe decribes three fluorescence states, high in the absence of ligands, intermediate in the presence of Mg^{2+} or ADP, low in the presence of vanadate or AMPPCP or following phosphorylation with P_i, ATP or acetyl phosphate.

Further conformational changes are detected by probes attached to the cysteines located at the beginning of the Hinge domain of SR Ca-ATPase. IAEDANS attached to C674 records ATP binding to the Ca-activated enzyme through a 14-23% quench in fluorescence (Suzuki et al., 1987; Obara et al., 1988; Suzuki et al., 1994). AMPPNP is just as effective showing it is a conformational change of nucleotide binding. The fluorescence is not changed by the E_1P to E_2P transition but rather is reversed on hydrolysis of E_2P. Interestingly, the fluorescence quench is biphasic with CaATP as substrate (Suzuki et al., 1994). The first phase is associated with nucleotide binding and the second with phosphorylation. The rate of the first phase is rapid irrespective of the metal ion in the metal. ATP complex whereas the phosphorylation step is sensitive to the metal ion. The conformational change on nucleotide binding is reported by the lowered mobility of an iodoacetamide spin-label attached to the same residue (see page 60 for references).

N-(1-anilinonaphth-4-yl)maleimide (ANM) probably attached to C344 or C364 of SR Ca-ATPase and IAEDANS at C674 show highly restricted mobility according to time resolved and steady-state fluorescence anisotropy measurements, but exhibited different rotational motions indicative of independent domain wobbling (Suzuki et al., 1989). Ca^{2+} binding selectively increases the faster wobbling of the phosphorylation domain to which ANM is attached.

The selective conformational changes detected by different probes have permitted the simultaneous observation of the appearance of two different E_1P species by placing fluorescent probes at different points on the protein. Derivatization of pig kidney Na–K-ATPase at C964 in M9 with N-[-(2-benzimidazolyl)phenyl]maleimide (BIPM) and at K501 with FITC still permits phosphorylation with acetyl phosphate (Taniguchi et al., 1988). The fluorescence of BIPM decreases simultaneously with the rise in phosphoenzyme level but the fluorescence change of the fluorescein was significantly delayed; this the authors attributed to the sequential formation of different phosphoenzyme intermediates, both of which were reactive to acetate and may be correlated with different Na^+ binding states. Multiple energy transfer changes between the two probes were measured during the catalytic cycle which could indicate changes in the distance between them during stages of the passage of the cations through the membrane section (Taniguchi and Mardh, 1993).

The dog kidney Na–K-ATPase is specifically labeled by IAF on C457 at the end of the second putative helix in the Connect region without inhibiting activity (Tyson et al., 1989). Its fluorescence increases with the binding of ATP, ADP, and Na^+ and the reverse occurs with K^+ or Rb^+ binding and therefore, like FITC, reports on the E_1 and E_2 forms (Kapakos and Steinberg, 1986). Again, labeling of the Na–K-ATPase with IAF and BIPM and their sensitivity to select conformations has allowed several different conformations of the enzyme induced by K^+, Na^+, and ATP to be resolved kinetically and these can be related to the binding and dissociation of individual cations at the transport sites (Pratap and Robinson, 1993).

The challenge for the future is to map with increasing detail the structural changes that occur differentially through the protein matrix with each catalytic intermediate and link these changes to their energetic requirements. Kinetic analyses of the cycles have shown that energy transduction can be divided into discrete steps by suitable choice of conditions and ligands, providing some hope that through-protein stabilization and destabilization mechanisms may be elucidated for each step.

VII. SUMMARY

P-type ATPases comprise a family of ATP-dependent ion pumps that are related through structural homology and catalytic mechanism. Most higher animal pumps have a high affinity for ATP and are regulated through a low-affinity ATP binding site, whereas bacterial, plant, and lower animal pumps exhibit a low affinity for ATP and are not regulated. The substrate specificity of the former group is low and large substitutions on the ATP molecule are tolerated. This may be because the principal mode of regulation of the cycle is by nucleotide binding close to or at the active site. Analysis of intron/exon boundaries of the large cytoplasmic loop of the genes of plant and animal ATPases suggests a five-domain structure interupted by antigenic elaborations in the larger pumps. Labeling and mutagenesis studies identify active site residues and lead to a speculative arrangement of residues binding ATP and tertiary structural model of the extramembranous portion of the protein. Probes located at the active site and the reactivity of active site residues have yielded information of changes in active site conformation that seem to reflect local disturbances rather than fundamental structural movements such as hinge bending or opening and closing of the site. An axis of energy coupling from the active site through the stalk helices S4 and S5 to the membrane region is beginning to be elucidated.

REFERENCES

Abbott, A. J., Amler, E., & Ball, W. J., Jr. (1991). Immunochemical and spectroscopic characterization of two fluorescein 5'-isothiocyanate labeling sites on Na^+,K^+-ATPase. Biochemistry 30, 1692-1701.

Abbott, A., & Ball, W. J., Jr. (1992). The inhibitory monoclonal antibody M7-PB-E9 stabilizes E_2 conformational states of Na^+,K^+-ATPase. Biochemistry 31, 11236-11243.

Abbott, A., & Ball, W. J., Jr. (1993). The Epitope for the inhibitory antibody M7-PB-E9 contains Ser-646 and Asp-652 of the the sheep Na^+,K^+-ATPase α-subunit. Biochemistry 32, 3511-3518.

Abrahams, J. P., Leslie, A. G. W., Lutter, R., & Walker, J. E. (1994). Structure at 2.8Å resolution of F_1-ATPase from bovine heart mitochondria. Nature 370, 621-628.

Aderem, A. A., McIntosh, D. B., & Berman, M. C. (1979). Occurrence and role of tightly bound adenine nucleotides in sarcoplasmic reticulum of rabbit skeletal muscle. Proc. Natl. Acad. Sci. USA 76, 3622-3626.

Ahlemeyer, B., Weintraut, H., Antolovic, R., & Schoner, W. (1992). Chick heart cells with high intracellular calcium concentration have a higher affinity for cardiac glycosides than those with low intracellular calcium concentration, as revealed by affinity labelling with a digoxigenin derivative. Eur. J. Biochem. 205, 269-275.

Amory, A., Goffeau, A., McIntosh, D. B., & Boyer, P. D. (1982). Exchange of oxygen between phosphate and water catalyzed by plasma membrane ATPase from the yeast *Schizosaccharomyces pombe*. J. Biol. Chem. 257, 12509-12516.

Andersen, J. P., Møller, J. V., & Jørgensen, P. L. (1982). The functional unit of sarcoplasmic reticulum Ca^{2+}-ATPase. J. Biol. Chem. 257, 8300-8307.

Andersen, J. P., Vilsen, B., Nielsen, H., & Møller, J. V. (1986). Characterization of detergent-solubilized sarcoplasmic reticulum Ca^{2+}-ATPase by high-performance liquid chromatography. Biochemistry 25, 6439-6447.

Andersen, J. P., Vilsen, B., Leberer, E., & MacLennan, D. H. (1989). Functional consequences of mutations in the β-strand sector of the Ca^{2+}-ATPase of sarcoplasmic reticulum. J. Biol. Chem. 264. 21018-21023.

Andersen, J. P., & Vilsen, B. (1992). Structural basis for the E_1/E_1P-E_2/E_2P conformation changes in the sarcoplasmic reticulum Ca^{2+}-ATPase studied by site-specific mutagenesis. Acta Physiol. Scand. 146, 151-159.

Andersen, J. P., & Vilsen, B. (1993). Functional consequences of substitution of the seven-residue segment LysIleArgAspGlnMetAla240 located in the stalk helix S3 of the Ca^{2+}-ATPase of sarcoplasmic reticulum. Biochemistry 32, 10015-10020.

Andersen, J. P. (1995). Functional consequences of alterations to amino acids at the M5S5 boundary of the Ca^{2+}-ATPase of sarcoplasmic reticulum. J. Biol. Chem. 270, 908-914.

Argüello, J. M., & Kaplan, J. H. (1990). *N*-acetylimidazole inactivates renal Na,K-ATPase by disrupting ATP binding to the catalytic site. Biochemistry 29, 5775-5782.

Asano, S., Kamiya, S., & Takeguchi, N. (1992). The energy transduction mechanism is different among P-type ion- transporting ATPases. J. Biol. Chem. 267, 6590-6595.

Askari, A., Huang, W-H., & McCormick, P. W. (1983). (Na^+ + K^+)-dependent adenosine triphosphatase. J. Biol. Chem. 258, 3453-3460.

Askari, A., & Huang, W-H. (1984). Reaction of (Na^+ + K^+)-dependent adenosine triphosphatase with inorganic phosphate. J. Biol. Chem. 259, 4169-4176.

Askari, A., Kakar, S. S., & Huang, W-H. (1988). Ligand binding sites of the Ouabain-complexed (Na^+ + K^+)-ATPase. J. Biol. Chem. 263, 235-242.

Ball, W. J. Jr. (1984). Immunochemical characterization of a functional site of (Na^+,K^+)-ATPase. Biochemistry 23, 2275-2281.

Barford, D., Flint, A. J., & Tonks, N. K. (1994). Crystal structure of human protein tyrosine phosphatase 1B. Science 263, 1397-1404.

Barrabin, H., Scofano, H. M., & Inesi, G. (1984). Adenosinetriphosphatase site stoichiometry in sarcoplasmic reticulum vesicles and purified enzyme. Biochemistry 23, 1542-1548.

Bayle, D., Robert, J. C., Bamberg, K., Benkouka, F., Cheret, A.M., Lewin, M. J. M., Sachs, G., & Soumarmon, A. (1992). Location of the cytoplasmic epitope for a K^+-competitive antibody of the (H^+,K^+)-ATPase. J. Biol. Chem. 267, 19060-19065.

Beaugé, L. A., & Glynn, I. M. (1980). The equilibrium between different conformations of the unphosphorylated sodium pump: Effects of ATP and of potassium ions, and their relevance to potassium transport. J. Physiol. 299, 367-383.

Berman, M. C., McIntosh, D. B., & Kench, J. E. (1977). Proton inactivation of Ca^{2+} transport by sarcoplasmic reticulum. J. Biol. Chem. 252, 994-1001.

Berman, M. C. (1986). Absorbance and fluorescence properties of 2'(3')-0-(2,4,6-trinitrophenyl)adenosine 5'-triphosphate bound to coupled and uncoupled Ca^{2+}-ATPase of skeletal muscle sarcoplasmic reticulum. J. Biol. Chem. 261, 16494-16501.

Bidwai, A. P., Morjana, N. A., & Scarborough, G. A. (1989). Studies on the active site of the *Neurospora crassa* plasma membrane H^+-ATPase with periodate-oxidized nucleotides. J. Biol. Chem. 264, 11790-11795.

Bigelow, D. J., & Inesi, G. (1991). Frequency-domain fluorescence spectroscopy resolves the location of maleimide-directed spectroscopic probes within the tertiary structure of the Ca-ATPase of sarcoplasmic reticulum. Biochemistry 30, 2113-2125.

Bigelow D. J., Squier, T. C., & Inesi, G. (1992). Phosphorylation-dependent changes in the spatial relationship between Ca-ATPase polypeptide chains in sarcoplasmic reticulum membranes. J. Biol. Chem. 267, 6952-6962.

Bishop, J. E., Johnson, J. D. & Berman, M. C. (1984). Transient kinetic analysis of turnover-dependent fluorescence of 2',3'-O-(2,4,6-trinitrophenyl)-ATP bound to Ca^{2+}-ATPase of sarcoplasmic reticulum. J. Biol. Chem. 259, 15163-15171.

Bishop, J. E., Nakamoto, R. K., & Inesi, G. (1986). Modulation of the binding characteristics of a fluorescent nucleotide derivative to the sarcoplasmic reticulum adenosinetriphosphatase. Biochemistry 25, 696-703.

Bishop, J. E., Al-Shawi, M. K., & Inesi, G. (1987). Relationship of the regulatory nucleotide site to the catalytic site of the sarcoplasmic reticulum Ca^{2+}-ATPase. J. Biol. Chem. 262, 4658-4663.

Bishop, J. E., Squier, T. C., Bigelow, D. J., & Inesi, G. (1988). (Iodoacetamido)fluorescein labels a pair of proximal cysteines on the Ca^{2+}-ATPase of sarcoplasmic reticulum. Biochemistry 27, 5233-5240.

Blake, C. C. F. (1978). Do genes-in-pieces imply proteins-in-pieces? Nature 273, 267.

Blanco, G., DeTomaso, A. W., Koster, J., Xie, Z. J., & Mercer, R. W. (1994). The α-subunit of the Na,K-ATPase has catalytic activity independent of the β-subunit. J. Biol. Chem. 269, 23420-23425.

Bodley, A. L., & Jencks, W. P. (1987). Acetyl phosphate as a substrate for the calcium ATPase of sarcoplasmic reticulum. J. Biol. Chem. 262, 13997-14004.

Brandl, C. J., Green, N. M., Korczak, B., & MacLennan, D. H. (1986) Two Ca^{++} ATPase genes: Homologies and mechanistic implication of the deduced amino acid sequences. Cell 44, 597-607

Briggs, F. N., Al-Jumaily, W., & Haley, B. E. (1980). Photoaffinity labeling of the (Ca+Mg)ATPase of skeletal and cardiac sarcoplasmic reticulum with [α ^{32}P]-8-azido-ATP. Cell Calcium 1, 205-215.

Brooker, R. J., & Slayman, C. W. (1983). Effects of Mg^{2+} ions on the plasma membrane [H^+]-ATPase of *Neurospora crassa*. J. Biol. Chem. 258, 8827-8832.

Brotherus, J. R., Møller, J. V., & Jorgensen, P. L. (1981). Soluble and active renal Na,K-ATPase with maximum protein molecular mass 170,000 +/-9,000 daltons; formation of larger units by secondary aggregation. Biochem. Biophys. Res. Commun. 100, 146-154.

Buxbaum, E., & Schoner, W. (1990). Blocking of Na^+/K^+ transport by the $MgPO_4$ complex analogue $Co(NH_3)_4PO_4$ leaves the Na^+/Na^+-exchange reaction of the sodium pump unaltered and shifts its high-affinity ATP-binding site to a Na^+-like form. Eur. J. Biochem. 193, 355-360.

Buxbaum, E., & Schoner, W. (1991). Phosphate binding and ATP-binding sites coexist in Na^+/K^+-transporting ATPase, as demonstrated by the inactivating $MgPO_4$ complex analogue $Co(NH_3)_4PO_4$. Eur. J. Biochem. 195, 407-419.

Campos, M., & Beaugé, L. (1994). Na^+-ATPase activity of Na^+,K^+-ATPase. J. Biol. Chem. 269, 18028-18036.

Capasso, J. M., Hoving, S., Tal, D. M., Goldshleger, R., & Karlish, S. J. D. (1992). Extensive digestion of Na^+,K^+-ATPase by specific and nonspecific proteases with preservation of cation occlusion sites. J. Biol. Chem. 267, 1150-1158.

Caride, A. J., Rega, A. F., & Garrahan, P. J. (1982). The role of the sites for ATP of the Ca^{2+}-ATPase from the human red cell membranes during Ca^{2+}-phosphatase activity. Biochim. Biophys. Acta 689, 421-428.

Caroni, P., & Carafoli, E. (1981). The Ca^{2+}-pumping ATPase of heart sarcolemma. J. Biol. Chem. 256, 3263-3270.

Champeil, P., Gingold, M. P., & Guillain, F. (1983). Effect of magnesium on the calcium-dependent transient kinetics of sarcoplasmic reticulum ATPase, studied by stopped flow fluorescence and phosphorylation. J. Biol. Chem. 258, 4453-4458.

Champeil, P., & Guillain, F. (1986). Rapid filtration study of the phosphorylation-dependent dissociation of calcium from transport sites of purified sarcoplasmic reticulum ATPase and ATP modulation of the catalytic cycle. Biochemistry 25, 7623-7633.

Champeil, P., Riollet, S., Orlowski, S., Guillain, F., Seebregts, C. J., & McIntosh, D. B. (1988). ATP regulation of sarcoplasmic reticulum Ca^{2+}-ATPase. J. Biol. Chem. 263, 12288-12294.

Chang, A., & Slayman, C. W. (1990). A structural change in the *Neurospora* plasma membrane $[H^+]$ATPase induced by *N*-ethylmaleimide. J. Biol. Chem. 265, 15531-15536.

Chen, Z., Coan, C., Fielding, L., & Cassafer, G. (1991). Interaction of CrATP with the phosphorylation site of the sarcoplasmic reticulum ATPase. J. Biol. Chem. 266, 12386-12394.

Chipman, D. M., & Jencks, W. P. (1988). Specificity of the sarcoplasmic reticulum calcium ATPase at the hydrolysis step. Biochemistry 27, 5707-5712.

Clarke, D. M., Maruyama, K., Loo, T. W., Leberer, E., Inesi, G., & MacLennan, D. H. (1989). Functional consequences of glutamate, aspartate, glutamine, and asparagine mutations in the stalk sector of the Ca^{2+}-ATPase of sarcoplasmic reticulum. J. Biol. Chem. 264, 11246-11251.

Clarke, D. M., Loo, T.W., & MacLennan, D. H. (1990a). Functional consequences of mutations of conserved amino acids in the β-strand domain of the Ca^{2+}-ATPase of sarcoplasmic reticulum. J. Biol. Chem. 265, 14088-14092.

Clarke, D. M., Loo, T.W., & MacLennan, D. H. (1990b). Functional consequences of alterations to amino acids located in the nucleotide binding domain of the Ca^{2+}-ATPase of sarcoplasmic reticulum. J. Biol. Chem. 265, 22223-22227.

Clore, G. M., Gronenborn, A. M., Mitchinson, C., & Green, N. M. (1982). ^{1}H-NMR studies on nucleotide binding to the sarcoplasmic reticulum Ca^{2+}-ATPase. Eur. J. Biochem. 128, 113-117.

Coan, C. R., & Inesi, G. (1977). Ca^{2+}-dependent effect of ATP on spin-labeled sarcoplasmic reticulum. J. Biol. Chem. 252, 3044-3049.

Coan, C., & Keating, S. (1982). Reactivity of sarcoplasmic reticulum adenosinetriphosphatase with iodoacetamide spin-label: Evidence for two conformational states of the substrate binding site. Biochemistry, 21, 3214-3220.

Coan, C., Amaral, J. A., Jr., & Verjovski-Almeida, S. (1993). Elimination of the hydroxyl groups in the ribose ring of ATP reduces its ability to phosphorylate the sarcoplasmic reticulum Ca^{2+}-ATPase. J. Biol. Chem. 268, 6917-6924.

Coll, R. J., & Murphy, A. J. (1991). Kinetic evidence for two nucleotide binding sites on the CaATPase of sarcoplasmic reticulum. Biochemistry 30, 1456-1461.

Coll, R. J., & Murphy, A. J. (1992). Fluoride-inhibited calcium ATPase of sarcoplasmic reticulum. J. Biol. Chem. 267, 21584-21587.

Collins, J. H., & Leszyk, J. (1987). The "γSubunit" of Na,K-ATPase: A small, amphiphilic protein with a unique amino acid sequence. Biochemistry 26, 8665-8668.

Colyer, J., Mata, A. M., Lee, A. G., & East, J. M. (1989). Effects on ATPase activity of monoclonal antibodies raised against (Ca^{2+} + Mg^{2+})-ATPase from rabbit skeletal muscle sarcoplasmic reticulum and their correlation with epitope location. Biochem. J. 262, 439-447.

Craig, W. S. (1982). Monomer of sodium and potassium ion activated adenosinetriphosphatase displays complete enzymatic function. Biochemistry 21, 5707-5717.

Daiho, T., Kubota, T., & Kanazawa, T. (1993). Stoichiometry of tight binding of magnesium and fluoride to phosphorylation and high-affinity binding of ATP, vanadate, and calcium in the sarcoplasmic reticulum Ca^{2+}-ATPase. Biochemistry 32, 10021-10026.

Daiho, T., & Kanazawa, T. (1994). Reduction of disulfide bonds in sarcoplasmic reticulum Ca^{2+}-ATPase by dithiothreitol causes inhibition of phosphoenzyme isomerization in catalytic cycle. J. Biol. Chem. 269, 11060-11064.

Davidson, G. A., & Berman, M. C. (1987). Phosphoenzyme conformational states and nucleotide-binding site hydrophobicity following thiol modification of the Ca^{2+}-ATPase of sarcoplasmic reticulum from skeletal muscle. J. Biol. Chem. 262, 7041-7046.

Davidson, G. A., & Varhol, R. J. (1995). Kinetics of thapsigargin-Ca^{2+}-ATPase (sarcoplasmic reticulum) interaction reveals a two-step binding mechanism and picomolar inhibition. J. Biol. Chem. 270, 11731-11734.

Davis, C. B., & Hammes, G. C. (1989). Topology of the yeast plasma membrane proton-translocating ATPase. J. Biol. Chem. 264, 370-374.

Davis, C. B., Smith, K. E., Campbell, B. N. Jr., & Hammes, G. G. (1990). The ATP binding site of the yeast plasma membrane proton-translocating ATPase. J. Biol. Chem. 265, 1300-1305.

DeJesus, F., Girardet, J-L., & Dupont, Y. (1993). Characterisation of ATP binding inhibition to the sarcoplasmic reticulum Ca^{2+}-ATPase by thapsigargin. FEBS Lett. 332, 229-232.

de Meis, L., & Fialho de Mello, M. C. (1973). Substrate regulation of membrane phosphorylation and of Ca^{2+} transport in the sarcoplasmic reticulum. J. Biol. Chem. 248, 3691-3701.

de Meis, L., Martins, O. B., Carvalho, M. G. C., and Alves, E. W. (1980). Role of water, hydrogen ion, and temperature on the synthesis of adenosine triphosphate by the sarcoplamsic reticulum adenosine triphosphatase in the absence of a calcium ion gradient. Biochemistry 19, 4252-4261.

de Meis, L. (1985). Role of water in processes of energy transduction: Ca^{2+}-transport ATPase and inorganic pyrophosphatase. Biochem. Soc. Symp. 50, 97-125.

Denu, J. M., Zhou, G., Guo, Y., & Dixon, J. E. (1995). The catalytic role of aspartic acid-92 in a human dual-specific protein-tyrosine-phosphatase. Biochemistry 34, 3396-3403.

de Souza, S. J., Long, M., Schoenbach, L., Roy, S. W., & Gilbert, W. (1996). Intron positions correlate with module boundaries in ancient proteins. Proc. Natl. Acad. Sci. USA 93, 14632-14636.

Dixon, D. A., & Haynes, D. H. (1989). Kinetic characterization of the Ca^{2+}-pumping ATPase of cardiac sarcolemma in four states of activation. J. Biol. Chem. 264, 13612-13622.

Doolittle, W. F. (1978). Genes in pieces: were they ever together? Nature 272, 581-582.

Dupont, Y., Chapron, Y., & Pougeois, R. (1982). Titration of the nucleotide binding sites of sarcoplasmic reticulum Ca^{2+}-ATPase with 2'3'-0-(2,4,6-trinitrophenyl) adenosine 5'-triphosphate and 5'-diphosphate. Biochem. Biophys. Res. Commun. 106, 1272-1279.

Dupont, Y., & Pougeois, R. (1983). Evaluation of H_2O activity in the free or phosphorylated catalytic site of Ca^{2+}-ATPase. FEBS Lett. 156, 93-98.

Dupont, Y., Pougeois, R., Ronjat, M., & Verjovsky-Almeida, S. (1985). Two distinct classes of nucleotide binding sites in sarcoplasmic reticulum Ca-ATPase revealed by 2',3'-O-(2,4,6-trinitrocyclohexadienylidene)-ATP. J. Biol. Chem. 260, 7241-7249.

Dux, L., Papp, S., & Martonosi, A. (1985). Conformational responses of the tryptic cleavage products of the Ca^{2+}-ATPase of sarcoplasmic reticulum. J. Biol. Chem. 260, 13454-13458.

Dzhandzhugazyan, K. N., Lutsenko, S. V., & Modyanov, N. N. (1988). Target-residues of the active site affinity modification are different in E_1 and E_2 forms. In: The Na^+,K^+-Pump Part A: Molecular Aspects (Skou, J. C., Nørby, J. G., Maunsbach, A. B., & Esmann, M., Eds.) pp. 181-188. New York, Alan R. Liss, Inc.

Ellis-Davies, G. C. R., & Kaplan, J. H. (1993). Modification of lysine 501 in Na,K-ATPase reveals coupling between cation occupancy and changes in the ATP binding domain. J. Biol. Chem. 268, 11622-11627.

Epstein, W., Walderhaug, M. O., Polarek, J. W., Hesse, J. E., Dorus, E., & Daniel, J. M. (1990). The bacterial Kdp K^+-ATPase and its relation to other transport ATPases, such as the Na^+/K^+-and Ca^{2+}-ATPases in higher oganisms. Phil. Trans. R. Soc. Lond. 326, 479-487.

Escalante, R., & Sastre, L. (1994). Structure of *Artemia franciscana* sarco/endoplasmic reticulum Ca-ATPase gene. J. Biol. Chem. 269, 13005-13012.

Esmann, M., & Skou, J. C. (1983). The effect of K^+ on the equilibrium between the E_2 and the K^+-occluded E_2 conformation of the ($Na^+ + K^+$)-ATPase. Biochim. Biophys. Acta 748, 413-417.

Esmann, M., & Skou, J. C. (1984). Kinetic properties of $C_{12}E_8$-solubilized ($Na^+ + K^+$)-ATPase. Biochim. Biophys. Acta 787, 71-80.

Esmann, M. (1994). Influence of Na^+ on conformational states in membrane-bound renal Na,K-ATPase. Biochemistry 33, 8558-8565.

Fagan, M. J., & Saier, M. H. (1994). P-type ATPases of eukaryotes and bacteria: sequence comparisons and construction of phylogenetic trees. J. Mol. Evol. 38, 57-99.

Faller, L. D. (1989). Competitive binding of ATP and the fluorescent substrate analogue 2',3'-O-(2,4,6-trinitrophenyl-cyclohexadienylidine)adenosine 5'-triphosphate to the gastric H^+,K^+-ATPase: Evidence for two classes of nucleotide sites. Biochemistry 28, 6771-6778.

Faller, L. D. (1990). Binding of the fluorescent substrate analogue 2'3'-O-(2,4,6-trinitrophenylcyclohexadienylidene)adenosine 5'triphosphate to the gastric H^+,K^+-ATPase: Evidence for cofactor-induced conformational changes in the enzyme. Biochemistry 29, 3179-3186.

Faller, L. D., Diaz, R. A., Scheiner-Bobis, G., & Farley, R. A. (1991). Temperature dependence of the rates of conformational changes reported by fluorescein 5'-isothiocyanate modification of H^+,K^+- and Na^+,K^+-ATPases. Biochemistry 30, 3503-3510.

Farley, R. A., Goldman, D. W., & Bayley, H. (1980). Identification of regions of the catalytic subunit of (Na-K)-ATPase embedded within the cell membrane. J. Biol. Chem. 255, 860-864.

Farley, R. A., Tran, C. M., Carilli, C. T., Hawke, D., & Shively, J. E. (1984). The amino acid sequence of a fluorescein-labeled peptide from the active site of (Na,K)-ATPase. J. Biol. Chem. 259, 9532-9535.

Farley, R. A., & Faller, L. D. (1985). The amino acid sequence of an active site peptide from the H,K-ATPase of gastric mucosa. J. Biol. Chem. 260, 3899-3901.

Ferreira, S. T., & Verjovski-Almeida, S. (1988). Stoichiometry and mapping of the nucleotide sites in sarcoplasmic reticulum ATPase with the use of UTP. J. Biol. Chem. 263, 9973-9980.

Ferreira-Pereira, A., Alves-Ferreira, M., & de Carvalho-Alves, P. C. (1994). *p*-Nitrophenylphosphatase activity of plasma membrane H^+-ATPase from yeast. J. Biol. Chem. 269, 12074-12079.

Filoteo, A. G., Gorski, J. P., & Penniston, J. T. (1987). The ATP-binding site of the erythrocyte membrane Ca^{2+} pump. J. Biol. Chem. 262, 6526-6530.

Forbush, B. III. (1987). Rapid release of ^{42}K and ^{86}Rb from an occluded state of the Na,K-Pump in the presence of ATP or ADP. J. Biol. Chem. 262, 11104-11115.

Forge, V., Mintz, E., & Guillain, F. (1993). Ca^{2+} binding to sarcoplasmic reticulum ATPase revisited. J. Biol. Chem. 268, 10961-10968.

Friedman, Z., & Makinose, M. (1970). Phosphorylation of skeletal muscle microsomes by acetylphosphate. FEBS Lett. 11, 69-72.

Froud, R. J., & Lee, A. G. (1986). Conformational transitions in the $Ca^{2+} + Mg^{2+}$-activated ATPase and the binding of Ca^{2+} ions. Biochem. J. 237, 197-206.

Gafni, A., & Boyer, P. D. (1984). Characterization of sarcoplasmic reticulum adenosinetriphosphatase purified by selective column adsorption. Biochemistry 23, 4362-4367.

Gantzer, M. L., Klevickis, C., & Grisham, C. M. (1982). Interactions of $Co(NH_3)_4ATP$ and $Cr(H_2O)_4ATP$ with Ca^{2+}-ATPase from sarcoplasmic reticulum and Mg^{2+}-ATPase and ($Na^+ + K^+$)-ATPase from kidney medulla. Biochemistry 21, 4083-4088.

Geering, K., Kraehenbuhl, J-P., & Rossier, B. C. (1987). Maturation of the catalytic α-subunit of NA,K-ATPase during intracellular transport. J. Cell Biol. 105, 2613-2619.

Gevondyan, N. M., Gevondyan, V. S.,& Modyanov, N. N. (1993). Analysis of disulfide bonds in the Na^+,K^+-ATPase α-subunits. Biol. Mol. Biol. Intl. 29, 327-337.

Ghislain, M., Schlesser, A., & Goffeau, A. (1987). Mutation of a conserved glycine residue modifies the vanadate sensitivity of the plasma membrane H^+-ATPase from *Schizosaccharomyces pombe*. J. Biol. Chem. 262, 17549-17555.

Ghosh, J., Ray, M., Sarkar, S., & Bhaduri, A. (1990). A high affinity Ca^{2+}-ATPase on the surface membrane of *Leishmania donovani* promastigote. J. Biol. Chem. 265, 11345-11351.

Gilbert, W. (1978). Why genes in pieces? Nature 271, 501.

Glynn, I. M., & Richards, D. E. (1982). Occlusion of rubidium ions by the sodium-potassium pump: Its implications for the mechanism of potassium transport. J. Physiol. 330, 17-43.

Go, M. (1983). Modular structural units, exons, and function in chicken lysozyme. Proc. Natl. Acad. Sci. USA 80, 1964-1968.

Goffeau, A., & de Meis, L. (1990). Effects of phosphate and hydrophobic molecules on two mutations in the β-strand sector of the H^+-ATPase from the yeast plasma membrane. J. Biol. Chem. 265, 15503-15505.

Green, N. M. (1989). ATP-driven cation pumps : alignment of sequences. Biochem. Soc. Trans. 17, 972-974.

Guillain, F., Champeil, P., Lacapère, J-J., & Gingold, M. P. (1981). Stopped flow and rapid quenching measurement of the transient steps induced by calcium binding to sarcoplasmic reticulum adenosine triphosphatase. J. Biol. Chem. 256, 6140-6147.

Gutowski-Eckel, Z., Karlheinz, M., & Bäumert, H. G. (1993). Identification of a cross-linked double-peptide from the catalytic site of the Ca-ATPase of sarcoplasmic reticulum formed by the Ca^{2+}- and pH-dependent reaction with ATP P-imidazolidate. FEBS Lett. 324, 314-318.

Hamer, E., & Schoner, W. (1993). Modification of the E_1ATP binding site of Na^+/K^+-ATPase by the chromium complex of adenosine 5'- [β, γ-methylene]triphosphate blocks the overall reaction but not the partial activities of the E_2 conformation. Eur. J. Biochem. 213, 743-748.

Hanel, A. M., & Jencks, W. P. (1990). Phosphorylation of the calcium-transporting adenosinetriphosphatase by lanthanum ATP: Rapid phosphoryl transfer following a rate-limiting conformational change. Biochemistry 29, 5210-5220.

Hardwicke, P. M. D., & Green, N. M. (1974). The effect of delipidation on the adenosine triphosphatase of sarcoplasmic reticulum. Eur. J. Biochem. 42, 183-193.

Harris, S. L., Perlin, D. S., Seto-Young, D., & Haber, J. E. (1991). Evidence for coupling between membrane and cytoplasmic domains of the yeast plasma membrane H^+-ATPase. J. Biol. Chem. 266, 24439-24445.

Hegyvary, C., & Post, R. L. (1971). Binding of adenosine triphosphate to sodium and potassium ion-stimulated adenosine triphosphate. J. Biol. Chem. 246, 5235-5240.

Helmich-de Jong, M. L., van Emst-de Vries, S. E., Swarts, H. G. P., Schuurmans Stekhoven, F. M. A. H., & de Pont, J. J. H. H. M. (1986a). Presence of a low-affinity nucleotide binding site on the (K^+ + H^+)-ATPase phosphoenzyme. Biochim. Biophys. Acta 860, 641-649.

Helmich-de Jong, M. L., van Duynhoven, J. P. M., Schuurmans Stekhoven, F. M. A. H., & de Pont, J. J. H. H. M. (1986b). Eosin, a fluorescent marker for the high-affinity ATP site of (K^+ + H^+)-ATPase. Biochim. Biophys. Acta 858, 254-262.

Herscher, C. J., Rega, A. F., & Garrahan, P. J. (1994). The dephosphorylation reaction of the Ca^{2+}-ATPase from plasma membranes. J. Biol. Chem. 269, 10400-10406.

Hesse, J. E., Wieczorek, L., Altendorf, K., Reicin, A. S., Dorus, E., & Epstein, W. (1984). Sequence homology between two membrane transport ATPases, the Kdp-ATPase of *Escherichia coli* and the Ca^{2+}-ATPase of sarcoplasmic reticulum. Proc. Natl. Acad. Sci. USA 81, 4746-4750.

Hiatt, A., McDonough, A. A., & Edelman, I. S. (1984). Assembly of the (Na^+ + K^+)-adenosine triphosphatase. J. Biol. Chem. 259, 2629-2635.

Highsmith, S. (1986). Solvent accessibility of the adenosine 5'-triphosphate catalytic site of sarcoplasmic reticulum CaATPase. Biochemistry 25, 1049-1054.

Hinz, H. R., & Kirley, T. L. (1990). Lysine 480 is an essential residue in the putative ATP site of lamb kidney (Na,K)-ATPase. J. Biol. Chem. 265. 10260-10265.

Horvath, L. I., Dux, L., Hankovsky, H. O., Hideg, K., & Marsh, D. (1990). Saturation transfer electron spin resonance of Ca^{2+}-ATPase covalently spin-labeled with beta substituted vinyl ketones and maleimide-nitroxide derivatives. Effects of segmental motion and labeling levels. Biophys. J. 58, 231-241.

Huang, W -H., & Askari, A. (1984). Simultaneous bindings of ATP and vanadate to (Na^+ + K^+)-ATPase. J. Biol. Chem. 259, 13287-13291.

Hugentobler, G., Heid, I., & Solioz, M. (1983). Purification of a putative K^+-ATPase from *Streptococcus faecalis*. J. Biol. Chem. 258, 7611-7617.

Inesi, G. (1971). p-Nitrophenyl phosphate hydrolysis and calcium ion transport in fragmented sarcoplasmic reticulum. Science 171, 901-903.

Inesi, G., Kurzmack, M., Nakamoto, R., de Meis, L., & Bernhard, S. A. (1980a). Uncoupling of calcium control and phosphohydrolase activity in sarcoplasmic reticulum vesicles. J. Biol. Chem. 255, 6040-6043.

Inesi, G., Kurzmack, M., Coan, C., & Lewis, D. E. (1980b). Cooperative calcium binding and ATPase activation in sarcoplasmic reticulum vesicles. J. Biol. Chem. 255, 3025-3031.

Jackson, R. J., Mendlein, J., & Sachs, G. (1983). Interaction of fluorescein isothiocyanate with the (H^+ + K^+)-ATPase. Biochim. Biophys. Acta 731, 9-15.

Jaisser, F., Canessa, C. M., Horisberger, J-D., & Rossier, B. C. (1992). Primary sequence and functional expression of a novel ouabain-resistant Na,K-ATPase. J. Biol. Chem. 267, 16895-16903.

Jencks, W. P. (1980). The utilization of binding energy in coupled vectorial processes. Adv. Enzym. 51, 75-106.

Jensen, J., & Nørby, J. G. (1971). On the specificity of the ATP-binding site of (Na^+ + K^+) activated ATPase from brain microsomes. Biochim. Biophys. Acta 233, 395-403.

Jensen, J., & Ottolenghi, P. (1983). The abolition of subunit-subunit interaction and the maximum weight of the nucleotide-binding unit. Biochim. Biophys. Acta 731, 282-289.

Jørgensen, P. L. (1975). Purification and characterization of (Na^+,K^+)-ATPase V. Conformational changes in the enzyme. Transitions between the Na-form and the K-form studied with tryptic digestion as a tool. Biochim. Biophys. Acta. 401, 399-415.

Jørgensen, P. L. (1977). Purification and characterization of (Na^+,K^+)-ATPase VI. Differential tryptic modification of catalytic functions of the purified enzyme in presence of NaCl and KCl. Biochim. Biophys. Acta. 466, 97-108.

Jørgensen, K. E., Lind, K. E., Røigaard-Petersen, H. & Møller, J. V. (1978). The functional unit of calcium-plus-magnesium-ion-dependent adenosine triphosphate from sarcoplasmic reticulum. Biochem. J. 169, 489-498.

Jørgensen, P. L., & Petersen, J. (1982). High-affinity ^{86} Rb-binding and structural changes in the α-subunit of Na^+,K^+-ATPase as detected by tryptic digestion and fluorescence analysis. Biochim. Biophys. Acta 705, 38-47.

Jørgensen, P. L., & Andersen, J. P. (1988). Structural basis for E_1-E_2 conformational transitions in Na,K-pump and Ca-pump proteins. J. Membrane Biol. 103, 95-120.

Kapakos, J. G., & Steinberg, M. (1986). Ligand binding to (Na,K)-ATPase labeled with 5-iodoacetamidofluorescein. J. Biol. Chem. 261, 2084-2089.

Karlish, S. J. D., Yates, D. W., & Glynn, I. M. (1978). Conformational transitions between Na^+-bound and K^+-bound forms of (Na^+ + K^+)-ATPase, studied with formycin nucleotides. Biochim. Biophys. Acta 525, 252-264.

Karlish, S. J. D. (1980). Characterization of conformational changes in (Na,K)ATPase labeled with fluorescein at the active site. J. Bioenerg. Biomembranes 12, 111-136.

Kawakita, M., Yasuoka, K., & Kaziro, Y. (1980). Selective modification of functionally distinct sulfhydryl groups of sarcoplasmic reticulum Ca^{2+},Mg^{2+}-adenosine triphosphatase with *N*-ethylmaleimide. J. Biochem. 87, 609-617.

Kawakita, M., & Yamashita, T. (1987). Reactive sulfhydryl groups of sarcoplasmic reticulum ATPase III. Identification of cysteine residues whose modification with *N*-ethylmaleimide leads to loss of the Ca^{2+}-transporting activity. J. Biochem. 102, 103-109.

Kaya, S., Tsuda, T., Hagiwara, K., Fukui, T., & Taniguchi, K. (1994). Pyridoxal 5'-phosphate probes at Lys-480 can sense the binding of ATP and the formation of phosphoenzymes in Na^+,K^+-ATPase. J. Biol. Chem. 269, 7419-7422.

Keeling, D. J., Taylor, A. G.,& Schudt, C. (1989). The binding of a K^+ competitive ligand, 2-methyl,8-(phenylmethoxy)imidazo(1,2-a)pyridine 3-acetonitrile, to the gastric (H^+ + K^+)-ATPase. J. Biol. Chem. 264, 5545-5551.

Kessler, F., Bennardini, F., Bachs, O., Serratosa, J., James, P., Caride, A. J., Gazzotti, P., Penniston, J. T., & Carafoli, E. (1990). Partial purification and characterization of the Ca^{2+}-pumping ATPase of the liver plasma membrane. J. Biol. Chem. 265, 16012-16019.

Kimura, M., Yamaguchi, Y., Takada, S., & Tanabe, K. (1993). Cloning of a Ca^{2+}-ATPase gene of *Plasmodium falciparum* and comparison with vertebrate Ca^{2+}-ATPases. J. Cell Science 104, 1129-1136.

Kirley, T. L., & Peng, M. (1991). Identification of cysteine residues in lamb kidney (Na,K)-ATPase essential for ouabain binding. J. Biol. Chem. 266, 19953-19957.

Kison, R., Meyer, H. E., & Schoner, W. (1989). Characterization of a cysteine-containing peptide after affinity labelling of Ca^{2+}-ATPase of sarcoplasmic reticulum with the disulfide of 3'(2')-*O*-biotinyl-thioinosine triphosphate. Eur. J. Biochem. 181, 503-511.

Klevickis, C., & Grisham, C. M. (1982). Phosphorus-31 nuclear magnetic resonance studies of the conformation of an adenosine 5'-triphosphate analogue at the active site of (Na^+ + K^+)-ATPase from kidney medulla. Biochemistry 21, 6979-6984.

Koepsell, H., Hulla, F. W., & Fritzsch, G. (1982). Different classes of nucleotide binding sites in the (Na^+ + K^+)-ATPase studied by affinity labeling and nucleotide-dependent SH-group modifications. J. Biol. Chem. 257, 10733-10741.

Koland, J. G., & Hammes, G. G. (1986). Steady state kinetic studies of purified yeast plasma membrane proton-translocating ATPase. J. Biol. Chem. 261, 5936-5942.

Korczak, B., Zarain-Herzberg, A., Brandl, C. J., Ingles, C. J., Green, N. M., & MacLennan, D. H. (1988). Structure of the rabbit fast-twitch skeletal muscle Ca^{2+}-ATPase gene. J. Biol. Chem. 263, 4813-4819.

Kosk-Kosicka, D., Kurzmack, M., and Inesi, G. (1983). Kinetic characterization of detergent-solubilized sarcoplasmic reticulum adenosinetriphosphatase. Biochemistry 22, 2559-2567.

Krishna, S., Cowan, G., Meade, J. C., Wells, R. A., Stringer, J. R, & Robson, K. J. (1993). A family of cation ATPase-like molecules from *Plasmodium falciparum*. J. Cell Biol. 120, 385-398.

Kubota, T., Daiho, T., & Kanazawa, T. (1993). Quasi-irreversible inactivation of the sarcoplasmic reticulum Ca^{2+}-ATPase by simultaneous tight binding of magnesium and fluoride to the catalytic site. Biochim. Biophys. Acta 1163, 131-143.

Kuby, S. A., Palmieri, R. H., Frischat, A., Fischer, A. H., Wu, L. H., Maland, L., & Manship, M. (1984). Studies on adenosine triphosphate transphosphorylases. Amino acid sequence of rabbit muscle ATP-AMP transphosphorylase. Biochemistry, 23, 2393-2399.

Lacapère, J-J., Bennett, N., Dupont, Y., & Guillain, F. (1990). pH and magnesium dependence of ATP binding to sarcoplasmic reticulum ATPase. J. Biol. Chem. 265, 348-353.

Lacapère, J-J., & Guillain, F. (1990). Reaction mechanism of Ca^{2+}-ATPase of sarcoplasmic reticulum. Equilibrium and transient study of phosphorylation with Ca-ATP as substrate. J. Biol. Chem. 265, 8583-8589..

Lacapère, J-J., & Guillain, F. (1993). The reaction mechanism of Ca^{2+}-ATPase of sarcoplasmic reticulum. Direct measurement of Mg.ATP dissociation constant gives similar values in the presence or absence of calcium. Eur. J. Biochem. 211, 117-126.

Lacapère, J-J., Garin, J., Trinnaman, B., & Green, N. M. (1993). Identification of amino acid residues photolabeled with 8-azidoadenosine 5'-diphosphate in the catalytic site of sarcoplasmic reticulum Ca-ATPase. Biochemistry 32, 3414-3421.

Lane, L. K., Feldmann, J. M., Flarsheim, C. E., & Rybczynski, C. L. (1993). Expression of Rat αl Na,K-ATPase containing substitutions of "essential" amino acids in the catalytic center. J. Biol. Chem. 268, 17930-17934.

Lane, L. K. (1993). Functional expression of rat alpha 1 Na,K-ATPase containing substitutions for cysteines 454, 458, 459, 513 and 551. Biochem. & Mol. Biol. Intl. 31, 817-822.

Le D. T. (1986). Purification of labeled cyanogen bromide peptides of the α polypeptide from sodium ion and potassium ion activated adenosinetriphosphatase modified with *N*-[^{3}H]ethylmaleimide. Biochemistry 25, 2379-2386.

Lee, C-P., Kao, M-C., French, B. A., Putney, S. D., & Chang, S. H. (1987). The rabbit muscle phosphofructokinase gene. J. Biol. Chem. 262, 4195-4199.

le Maire, M., Lund, S., Viel, A., Champeil, P., and Moller, J. V. (1990). Ca^{2+}-induced conformational changes and location of Ca^{2+} transport sites in sarcoplasmic reticulum Ca^{2+}-ATPase as detected by the use of proteolytic enzyme (V8). J. Biol. Chem. 265, 1111-1123.

Lesniak, W., Modyanov, N., & Chiesi, M. (1994). Reversible inactivation of the sarcoplasmic reticulum Ca^{2+}-ATPase coupled to rearrangement of cytoplasmic protein domains as revealed by changes in trypsinization pattern. Biochemistry 33, 13678-13683.

Levy, D., Seigneuret, M., Bluzat, A., & Rigaud, J-L. (1990). Evidence for proton countertransport by the sarcoplasmic reticulum Ca^{2+}-ATPase during calcium transport in reconstituted proteoliposomes with low ionic permeability. J. Biol. Chem. 265, 19524-19534.

Lewis, S. M., & Thomas, D. D. (1992). Resolved conformational states of spin-labeled Ca-ATPase during the enzymatic cycle. Biochemistry 31, 7381-7389.

Lin, S-H., & Faller, L. D. (1993). Time resolution of fluorescence changes observed in titrations of fluorescein 5'-isothiocyanate-modified Na,K-ATPase with monovalent cations. Biochemistry 32, 13917-13924.

Lu, Y-Z., Xu, Z-C., & Kirchberger, M. A. (1993). Evidence for an effect of phospholamban on the regulatory role of ATP in calcium uptake by the calcium pump of the cardiac sarcoplasmic reticulum. Biochemistry 32, 3105-3111.

MacLennan, D. H., Brandl, C. J., Korczak, B., & Green, N. M. (1985). Amino-acid sequence of a Ca^{2+} + Mg^{2+}-dependent ATPase from rabbit muscle sarcoplasmic reticulum, deduced from its complementary DNA sequence. Nature 316, 696-700.

MacLennan, D. H., Clarke, D. M., Loo, T. W., & Skerjanc, I. S. (1992). Site-directed mutagenesis of the Ca^{2+} ATPase of sarcoplasmic reticulum. Acta Physiol. Scand. 146, 141-150.

Maeda, M., Tagaya, M., & Futai, M. (1988). Modification of gastric (H^+ + K^+)-ATPase with pyridoxal 5'-phosphate. J. Biol. Chem. 263, 3652-3656.

Mahaney, J. E., Froehlich, J. P., & Thomas, D. D. (1995). Conformational transitions of the sarcoplasmic reticulum Ca- ATPase studied by time-resolved EPR and quenched-flow kinetics. Biochemistry 34, 4864-4879.

Maldonado, A. M., & Portillo, F. (1995). Genetic analysis of the fluorescein isothiocyanate binding site of the yeast plasma membrane H^+-ATPase. J. Biol. Chem. 270, 8655-8659.

Malik, B., Jamieson, G. A. Jr., & Ball, W. J. Jr. (1993). Identification of the amino acids comprising a surface-exposed epitope within the nucleotide-binding domain of the Na^+,K^+- ATPase using a random peptide library. Protein Science 2, 2103-2111.

Maruyama, K., & MacLennan, D. H. (1988). Mutation of aspartic acid-351, lysine-352, and lysine-515 alters the Ca^{2+} transport activity of the Ca^{2+}-ATPase expressed in COS-1 cells. Proc. Natl. Acad. Sci. USA 85, 3314-3318.

Maruyama, K., Clarke, D. M., Fujii, J., Inesi, G., Loo, T. W., & MacLennan, D. H. (1989). Functional consequences of alterations to amino acids located in the catalytic center (Isoleucine 348 to Threonine 357) and nucleotide-binding domain of the Ca^{2+}-ATPase of sarcoplasmic reticulum. J. Biol. Chem. 264, 13038-13042.

McIntosh, D. B., & Berman, M. C. (1978). Calcium ion stabilization of the calcium transport system of sarcoplasmic reticulum. J. Biol. Chem. 253, 5140-5146.

McIntosh, D. B., & Boyer, P. D. (1983). Adenosine 5'-triphosphate modulation of catalytic intermediates of calcium ion activated adenosinetriphosphatase of sarcoplasmic reticulum subsequent to enzyme phosphorylation. Biochemistry 22, 2867-2875.

McIntosh, D. B., & Ross, D. C. (1988). Reaction cycle of solubilized monomeric Ca^{2+}-ATPase of sarcoplasmic reticulum is the same as that of the membrane form. J. Biol. Chem. 263, 12220-12223.

McIntosh, D. B., Ross, D. C., Champeil, P., & Guillain, F. (1991). Crosslinking the active site of sarcoplasmic reticulum Ca^{2+}-ATPase completely blocks Ca^{2+} release to the vesicle lumen. Proc. Natl. Acad. Sci. USA 88, 6437-6441.

McIntosh, D. B., Woolley, D. G., & Berman, M. C. (1992). 2',3'-0-(2,4,6-trinitrophenyl)-8-azido-AMP and -ATP photolabel Lys-492 at the active site of sarcoplasmic reticulum Ca^{2+}-ATPase. J. Biol. Chem. 267, 5301-5309.

McIntosh, D. B. (1992). Glutaraldehyde cross-links Lys-492 and Arg-678 at the active site of sarcoplasmic reticulum Ca^{2+}-ATPase. J. Biol. Chem. 267, 22328-22335.

McIntosh, D. B., & Woolley, D. G. (1994). Catalysis of an ATP analogue untethered and tethered to lysine 492 of sarcoplasmic reticulum Ca^{2+}-ATPase. J. Biol. Chem. 269, 21587-21595.

Meissner, G. (1973). ATP and Ca^{2+} binding by the Ca^{2+} pump protein of sarcoplasmic reticulum. Biochim. Biophys. Acta 298, 906-926.

Mendlein, J., Ditmars, M. L., & Sachs, G. (1990). Calcium binding to the H^+,K^+-ATPase. J. Biol. Chem. 265, 15590-15598.

Michelson, A. M., Blake, C. C. F., Evans, S. T., & Orkin, S. H. (1985). Structure of the human phosphoglycerate kinase gene and the intron-mediated evolution and dispersal of the nucleotide-binding domain. Proc. Natl. Acad. Sci. *USA*. 82, 6965-6969.

Mitchinson, C., Wilderspin, A. F., Trinnaman, B. J., & Green, N. M. (1982). Identification of a labeled peptide after stoichiometric reaction of fluorescein isothiocyanate with the Ca^{2+}-dependent adenosine triphosphatase of sarcoplasmic reticulum. FEBS Lett. 146, 87-92.

Moczydlowski, E. G. & Fortes, P. A. G. (1981a). Characterization of 2',3'-O-(2,4,6-trinitrocyclohexadienylidine)adenosine 5'triphosphate as a fluorescent probe of the ATP site of sodium and potassium transport adenosine triphosphatase. J. Biol. Chem. 256, 2346-2356.

Moczydlowski, E. G. & Fortes, P. A. G. (1981b). Inhibition of sodium and potassium adenosine triphosphatase by 2'3'-O-(2,4,6-trinitrocyclohexadienylidene) adenine nucleotides. J. Biol. Chem. 256, 2357-2366.

Modyanov, M., Lutsenko, S., Chertova, E., & Efremov, R. (1991). Architecture of the sodium pump molecule: probing the folding of the hydrophobic domain. Soc. Gen. Physiol. 46, 99-115.

Moutin, M. J., Cuillel, M., Rapin, C., Miras, R., Anger, M., Lompre, A. M., & Dupont, Y. (1994). Measurements of ATP binding on the large cytoplasmic loop of the sarcoplasmic reticulum Ca^{2+}-ATPase overexpressed in Escherichia coli. J. Biol. Chem. 269, 11147-11154.

Muallem S., & Karlish, S. J. D. (1983). Catalytic and regulatory ATP-binding sites of the red cell Ca^{2+} pump studied by irreversible modification with fluorescein isothiocyanate. J. Biol. Chem. 258, 169-175.

Munson, K. B. (1981). Light-dependent inactivation of (Na^+ + K^+)-ATPase with a new photoaffinity reagent, chromium arylazido-β-alanyl ATP. J. Biol. Chem. 256, 3223-3230.

Murphy, A. J. (1977). Sarcoplasmic reticulum adenosine triphosphatase: Labeling of an essential lysyl residue with pyridoxal-5'-phosphate. Arch. Biochem. Biophys. 180, 114-120.

Murphy, A. J. (1978). Effects of divalent cations and nucleotides on the reactivity of the sulfhydryl groups of sarcoplasmic reticulum membranes. J. Biol. Chem. 253, 385-389.

Murphy, A. J. (1988). Affinity labeling of the active site of the Ca^{2+}-ATPase of sarcoplasmic reticulum. Biochim. Biophys. Acta 946, 57-65.

Murphy, A. J. (1990). Reaction of a carbodiimide adduct of ATP at the active site of sarcoplasmic reticulum calcium ATPase. Biochemistry 29, 11236-11242

Murphy, A. J., & Coll, R. J. (1992). Fluoride is a slow, tight-binding inhibitor of the calcium ATPase of sarcoplasmic reticulum. J. Biol. Chem. 267, 5229-5235.

Nakamoto, R. K., & Inesi, G. (1984). Studies of the interactions of 2'3'-O-(2,4,6-(trinitrocyclohexyldienylidine)adenosine nucleotides with the sarcoplasmic reticulum (Ca^{2+} + Mg^{2+})-ATPase active site. J. Biol. Chem. 259, 2961-2970.

Nakamura, Y., & Tonomura, Y. (1978). Reaction mechanism of ρ-nitrophenylphosphatase of sarcoplasmic reticulum. J. Biochem. 83, 571-583.

Navarre, C., Ghislain, M., Leterme, S., Ferroud, C., Dufour, J- P., & Goffeau, A. (1992). Purification and complete sequence of a small proteolipid associated with the plasma membrane H^+- ATPase of *Saccharomyces cerevisiae*. J. Biol. Chem. 267, 6425- 6428.

Nediani, C., Fiorillo, C., Marchetti, E., Bandinelli, R., Degl' Innocenti, D., & Nassi, P. (1995). Acylphosphatase: A potential modulator of heart sarcolemma Na^+,K^+ pump. Biochemistry 34, 6668-6674.

Nicholas, R. A. (1984). Purification of the membrane-spanning tryptic peptides of the polypeptide from sodium and potassium ion activated adenosinetriphosphatase labeled with 1-tritiospiro[adamantane-4,3'-diazirine]. Biochemistry 23, 888-898.

Noumi, T., Maeda, M., & Futai, M. (1988). A homologous sequence between H^+-ATPase (F_oF_1) and cation-transporting ATPases. J. Biol. Chem. 263, 8765-8770.

Obara, M., Suzuki, H., & Kanazawa, T. (1988). Conformational changes in the vicinity of the *N*-iodoacetyl-*N'*-(5-sulfo-1-naphthyl)ethylenediamine attached to the specific thiol of sarcoplasmic reticulum Ca^{2+}-ATPase throughout the catalytic cycle. J. Biol. Chem. 263, 3690-3697.

Odermatt, A., Suter, H., Krapf, R., & Solioz, M. (1993). Primary structure of two P-type ATPases involved in copper homeostasis in *Enterococcus hirae*. J. Biol. Chem. 268, 12775-12779.

Odom, T. A., Chipman, D. M., Betts, G., & Bernhard, S. A. (1981). Transient and steady-state kinetic studies of sodium-potassium adenosine triphosphatase using β-(3-furyl)acryloyl phosphate as chromophoric substrate assay. Biochemistry 20, 480-486.

Ohta, T., Morohashi, M, Kawamura, M., & Yoshida, M. (1985). The amino acid sequence of the fluorescein-labeled peptides of electric ray and brine shrimp (Na,K)-ATPase. Biochem. Biophys. Res. Commun. 130, 221-228.

Ohta, T., Nagano, K., & Yoshida, M. (1986). The active site structure of Na^+/K^+-transporting ATPase: Location of the 5'-(*p*-fluorosulfonyl)benzoyladenosine binding site and soluble peptides released by trypsin. Proc. Natl. Acad. Sci. USA 83, 2071-2075.

Oliveira, C. R. G., Coan, C., & Verjovski-Almeida, S. (1988). Utilization of arylazido-ATP by sarcoplasmic reticulum ATPase in the absence of calcium. J. Biol. Chem. 263, 1861-1867.

Ottolenghi, P., & Jensen, J. (1983). The K^+-induced apparent heterogeneity of high-affinity nucleotide-binding sites in (Na^+ + K^+)-ATPase can only be due to the oligomeric structure of the enzyme. Biochim. Biophys. Acta 727, 89-100.

Ovchinnikov, Y. A., Monastyrskaya, G. S., Broude, N. E., Ushkaryov, Y. A., Melkov, A. M., Smirnov, Y. V., Malyshev, I. V., Allikmets, R. L., Kostina, M. B., Dulubova, I. E., Kiyatkin, N. I., Grishin, A. V., Modyanov, N. N., & Sverdlov, E. D. (1988). Family of human Na^+,K^+-ATPase genes. FEBS Lett. 233, 87-94.

Pardo, J. P., & Slayman, C. W. (1988). The fluorescein isothiocyanate-binding site of the plasma-membrane H^+-ATPase of *Neurospora crassa*. J. Biol. Chem. 263, 18664-18668.

Pardo, J. P., & Slayman, C. W. (1989). Cysteine 532 and cysteine 545 are the *N*-ethylmaleimide-reactive residues of the *Neurospora* plasma membrane H^+-ATPase. J. Biol. Chem. 264, 9373-9379.

Patzelt-Wenczler, R., & Schoner, W. (1981). Evidence for two different reactive sulfhydryl groups in the ATP-binding sites of (Na^+ + K^+)-ATPase. Eur. J. Biochem. 114, 79-87.

Pauls, H., Bredenbröcker, B., & Schoner, W. (1980). Inactivation of (Na^+ + K^+)-ATPase by chromium(III) complexes of nucleotide triphosphates. Eur. J. Biochem. 109, 523-533.

Pauls, H., Serpersu, E. H., Kirch, U., & Schoner, W. (1986). Chromium(III)ATP activating (Na^+ + K^+)-ATPase supports Na^+-Na^+ and Rb^+-Rb^+ exchanges in everted red blood cells but not Na^+,K^+ transport. Eur. J. Biochem. 157, 585-595.

Pedemonte, C. H., Kirley, T. L., Treuheit, M. J., & Kaplan, J. H. (1992). Inactivation of the Na,K-ATPase by modification of lys-501 with 4-acetamido-4'-isothiocyanatostilbene-2,2'acid (SITS). FEBS Lett. 314, 97-100.

Peluffo, R. D., Garrahan, P. J., & Rega A. F. (1992). Low affinity superphosphorylation of the Na,K-ATPase by ATP. J. Biol. Chem. 267, 6596-6601.

Peluffo, R. D., Rossi, R. C., Garrahan, P. J., & Rega, A. F. (1994). Low affinity acceleration of the phosphorylation reaction of the Na,K-ATPase by ATP. J. Biol. Chem. 269, 1051-1056.

Perez, C., Michelet, B., Ferrant, V., Bogaerts, P., & Boutry, M. (1992). Differential expression within a three-gene subfamily encoding a plasma membrane H^+-ATPase in *Nicotiana plumbaginifolia*. J. Biol. Chem. 267, 1204-1211.

Perlin, D. S., Harris, S. L., Seto-Young, D., & Haber, J. E. (1989). Defective H^+-ATPase of hygromycin B-resistant *pma1* mutants from *Saccharomyces cerevisiae*. J. Biol. Chem. 264, 21857-21864.

Petithory, J. R., & Jencks, W. P. (1986). Phosphorylation of the calcium adenosinetriphosphatase of sarcoplasmic reticulum: Rate-limiting conformational change followed by rapid phosphoryl transfer. Biochemistry 25, 4493-4497.

Pick, U., & Karlish, S. J. D. (1980). Indications for an oligomeric structure and for conformational changes in sarcoplasmic reticulum Ca^{2+}-ATPase labelled selectively with fluorescein. Biochim. Biophys. Acta 626, 255-261.

Pick, U. (1981a). Interaction of fluorescein isothiocyanate with nucleotide-binding sites of the Ca-ATPase from sarcoplasmic reticulum. Eur. J. Biochem. 121, 187-195.

Pick, U. (1981b). Dynamic interconversions of phosphorylated and non-phosphorylated intermediates of the Ca-ATPase from sarcoplasmic reticulum followed in a fluorescein-labeled enzyme. FEBS Lett. 123, 131-136.

Pick, U., & Bassilian, S. (1981). Modification of the ATP binding site of the Ca^{2+}-ATPase from sarcoplasmic reticulum by fluorescein isothiocyanate. FEBS Lett. 123, 127-130.

Pick, U. (1982). The interaction of vanadate ions with the Ca-ATPase from sarcoplasmic reticulum. J. Biol. Chem. 257, 6111-6119.

Pick, U., & Karlish, S. J. D. (1982). Regulation of the conformational transition in the Ca-ATPase from sarcoplasmic reticulum by pH, temperature, and calcium ions. J. Biol. Chem. 257, 6120-6126.

Pickart, C. M., & Jencks, W. P. (1982). Slow dissociation of ATP from the calcium ATPase. J. Biol. Chem. 257, 5319-5322.

Pickart, C. M., & Jencks, W. P. (1984). Energetics of the calcium-transporting ATPase. J. Biol. Chem. 259, 1629-1643.

Portillo, F., & Serrano, R. (1988). Dissection of functional domains of the yeast proton-pumping ATPase by directed mutagenesis. EMBO J. 7, 1793-1798.

Portillo, F., & Serrano, F. (1989). Growth control strength and active site of yeast plasma membrane ATPase studied by site-directed mutagenesis. Eur. J. Biochem. 186, 501-507.

Post, R. L., Hegyvary, C., & Kume, S. (1972). Activation by adenosine triphosphate in the phosphorylation kinetics of sodium and potassium ion transport adenosine triphosphatase. J. Biol. Chem. 247, 6530-6540.

Pratap, P. R., & Robinson, J. D. (1993). Rapid kinetic analyses of the Na^+/K^+-ATPase distinguish among different criteria for conformational change. Biochim. Biophys. Acta 1151, 89-98.

Rabon, E. C., Smillie, K., Seru, V., & Rabon, R. (1993). Rubidium occlusion within tryptic peptides of the H,K-ATPase. J. Biol. Chem. 268, 8012-8018.

Reinstein, J., & Jencks, W. P. (1993). The Binding of ATP and Mg^{2+} to the calcium adenosinetriphosphatase of sarcoplasmic reticulum follows a random mechanism. Biochemistry 32, 6632-6642

Renaud, K. J., Inman, E. M., & Fambrough, D. M. (1991). Cytoplasmic and transmembrane domain deletions of Na,K-ATPase β-subunit. J. Biol. Chem. 266, 20491-20497.

Rephaeli, A., Richards, D. E., & Karlish, S. J. D. (1986). Electrical potential accelerates the E_1P(Na)→ E_2P conformational transition of (Na,K)-ATPase in reconstituted vesicles. J. Biol. Chem. 261, 12437-12440.

Robinson, J. D. (1970). Phosphate activity stimulated by Na^+ plus K^+: Implications for the (Na^+ plus K^+)-dependent adenosine triphosphatase. Arch. Biochem. Biophys. 139, 164-171.

Rogers, J. (1985). Exon shuffling and intron insertion in serine protease genes. Nature 315, 458-459.

Ross, D. C., & McIntosh, D. B. (1987a). Intramolecular cross- linking of domains at the active site links A_1 and B subfragments of the Ca^{2+}-ATPase of sarcoplasmic reticulum. J. Biol. Chem. 262, 2042-2049.

Ross, D. C., & McIntosh, D. B. (1987b). Intramolecular cross-linking at the active site of the Ca^{2+}-ATPase of sarcoplasmic reticulum. J. Biol. Chem. 262, 12977-12983.

Ross, D. C., Davidson, G. A., & McIntosh, D. B. (1991). Mechanism of inhibition of sarcoplasmic reticulum Ca^{2+}-ATPase by active site cross-linking. J. Biol. Chem. 266, 4613-4621.

Rossi, R. C., & Nørby, J. G. (1993). Kinetics of K^+-stimulated dephosphorylation and simultaneous K^+ occlusion by Na,K-ATPase, studied with the K^+ congener Tl^+. J. Biol. Chem. 268, 12579-12590.

Sagara, Y., & Inesi, G. (1991). Inhibition of the sarcoplasmic reticulum Ca^{2+} transport ATPase by thapsigargin at subnanomolar concentrations. J. Biol. Chem. 266, 13503-13506.

Saito-Nakatsuka, K., Yamashita, T., Kubota, I., & Kawakita, M. (1987). Reactive sulfhydryl groups of sarcoplasmic reticulum ATPase. I. Location of a group which is most reactive with *N*-ethylmaleimide. J. Biochem. 101, 365-376.

Sasaki, T., Inui, M., Kimura, Y., Kuzuya, T., & Tada, M. (1992). Molecular mechanism of regulation of Ca^{2+} pump ATPase by phospholamban in cardiac sarcoplasmic reticulum. J. Biol. Chem. 267, 1674-1679.

Scheiner-Bobis, G., & Schoner, W. (1985). Demonstration of an Mg^{2+}-induced conformational change by photoaffinity labelling of the high-affinity ATP-binding site of (Na^+ + K^+)-ATPase with 8-azido-ATP. Eur. J. Biochem. 152, 739-746.

Scheiner-Bobis, G., Fahlbusch, K., & Schoner, W. (1987). Demonstration of cooperating α subunits in working (Na^+ + K^+)-ATPase by the use of MgATP complex analogue cobalt tetrammine ATP. Eur. J. Biochem. 168, 123-131.

Scheiner-Bobis, G., Antonipillai, J., & Farley, R. A. (1993). Simultaneous binding of phosphate and TNP-ADP to FITC-modified Na^+,K^+-ATPase. Biochemistry 32, 9592-9599.

Scofano, H. M., Vieyra, A., & de Meis, L. (1979). Substrate regulation of the sarcoplasmic reticulum ATPase. J. Biol. Chem. 254, 10227-10231.

Seebregts, C. J. & McIntosh, D. B. (1989). 2',3'-O-(2,4,6-trinitrophenyl)-8-azido-adenosine mono-, di-, and triphosphates as photoaffinity probes of the Ca^{2+}ATPase of sarcoplasmic reticulum. J. Biol. Chem. 264, 2043-2052.

Sen, P. C., Kapakos, J. G., & Steinberg, M. (1981). Modification of (Na^+ + K^+)-dependent ATPase by fluorescein isothiocyanate : Evidence for the involvement of different amino groups at different pH values. Arch. Biochem. Biophys. 211, 652-661.

Serpersu, E. H., Kirch, U., & Schoner, W. (1982). Demonstration of a stable occluded form of Ca^{2+} by the use of the chromium complex of ATP in the Ca^{2+}-ATPase of sarcoplasmic reticulum. Eur. J. Biochem. 122, 347-354.

Serpersu, E. H., Bunk, S., & Schoner, W. (1990). How do MgATP analogues differentially modify high-affinity and low-affinity ATP binding site of Na^+/K^+-ATPase? Eur. J. Biochem. 191, 397-404.

Serrano, R., & Portillo, F. (1990). Catalytic and regulatory sites of yeast plasma membrane H^+-ATPase studied by directed mutagenesis. Biochim. Biophys. Acta 1018, 195-199.

Serrano, R., Villalba, J. M., Palmgren, M. G., Portillo, F., Parets-Soler, A., Roldan, M., Ferguson, C., & Montesinos, C. (1992). Studies of the plasma H^+-ATPase of yeast and plants. Biochem. Soc. Trans. 20, 562-566.

Shaffer, E., Azari, J., & Dahms, A. S. (1978). Properties of the P_i-Oxygen exchange reaction catalyzed by (Na^+,K^+)-dependent adenosine triphosphatise. J. Biol. Chem. 253, 5696-5706.

Shigekawa, M., Dougherty, J. P., & Katz, A. M. (1978). Reaction mechanism of Ca^{2+}-dependent ATP hydrolysis by skeletal muscle sarcoplasmic reticulum in the absence of added alkali metal salts. J. Biol. Chem. 253, 1442-1450.

Shigekawa, M., Wakabayashi, S., & Nakamura, H. (1983). Reaction mechanism of Ca^{2+}-dependent adenosine triphosphatase of sarcoplasmic reticulum. J. Biol. Chem. 258, 8698-8707.

Shull, G. E., & Greeb, J. (1988). Molecular cloning of two isoforms of the plasma membrane Ca^{2+}-transporting ATPase from rat brain. J. Biol. Chem. 263, 8646-8657.

Skou, J. C., & Esmann, M. (1983). Effect of magnesium ions on the high-affinity binding of eosin to the $(Na^+ + K^+)$-ATPase. Biochim. Biophys. Acta 727, 101-107.

Skriver, E., Maunsbach, A. B., Herbert, H., Scheiner-Bobis, G., & Schoner, W. (1989). Two-dimensional crystalline arrays of Na,K-ATPase with new subunit interactions induced by cobalt-tetrammine-ATP. J. Ultrastr. Mol. Struct. Res. 102, 189-195.

Solioz, M., Mathews, S., & Fürst, P. (1987). Cloning of the K^+-ATPase of *Streptococcus faecalis*. J. Biol. Chem. 262, 7358-7362.

Stahl, N., & Jencks, W. P. (1984). Adenosine 5'-triphosphate at the active site accelerates binding of calcium to calcium adenosinetriphosphatase. Biochemistry 23, 5389-5392.

Stefanova, H. I., East, J. M., Gore, M. G., & Lee, A. G. (1992). Labeling the $(Ca^{2+}$-$Mg^{2+})$-ATPase of sarcoplasmic reticulum with 4-(Bromomethyl)-6,7-dimethoxycoumarin: Detection of conformational changes. Biochemistry 31, 6023-6031.

Stefanova, H. I., Mata, A. M., East, J. M., Gore, M. G., & Lee, A. G. (1993a). Reactivity of lysyl residues on the $(Ca^{2+}$-$Mg^{2+})$-ATPase to 7-amino-4-methylcoumarin-3-acetic acid succinimidyl ester. Biochemistry 32, 356-362.

Stefanova, H. I., Mata, A. M., Gore, M. G., East, J. M., & Lee, A. G. (1993b). Labeling the $(Ca^{2+}$-$Mg^{2+})$-ATPase of sarcoplasmic reticulum at glu-439 with 5-(bromomethyl)fluorescein. Biochemistry 32, 6095-6103.

Stein, W. D. (1995). The sodium pump in the evolution of animal cells. Phil. Trans. Roy. Soc. (in press).

Stewart, J. M. MacD., & Grisham, C. M. (1988). ^1H nuclear magnetic resonance studies of the conformation of an ATP analogue at the active site of Na,K-ATPase from kidney medulla. Biochemistry 27, 4840-4848.

Stewart, J. M. MacD., Jørgensen, P. L., & Grisham, C. M. (1989). Nuclear Overhauser effect studies of the conformation of $Co(NH_3)_4ATP$ bound to kidney Na,K-ATPase. Biochemistry 28, 4695-4701.

Stokes, D. L., Taylor, W. R., & Green, N. M. (1994). Structure, transmembrane topology and helix packing of P-type ion pumps. FEBS Lett. 346, 32-38.

Stoltzfus, A., Spencer, D. F., Zuker, M., Logsdon Jr, J. M., & Doolittle, W. F. (1994). Testing the Exon theory of genes: the evidence from protein structure. Science 265, 202-207.

Story, R. M., & Steitz, T. A. (1992). Structure of the recA protein-ADP complex. Nature 355, 374-376.

Stuckey, J. A., Schubert, H. L., Fauman, E. B., Zhang, Z-Y., Dixon, J. E., & Saper, M. A. (1994). Crystal structure of *Yersinia* protein tyrosine phosphatase at 2.5Å and the complex with tungstate. Nature 370, 571-575.

Su, X-D., Taddel, N., Stefani, M., Ramponi, G., & Nordlund, P. (1994). The crystal structure of a low-molecular-weight phosphotyrosine protein phosphatase. Nature 370, 575-578.

Suzuki, H., Obara, M., Kuwayama, H., & Kanazawa, T. (1987). A conformational change of *N*-iodoacetyl-*N'*-(5-sulfo-1- naphthyl)ethylenediamine-labeled sarcoplasmic reticulum Ca^{2+}-ATPase upon ATP binding to the catalytic site. J. Biol. Chem. 262, 15448-15456.

Suzuki, S., Kawato, S., Kouyama, T., Kinosita, K., Ikegami, A., & Kawakita, M. (1989). Independent flexible motion of submolecular domains of the Ca^{2+},Mg^{2+}-ATPase of sarcoplasmic reticulum measured by time-resolved fluorescence depolarization of site-specifically attached probes. Biochemistry 28, 7734-7740.

Suzuki, H., Kubota, T., Kubo, K., & Kanazawa, T. (1990). Existence of a low-affinity ATP-binding site in the unphosphorylated Ca^{2+}-ATPase of sarcoplasmic reticulum vesicles: Evidence from binding of 2'3'-O-(2,4,6-trinitrocyclohexadienylidene)-[^{3}H]AMP and -[^{3}H]ATP. Biochemistry 29, 7040-7045.

Suzuki, H., Nakamura, S., & Kanazawa, T. (1994). Effects of divalent cations bound to the catalytic site on ATP-induced conformational changes in the sarcoplasmic reticulum Ca^{2+}-ATPase: Stopped-flow analysis of the fluorescence of *N*-acetyl-*N'*-(5-sulfo-1-naphthyl)ethylenediamine attached to cysteine-674. Biochemistry 33, 8240-8246.

Sweadner, K. J. (1979). Two molecular forms of ($Na^+ + K^+$)-stimulated ATPase in brain. J. Biol. Chem. 254, 6060-6067.

Sweadner, K. J., & Farshi, S. K. (1987). Rat cardiac ventricle has two Na^+,K^+-ATPases with different affinities for ouabain: Developmental changes in immunologically different catalytic subunits. Proc. Natl. Acad. Sci. USA 84, 8404-8407.

Tada, M., Kirchberger, M. A., Repke, D. I., & Katz, A. M. (1974). The stimulation of calcium transport in cardiac sarcoplasmic reticulum by adenosine 3':5'-monophosphate-dependent protein kinase. J. Biol. Chem. 249, 6174-6180.

Tamura, S., Tagaya, M., Maeda, M., & Futai, M. (1989). Pig gastric ($H^+ + K^+$)-ATPase. J. Biol. Chem. 264, 8580-8584.

Taniguchi, K., Tosa, H., Suzuki, K., & Kamo, Y. (1988). Microenvironment of two different extrinsic fluorescence probes in $Na,^+$,K^+-ATPase changes out of phase during sequential appearance of reaction intermediates. J. Biol. Chem. 263, 12943-12947.

Taniguchi, K., & Mardh, S. (1993). Reversible changes in the fluorescence energy transfer accompanying formation of reaction intermediates in probe-labeled (Na^+,K^+)-ATPase. J. Biol. Chem. 268, 15588-15594.

Tate, C. A., Bick, R. J., Chu, A., Van Winkle, W. B., & Entman, M. L. (1985). Nucleotide specificity of cardiac sarcoplasmic reticulum. J. Biol. Chem. 260, 9618-9623.

Tate, C. A., Bick, R. J., Blaylock, S. L., Youker, K. A., Scherer, N. M., & Entman, M. L. (1989). Nucleotide specificity of canine cardiac sarcoplasmic reticulum. J. Biol. Chem. 264, 7809-7813.

Tate, C. A., Shin, G., Walseth, T. F., Taffet, G. E., Bick, R. J., & Entman, M. L. (1991). Nucleotide specificity of cardiac sarcoplasmic reticulum. J. Biol. Chem. 276, 16165-16170.

Teruel, J. A., & Inesi, G. (1988). Roles of phosphorylation and nucleotide binding domains in calcium transport by sarcoplasmic reticulum adenosinetriphosphatase. Biochemistry 27, 5885-5890.

Toyofuku, T., Kurzydlowski, K., Tada, M., & MacLennan, D. H. (1994). Amino acids Lys-Asp-Asp-Lys-Pro-Val[402] in the Ca^{2+}-ATPase of cardiac sarcoplasmic reticulum are critical for functional association with phospholamban. J. Biol. Chem. 269, 22929-22932.

Toyoshima, C., Sasabe, H., & Stokes, D. L. (1993). Three-dimensional cryo-electron microscopy of the calcium ion pump in the sarcoplasmic reticulum membrane. Nature 362, 469-471.

Tran, C. M., Huston, E. E., & Farley, R. A. (1994a). Photochemical labeling and inhibition of Na,K-ATPase by 2-azido-ATP. J. Biol. Chem. 269, 1-8.

Tran, C. M., Scheiner-Bobis, G., Schoner, W., & Farley, R. A. (1994b). Identification of an amino acid in the ATP binding site of Na^+/K^+-ATPase after photochemical labeling with 8-azido-ATP. Biochemistry 33, 4140-4147.

Tunwell, R. E. A., Conlan, J. W., Matthews, I., East, J. M., & Lee, A. G. (1991). Definition of surface-exposed epitopes on the (Ca^{2+}-Mg^{2+})-ATPase of sarcoplasmic reticulum. Biochem. J. 279, 203-212.

Tyson, P. A., Steinberg, M., Wallick, E. T., & Kirley, T. L. (1989). Identification of the 5-Iodoacetamidofluorescein reporter site on the Na,K-ATPase. J. Biol. Chem. 264, 726-734.

Ulaszewski, S., Van Herck, J-C., Dufour, J-P., Kulpa, J., Nieuwenhuis, B., Goffeau, A. (1987). A single mutation confers vanadate resistance to the plasma membrane H^+-ATPase from the yeast *Schizosaccharomyces pombe*. J. Biol. Chem. 262, 223-228.

Van Uem, T. J. F., Peters, W. H. M., & De Pont, J. J. H. H. M. (1990). A monoclonal antibody against pig gastric H^+/K^+-ATPase, which binds to the cytosolic $E_1.K^+$ form. Biochim. Biophys. Acta 1023, 56-62.

Van Uem, T. J. F., Swarts, A.G., & De Pont, J. J. H. H. M. (1991). Determination of the epitope for the inhibitory monoclonal antibody 5-B6 on the catalytic subunit of gastric Mg^{2+}-dependent H^+ transporting and K^+ stimulated ATPase. Biochem. J. 280, 243-248.

Vilsen, B., & Andersen, J. P. (1986). Occlusion of Ca^{2+} in soluble monomeric sarcoplasmic reticulum Ca^{2+}-ATPase. Biochim. Biophys. Acta 855, 429-431.

Vilsen, B., Andersen, J. P., Petersen, J., & Jørgensen, P. L. (1987). Occlusion of $^{22}Na^+$ and $^{86}Rb^+$ in membrane-bound and soluble protomeric $\alpha\beta$-units of Na,K-ATPase. J. Biol. Chem. 262, 10511-10517.

Vilsen, B., & Andersen, J. P. (1992). Interdependence of Ca^{2+} occlusion sites in the unphosphorylated sarcoplasmic reticulum Ca^{2+}-ATPase complex with CrATP. J. Biol. Chem. 267, 3539-3550.

Vilsen, B., Andersen, J. P., & MacLennan, D. H. (1991). Functional consequences of alterations to amino acids located in the hinge domain of the Ca^{2+}-ATPase of sarcoplasmic reticulum. J. Biol. Chem. 266, 16157-16164.

Vulpe, C., Levinson, B., Whitney, S., Packman, S., & Gitschier, J. (1993). Isolation of a candidate gene for Menkes disease and evidence that it encodes a copper-transporting ATPase. Nature Genetics 3, 7-13.

Wakabayashi, S., Ogurusu, T., & Shigekawa, M. (1986). Factors influencing calcium release from the ADP-sensitive phosphoenzyme intermediate of the sarcoplasmic reticulum ATPase. J. Biol. Chem. 261, 9762-9769.

Wakabayashi, S., & Shigekawa, M. (1987). Effect of metal bound to the substrate site on calcium release from the phosphoenzyme intermediate of sarcoplasmic reticulum ATPase. J. Biol. Chem. 262, 11524-11531.

Wakabayashi, S., Imagawa, T., & Shigekawa, M. (1990). Does fluorescence of 4-nitrobenzo-2-oxa-1,3-diazole incorporated into sarcoplasmic reticulum ATPase monitor putative E_1-E_2 conformational transition? J. Biochem. 107, 563-571.

Wakabayashi, S., & Shigekawa, M. (1990). Mechanism for activation of the 4-nitrobenzo-2-oxa-1,3-diazole-labeled sarcoplasmic reticulum ATPase by Ca^{2+} and its modulation by nucleotides. Biochemistry 29, 7309-7318.

Walker, J. E., Saraste, M., Runswick, M. J., & Gay, N. J. (1982). Distantly related sequences in the α- and β-subunits of ATP synthase, myosin, kinases and other ATP-requiring enzymes and a common nucleotide binding fold. EMBO J. 1, 945-951.

Wang, K., & Farley, R. A. (1992). Lysine 480 is not an essential residue for ATP binding or hydrolysis by Na,K-ATPase. J. Biol. Chem. 267, 3577-3580.

Ward, D. G., & Cavieres, J. E. (1993). Solubilized Na,K-ATPase remains protomeric during turnover yet shows apparent negative cooperativity toward ATP. Proc. Natl. Acad. Sci. USA. 90, 5332-5336.

Watanabe, T., & Inesi, G. (1982). The use of 2',3'-O-(2,4,6-trinitrophenyl) adenosine 5'-triphosphate for studies of nucleotide interaction with sarcoplasmic reticulum vesicles. J. Biol. Chem. 257, 11510-11516.

Wawrzynow, A., Theibert, J. L., Murphy, C., Jona, I., Martonosi, A., & Collins, J. H. (1992). Sarcolipin, the "Proteolipid" of skeletal muscle sarcoplasmic reticulum, is a unique, amphipathic, 31-residue peptide. Arch. Biochem. Biophys. 298, 620-623.

Wawrzynow, A., Collins, J. H., & Coan, C. (1993). An iodoacetamide spin-label selectively labels a cysteine side chain in an occluded site on the sarcoplasmic reticulum Ca^{2+}-ATPase. Biochemistry 32, 10803-10811.

Xu, K., & Kyte, J. (1989). Nucleophilic behavior of lysine-501 of the α-polypeptide of sodium and potassium ion activated adenosinetriphosphatase consistent with a role in binding adenosine triphosphate. Biochemistry 28, 3009-3017.

Xu, K. (1989). Acid dissociation constant and apparent nucleophilicity of lysine-501 of the α-polypeptide of sodium and potassium ion activated adenosinetriphosphatase. Biochemistry 28, 6894-6899.

Yamada, S., & Ikemoto, N. (1980). Reaction mechanism of calcium-ATPase of sarcoplasmic reticulum. J. Biol. Chem. 255, 3108-3119.

Yamagata, K., Daiho, T., & Kanazawa, T. (1993). Labeling of lysine 492 with pyridoxal 5'-phosphate in the sarcoplasmic reticulum Ca^{2+}-ATPase. J. Biol. Chem. 268, 20930-20936.

Yamaguchi, M., & Kanazawa, T. (1984). Protonation of the sarcoplasmic reticulum Ca-ATPase during ATP hydrolysis. J. Biol. Chem. 259, 9526-9531.

Yamaguchi, M., & Kanazawa, T. (1985). Coincidence of H^+ binding and Ca^{2+} dissociation in the sarcoplasmic reticulum Ca-ATPase during ATP hydrolysis. J. Biol. Chem. 260, 4896-4900.

Yamamoto, H., Tagaya, M. Fukui, T., & Kawakita, M. (1988). Affinity labeling of the ATP-binding site of Ca^{2+}-transporting ATPase of sarcoplasmic reticulum by adenosine triphosphopyridoxal: Identification of the reactive lysyl residue. J. Biochem. 103, 452-457.

Yamamoto, H., Imamura, Y., Tagaya, M., Fukui, T., & Kawakita, M. (1989). Ca^{2+}-dependent conformational change of the ATP-binding site of Ca^{2+}-transporting ATPase of sarcoplasmic reticulum as revealed by an alteration of the target-site specificity of adenosine triphosphopyridoxal. J. Biochem. 106, 1121-1125.

Yamasaki, K., Sano, N., Ohe, M., & Yamamoto, T. (1990). Determination of the primary structure of intermolecular cross-linking sites on the Ca^{2+}-ATPase of sarcoplasmic reticulum using ^{14}C-labeled N,N'-(1,4-phenylene)bismaleimide or *N*-ethylmaleimide. J. Biochem. 108, 918-925.

Yamasaki, K., Daiho, T., & Kanazawa, T. (1994). 3'-O-(5-fluoro-2,4-dinitrophenyl)-ATP exclusively labels Lys-492 at the active site of the sarcoplasmic reticulum. J. Biol. Chem. 269, 4129-4134.

Zhang, Z-Y., Wang, Y., & Dixon, J. E. (1994). Dissecting the catalytic mechanism of protein-tyrosine phosphatases. Proc. Natl. Acad. Sci. USA 91, 1624-1627.

Zolotarjova, N., Periyasamy, S. M., Huang, W-H., & Askari, A. (1995). Functional coupling of phosphorylation and nucleotide binding sites in the proteolytic fragments of Na^+/K^+-ATPase. J. Biol. Chem. 270, 3989-3995.

THE GASTRIC H^+-K^+-ATPase

Jai Moo Shin, Denis Bayle, Krister Bamberg, and George Sachs

Advances in Molecular and Cell Biology
Volume 23A, pages 101-142.

ISBN: 0-7623-0287-9

I. INTRODUCTION

This review of the gastric H^+-K^+-ATPase focuses on the molecular and biochemical properties of the enzyme. This enzyme is responsible for the elaboration of HCl by the parietal cell of the gastric mucosa. It catalyzes an electroneutral exchange of cytoplasmic protons for extracytoplasmic potassium. The extracytoplasmic potassium derives from cellular K^+ by K^+, Cl^- efflux into the lumen of the secretory canaliculus of the parietal cell (an infolding of the apical surface). It is a member of an ever-growing family of P type ion-motive ATPases that catalyze transport by means of conformational changes driven by cyclic phosphorylation and dephosphorylation of the catalytic subunit of the ATPase. This family of enzymes is represented in both prokaryotes and eukaryotes and are primary ion transporters.

Structurally, these membrane-embedded proteins can be divided into cytoplasmic, membrane and extracytoplasmic domains. Since ion transport must occur across the membrane domain, much of this review is devoted to an analysis of the nature of the membrane domain of the H,K ATPase which is compared to the membrane domains of the Na^+-K^+ and Ca^{2+}-ATPases. The membrane domain is traversed by the ions transported by these ATPases and this or the adjacent extracytoplasmic domain is the site of action of inhibitors such as ouabain in the case of the Na^+-K^+-ATPase and the substituted benzimidazoles and the imidazopyridines in the case of the H^+-K^+-ATPase.

II. CLASSES OF ION MOTIVE ATPases

There are three classes of ion transport ATPases: the F-, V-, and P-type ATPases. Their function in the case of the F type is ATP synthesis, in the case of the V type, transport of protons and in the case of the P type, transport of various cations such as Na^+, Ca^{2+}, H^+, Mg^{2+}, Cd^{2+}, Cu^{2+} and K^+.

The F1/F0 ATPase is present in mitochondria and synthesizes ATP at the expense of the proton motive force generated by electron transport (Futai et al., 1989). This enzyme shows a very similar organization and mechanism as compared to the V ATPase. The F ATPase has eight different subunits with a stoichiometry of, for F1 3α, 3β, 1γ, 1δ, 1ε, and for F_o, 1 a, 2 b, and 10-12 c subunits. The α subunit of the F1 ATPase has homology with the B subunit of the V ATPase, and the β subunit of F ATPase is homologous to the A subunit of V ATPase. The DCCD binding c subunit of the F_0 appears to be responsible for the proton transport (Hermolin and

Fillingame, 1989). This protein complex is responsible for ATP synthesis in mitochondria, bacteria, and chloroplasts.

The V ATPase generates the proton-motive force for accumulating neurotransmitters in storage vesicles and energizing the vacuolar system of eukaryotes, among others (Nelson, 1991). The V ATPase is therefore responsible for acidification of the interior of organelles inside the cell. Such an enzyme is also involved in pH homeostasis by the kidney and in bone resorption by the osteoclast. In these instances the enzyme is present in the plasma membrane. The V ATPase contains seven to ten subunits organized into two distinct domains. The cytoplasmic domain contains five different subunits with molecular weights between 72 and 26 kDa, and the membrane domain contains two different subunits with molecular masses of about 20 and 16 kDa. The A cytoplasmic subunit (molecular mass of 72 kDa) in cooperation with the B cytoplasmic subunit (molecular mass of 57 kDa) contains the catalytic site and the membrane subunits a and c conduct protons across the membrane.

The P type ATPase family in mammals contains the Ca^{2+}-ATPases, the Na^+-K^+-ATPases, and the H^+-K^+-ATPases. These ATPases are membrane bound enzymes with similar structural motifs (Jorgensen and Andersen, 1988; Glynn and Karlish, 1990; Rabon and Reuben, 1990). The Na^+-K^+-ATPase is an ubiquitous electrogenic pump mainly responsible for the maintenance of Na and K gradients between cells and their environment. The H^+-K^+-ATPase is the electroneutral pump responsible for producing gastric acid. The Ca^{2+}-ATPases encoded by the SERCA genes (Lytton et al., 1992) maintain the Ca^{2+} gradient necessary for this ion's ability to act as a second messenger or to initiate muscular contraction. The Na^+-K^+ and H^+-K^+-ATPases are composed of two subunits. The α subunits with molecular mass of about 100 kDa have the catalytic site and the β subunits with peptide mass of 35 kDa are noncovalently bound to the α subunit and are glycoproteins with most of their mass exposed to the extra-cytoplasmic surface. The Ca^{2+}-ATPases have a single catalytic protein of mass of about 100 kDa, which pumps Ca^{2+} in exchange for H^+(Yamaguchi and Kanazawa, 1984). There are also several isoforms of the plasma membrane Ca^{2+} pump which contains a calmodulin binding domain absent from the SERCA family (Lytton et al., 1992; Carafoli, 1992). There is about 60% sequence homology between the Na^+-K^+-ATPase and the H^+-K^+-ATPase α subunit, while the Ca^{2+}-ATPase of sarcoplasmic reticulum shows only about 15% overall homology with the H^+-K^+-ATPase. The hydropathy profile is, however, very similar. There is about 40% sequence homology between the β subunit of the Na^+-K^+-ATPase and the similar subunit of the H^+-K^+-ATPase (Canfield et al., 1990).

The global structural motif in both the F and V type ATPases is a separable catalytic cytoplasmic domain and an ion transporting membrane domain. Although the membrane domain contains few subunits, there are multiple copies of the proton transporting subunits suggesting that multiple membrane spanning helices are necessary for ion transport by these pumps. The ion selectivity of the membrane domain appears to determine the overall selectivity of transport. For example, whereas the F1/Fo ATPase of *E. coli* transports protons, and the F1/F0 ATPase of *Propionigenium modestum* transports Na^+, chimeric pumps constructed of the cytoplasmic domain of one F1/Fo pump and the membrane domain of the other transport the ion determined by the membrane domain rather than the cytoplasmic domain (Laubinger et al., 1990). In the case of the P type ATPases, they have several membrane spanning segments, many of them containing amino acids essential for ion transport. Chimeric constructs of these pumps appear to conform to what has been found for the F type pump, namely that it is the membrane domain that determines general ion selectivity and the cytoplasmic domain energizes the transport of the ion (Blostein et al., 1993). The hydropathy plot of the P type ATPases predicts that the membrane domain contains several transmembrane helices presumably enclosing the ion transport pathway.

The gastric H^+-K^+-ATPase, along with the Na^+-K^+-ATPase, have provided a therapeutic target in the treatment of disease. Digitalis, which inhibits the Na^+-K^+-ATPase is probably the oldest prescription drug still in use today, acting to selectively increase cell Ca^{2+} in cardiac muscle by allowing Na^+ loading and thereby decreasing Na^+/Ca^{2+} exchange. The substituted benzimidazoles which inhibit the H^+-K^+-ATPase are the newest class of anti-ulcer drugs used in the treatment of a variety of diseases of the upper GI tract. Imidazopyridines and arylquinolines are being investigated as an alternative means of inhibiting gastric acid secretion by inhibition of the H^+-K^+-ATPase.

III. STRUCTURE OF THE H^+-K^+-ATPase

A. General

The gastric H^+-K^+-ATPase consists of two subunits, a catalytic α subunit and a heavily glycosylated β subunit (Hall et al., 1990). While the Na^+-K^+-ATPase β subunit was identified soon after the discovery of the enzyme, the H^+-K^+-ATPase β subunit was identified some years after discovery of this enzyme because the β subunit was not easy to detect on SDS gel electrophoresis using the Laemmli buffer system due to smearing on the gel. The β subunit was eventually identified in work analyzing the carbohydrates of gastric membranes. Postembedding electron microscopic staining techniques showed that wheat germ agglutinin staining occurred on the extracellular face of the inside-out gastric vesicles. Then the β subunit was partially sequenced after deglycosylation followed by protease digestion (Hall et al., 1990). The partial amino acid sequence showed strong homology with the β subunit of the Na^+-K^+-ATPase. This led to the sequencing of the gastric H^+-K^+-ATPase β subunit from a gastric cDNA library (Reuben et al., 1990; Shull, 1990). The same subunit was found by determining the nature of the antigen in autoimmune gastritis (Toh et al., 1990). Using lectin affinity chromatography, the H^+-K^+-ATPase α subunit was co-purified with the β subunit, showing that the α subunit interacts with the β subunit (Okamoto et al., 1990; Callaghan et al., 1990; Callaghan et al., 1992). By crosslinking with low concentrations of glutaraldehyde, the α subunit was shown to be closely associated with the β subunit (Rabon et al., 1990a; Hall et al., 1991).

The only ion pump for which a detailed structure is available is bacteriorhodopsin which has been diffracted to a resolution of 2.8 Å. There are 7 transmembrane helices, and the crucial all trans retinal group that isomerizes to 13 cis retinal with light is bound to lys^{216} in the middle of the membrane. Asp^{85} and glu^{96} are on the cytoplasmic and extra-cytoplasmic side of the retinal group. Isomerization of the retinal group alters the directionality and pK_a of the $lysine^{216}$ Schiff base allowing sided deprotonation. There appears to be tilting of the transmembrane helices which form a narrow channel on the cytoplasmic side and a broader channel on the extra-cytoplasmic side of the retinal moiety (Baldwin et al., 1988; Ceska and Henderson, 1990). All the helices participate on one side or the other of the ion pathway across bacterial rhodopsin. Presumably protonation from the cytoplasmic side and deprotonation of the Schiff base to the extra-cytoplasmic face depends on the change in orientation of the Schiff base and a change in the peptide conformation on either side of the base.

Ion pumping by bacterial rhodopsin involves a change of conformation in the central part of the membrane domain which acts like a switch sending protons from one side of the membrane to the other. This change in conformation is driven by the isomerization of a bound photoabsorptive pigment. Changes in binding energy are presumably driven by the change in tilt of a significant number of transmembrane helices. EP type pumps lack photoabsorptive pigments but perhaps the cycle of phosphorylation and dephosphorylation determines both the switching and the conformational change in these pumps and is transmitted from the cytoplasmic to the membrane domain. The cytoplasmic domains of P type ATPases are much larger than those of bacterial rhodopsin, reflecting the fact that ATP, not light, is used as an energy source for vectorial ion transport.

Whereas two-dimensional crystals of the Ca^{2+}-ATPase have been diffracted to a resolution of 14 Å (Toyoshima et al., 1993) which allows visualization of the cytoplasmic, membrane and extra-cytoplasmic domain, crystals of the H^+-K^+-ATPase have only been diffracted to a resolution of 25 Å (Rabon et al., 1986). The Ca^{2+}-ATPase data show the presence of a bird-like head, a stalk, a transmembrane region which may contain 8 or 10 membrane-spanning helices significantly tilted with respect to each other and a small extra-cytoplasmic domain. The combination of electron diffraction, transmission electron microscopy using tannic acid staining and freeze fracture have allowed definition of the general dimensions of the H^+-K^+-ATPase showing a large cytoplasmic domain, a smaller membrane domain and an even smaller extra-cytoplasmic domain (Rabon et al., 1986; Mohraz et al., 1990). The dimensions calculated in this way are similar to those observed for the Ca^{2+}-ATPase.

B. The α Subunit of H^+-K^+-ATPase

General Aspects

The primary sequences of the α subunits deduced from cDNA have been reported for pig (Maeda et al., 1988a), rat (Shull and Lingrel, 1986) and rabbit (Bamberg et al., 1992). The hog gastric H^+-K^+-ATPase α subunit sequence deduced from its cDNA consists of 1034 amino acids and has a Mr of 114,285 (Maeda et al., 1988a). The sequence based on the known N-terminal amino acid sequence is one less than the cDNA derived sequence (Lane et al., 1986). The rat gastric H^+-K^+-ATPase consists of 1033 amino acids and has a Mr of 114,012 (Shull and Lingrel, 1986), and the rabbit gastric H^+-K^+-ATPase consists of 1,035 amino acids, showing a Mr of 114,201 (Bamberg et al., 1992). The degree of conservation among the α subunits is extremely high (over 97% identity). In addition, the gene sequence for human and the 5' part of the rat H^+-K^+-ATPase α subunits have been determined (Maeda et al., 1990; Newman et al., 1990; Oshiman et al., 1991). The human gastric H^+-K^+-ATPase gene has 22 exons and encodes a protein of 1035 residues including the initiator methionine residue (Mr = 114,047). These H^+-K^+-ATPase α subunits show high homology (~60% identity) with the Na^+-K^+-ATPase catalytic α subunit (Maeda et al., 1990). The putative distal colon H^+-K^+-ATPase α subunit has also been sequenced and shares 75% homology with both the H^+-K^+- and Na^+-K^+-ATPases (Crowson and Shull, 1992).

The gastric α subunit has conserved sequences along with the other P-type ATPases for the ATP binding site, the phosphorylation site, the pyridoxal 5'-phosphate binding site and the fluorescein isothiocyanate binding site. These sites are thought to be within the ATP binding domain in the large cytoplasmic loop between membrane spanning segments 4 and 5.

In the case of the hog gastric H^+-K^+-ATPase, pyridoxal 5'-phosphate is bound at lys^{497} of the α subunit in the absence but not the presence of ATP (Tamura et al., 1989), suggesting that lys^{497} is present in the ATP binding site or in its vicinity (Maeda et al., 1988b). The phosphorylation site was observed to be at asp^{386} (Walderhaug et al., 1985). Fluorescein isothiocyanate (FITC) covalently labels the gastric H^+-K^+-ATPase in the absence of ATP (Jackson et al., 1983). The binding site of FITC was at lys^{518} (Farley and Faller, 1985). However, several additional lysines such as those at positions 497 and 783, were shown to react with FITC during the inactivation of the Na^+-K^+-ATPase and to be protected from reaction with FITC when ATP was present in the incubation (Xu, 1989). Based on these data similar lysines of the H^+-K^+-ATPase could be near or in the ATP binding site.

Membrane Topology of the α Subunit

There are various methods available for determining membrane topology of a membrane inserted protein. The most commonly used is to determine the location of membrane-spanning α helices using hydropathy plots (Kyte and Doolittle, 1982). These are based on determining a moving average of hydrophobic, neutral, and hydrophilic amino acids using a variety of scales. When the average is greater than a predetermined value of 11 or more amino acids, then this sector is considered to be potentially membrane-spanning. In the absence of crystal information, hydropathy plot information has to be buttressed by specific biochemical information, as discussed below. The potential membrane-spanning segments are indicated in the text below as H1, H2 and so forth.

Biochemical methods use sites of labeling with cytoplasmic, extracytoplasmic, or membrane-directed chemical probes, determination of the peptides remaining in the membrane following cleavage of the extramembranal domain (usually cytoplasmic), determination of the sidedness of epitopes for antibodies and site-directed mutagenesis.

The hydrophobicity plots of the α subunit of the H^+-K^+-ATPase based on the primary amino acid sequence have suggested an 8- or 10-membrane spanning segment model for the secondary structure, as for the other mammalian P type ATPases. There is agreement on the first 4 membrane spanning segments in the N-terminal one-third of the α subunit. However, in the C-terminal one-third of the protein, a prediction from the hydropathy analysis is more difficult and therefore controversial in its interpretation.

The gastric enzyme is readily isolated as intact, ion-tight, uniformly (94%) inside-out vesicles. With such a sided preparation, a set of biochemical techniques can be used to establish the location of each region of the enzyme. More than one approach is necessary. The techniques that have been used include protease cleavage of the cytoplasmic domain, and allocation of the sites of reaction of cytoplasmic ligands such as ATP, FITC, or pyridoxal phosphate. Extra-cytoplasmic reagents such as the substituted benzimidazoles or photoaffinity imidazopyridine analogs have also provided information as to the secondary structure of the enzyme. The topology of epitopes for either polyclonal antibodies directed against specific peptides or of monoclonal antibodies generated against the enzyme or even intact parietal cells has also provided topological data. Finally the fact that the C terminal amino acids are tyr-tyr allows iodination in combination with carboxypeptidase cleavage to be used to determine the sidedness of the C-terminal domain.

An alternative approach is to determine the capability of putative membrane segments to act as signal anchor or stop transfer sequences in *in vitro* translation. This method has the additional advantage that the boundaries and determinant amino acids can be more closely defined for each segment found either by direct experimentation or from hydropathy analysis. Some insight can also be obtained as to the mechanisms involved in the folding of the enzyme.

Cytoplasmic Ligand Binding Sites. The phosphorylation site, FITC binding sites, and the pyridoxal phosphate binding site are cytoplasmic. Asp-386 is the phosphorylation site (Walderhaug et al., 1985), lys-518 is the FITC binding site

(Farley and Faller, 1985) and lys-497 the pyridoxal phosphate binding site (Tamura et al., 1989). These are all predicted to be present in the loop between H4 and H5, consistent with their cytoplasmic location and interaction with ATP.

Tryptic Cleavage of the α Subunit. The key assumption in the use of protease digestion as a topological probe is that the protease cleaves only cytoplasmic residues while maintaining vesicle integrity. The digestion pattern is thus simplified in that fewer bonds are accessible. If the cytoplasmic fragments are removed by washing, the pattern is simplified even further. Thus, instead of 97 possible tryptic fragments of the H^+-K^+-ATPase α subunit, only 4 or 5 membrane segment pairs should be left, along with the partially cleaved β subunit.

In either an 8- or 10-transmembrane segment model, each transmembrane pair connecting the luminal loop has at least one cysteine, which allows fluorescent labeling of the cysteines left after complete cleavage of the cytoplasmic domain. The N-terminal position of each fluorescent peptide fragment is then defined by N-terminal sequencing. The size of the fragment is determined from M_r measurements in the tricine gradient gels used for the separation. This then allows the C termination of the peptide to be identified using the lys or arg cleavage sites at this end. Although it is thought that there might be some chymotryptic activity in even the purest preparations of trypsin, in all the tryptic sequences we have analyzed, no evidence was obtained for chymotryptic cleavage sites.

Four transmembrane pairs connected by their luminal loop were detected in the hog gastric H^+-K^+-ATPase digest (Besancon et al., 1993; Shin et al., 1993). A tryptic peptide fragment beginning at gln^{104} represents the H1/loop/H2 sector. The H3/loop/H4 sector was found at a single peptide beginning at thr^{291}, and the H5/loop/H6 sector at a peptide beginning at leu^{776}. The H7/loop/H8 region was found in a single peptide fragment of 11 kDa, beginning at leu^{853}. These represent the first four transmembrane segment pairs of the proposed 10-transmembrane segment model (Bamberg et al., 1992). A typical digest result is shown in Figure 1 , to be compared with a digest result using intact sarcoplasmic reticulum vesicles.

From these studies, 4 membrane segment/loop/membrane segment sectors were identified, corresponding to H1/loop/H2 through H7/loop/H8. Although the hydropathy plot predicts two additional membrane spanning segments (H9, H10) at the C-terminal region of the enzyme, no evidence was obtained for this pair in spite of the fact that H9 is predicted to have 4 cysteine residues that are in principle able to bind fluorescein 5-maleimide (F-MI). The absence of F-MI labeling of these hydrophobic segments may suggest that the cysteines in H9 are post-translationally modified. Reduction, ester cleavage with hydroxylamine and chelating agents have not been successful in identifying this putative transmembrane pair after tryptic cleavage. The nature of the post-translational modification, if this is the reason, is at present, obscure.

Iodination. The C-terminal amino acids of the α subunit are tyr-tyr, which therefore can be iodinated with peroxidase-H_2O_2-^{125}I on the cytoplasmic side of in-

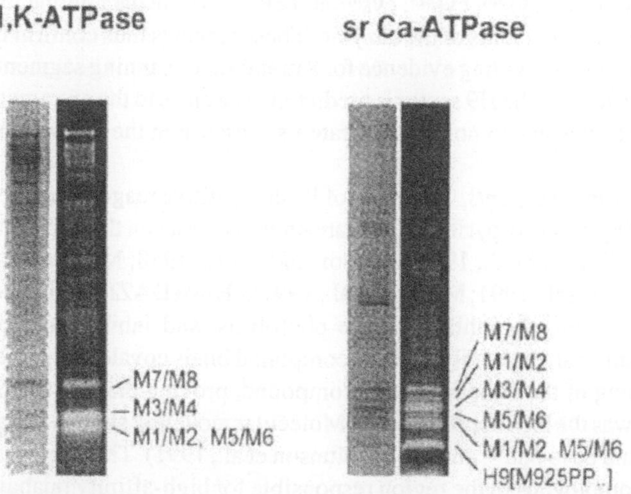

Figure 1. A tryptic digest of intact membrane vesicles of hog gastric H,K ATPase and rabbit SR vesicles was performed (trypsin/protein 1/4). The membrane fraction was isolated, solubilized in SDS, labeled with flourescein maleimide and separated on a SDS tricine gradient gel (Besancon et al., 1993; Shin et al., 1994). The bands were identified by N-terminal sequencing.

tact hog gastric vesicles. Digestion with carboxypeptidase Y then released about 28% of the counts incorporated into the α subunit, as would be predicted from a cytoplasmic location of the C terminal tyrosines (Scott et al., 1992). These data show that there is an even number of transmembrane segments in the α subunit.

Sided Reagents. Ouabain is the best known sided reagent for the Na^+-K^+-ATPase. High-affinity binding depends on the boundary amino acids between H1 and H2 and their connecting extra-cytoplasmic loop (Price and Lingrel, 1988). In all likelihood, given the large size of ouabain and the compact nature of the membrane domain, more regions are involved in ouabain binding. In the case of the gastric H^+-K^+-ATPase, two sets of reagents are known that also react exclusively with the extra-cytoplasmic surface and hence are useful in mapping the membrane disposition of these sites.

One class of reagents available as extra-cytoplasmic molecular probes are omeprazole, lansoprazole, and pantoprazole, which are all substituted benzimidazoles. These reagents are weak base, acid-activated compounds, which form cationic sulfenamides in acidic environments. The sulfenamides formed react with the SH group of cysteines in proteins to form relatively stable disulfides. Since the pump generates acid on its extra-cytoplasmic surface, only those cysteines available from that surface are accessible to these sulfenamides if labeling is carried out under acid transporting conditions. The cysteines that are labeled, depending on the reagent used (see below) are cys^{321}, cys^{813}, cys^{822} and cys^{892}. These data define the H3/H4, H5/H6 and H7/H8 segments of the enzyme. These reagents thus confirm the results of tryptic digestion providing evidence for 8 membrane spanning segments. Again, although a cysteine in the H9 sector is predicted to be close to the extra-cytoplasmic domain, these reagents do not demonstrate its presence in the mature pump.

K^+ Competitive Reagents. A series of K^+ competitive reagents has been developed, including imidazo-pyridines, that are known to react on the outside surface of the pump (Wallmark et al., 1987; Munson and Sachs, 1988; Mendlein and Sachs, 1990; Kaminski et al., 1991; Munson et al., 1991). ^{3}H-MeDAZIP is an imidazopyridine that competitively inhibits before photolysis, and inhibits and binds in a saturable manner after photolysis. This compound binds covalently in the same region of binding of the nonphotolyzed compound, proving that the sector binding this reagent was the H1/loop/H2 sector. Molecular modeling studies suggested that binding should be to phe^{124} and asp^{136} (Munson et al., 1991). This region of the H,K ATPase is homologous to the region responsible for high-affinity ouabain binding to the H^+-K^+-ATPase (Price et al., 1990). These data verify the presence of the first two segments spanning the membrane in the α subunit of the H^+-K^+-ATPase. A fluorescent K^+ competitive arylquinoline, 1-(2-methylphenyl)-4-methylamino-6--methyl-2,3-dihydropyrrolo[3,2-c]quinoline (MDPQ), shows enhanced hydrophobicity of its environment with formation of the E2-P[I] conformer of the enzyme, as if the H1/loop/H2 segment moves further into the membrane in the E2 conformation (Rabon et al., 1991).

Substituted Benzimidazoles. The substituted benzimidazole compounds available clinically with inhibitory activity on the H^+-K^+-ATPase are omeprazole, lansoprazole, and pantoprazole.

Omeprazole shows inhibitory activity only under the acidic conditions generated by the pump (Wallmark et al., 1984; Im et al., 1985; Lorentzon et al., 1985; Lindberg et al., 1986; Lorentzon et al., 1987; Keeling et al., 1987). The action of omeprazole following acidification is to react with available cysteine-SH groups. The reaction with the cysteine-SH produces a disulfide, cys-S-S-(Inhibitor), which covalently inhibits the enzyme in terms of ATPase activity and acid transport.

The enzyme was labeled using ^{3}H- omeprazole under acid transporting conditions. The cysteines reacting in the ATPase were determined by trypsinolysis followed by SDS-PAGE electrophoretic separation of the tryptic fragments and sequencing. The 2 cysteines reacting were found to be in position 813 or 822 and 892. These should be on or close to the extra-cytoplasmic face of the enzyme and are predicted to be in the H5/loop/H6 and H7/loop/H8 sector of the enzyme, respectively (Besancon et al., 1993). This reagent therefore defines two pairs of membrane spanning segments. Since it does not react with the β subunit, the 6 cysteines on this subunit predicted to be extra-cytoplasmic must be disulfide linked.

Lansoprazole, another substituted benzimidazole, labels at three positions on the enzyme, at cysteine 321, at cysteine 813 or 822, and at cysteine 892 (Sachs et al., 1993). Pantoprazole labels at both cys 813 and 822, with no labeling at either 892 or 321 (Shin et al., 1993). These cysteines are predicted to be in the H3/loop/H4 sector (cys 321) and in the H5/loop/H6 sector (cys 813,822) and in the H7/loop/H8 sector (cys 892). The labeling by lansoprazole provides additional evidence for the existence of the H3/H4 membrane spanning pair.

Antibody Epitope Studies on the α Subunit. There are many monoclonal antibodies that have been generated reacting with the H^+-K^+-ATPase. With the sequence of the α subunit deduced from cDNA, it is now possible to define the epitopes, and to determine the sidedness of these epitopes by staining intact or permeable cells with either fluorescent or immunogold techniques.

Antibody 95 inhibits ATP hydrolysis in the intact vesicles, and appears to be K^+ competitive (Bayle et al., 1992). Its epitope was identified by Western blotting of an *E. coli* expression library of fragments of the cDNA encoding the α subunit, as well as by Western analysis of tryptic fragments. The sequence recognized by this antibody was between amino acid positions 529 and 561. Since it inhibits intact vesicles this epitope must be cytoplasmic. Its epitope is close to the region known to bind the cytoplasmic reagent, FITC, namely in the loop between H4 and H5. The ability of this antibody to immunoprecipitate intact gastric vesicles showed that it was on the cytoplasmic surface of the pump.

Antibody 1218 was shown by Western analysis of tryptic fragments and by recombinant methodology to have its major epitope between amino acid positions 665 and 689 (Mercier et al., 1993). This epitope which is on the cytoplasmic surface of the enzyme is also in the loop between H4 and H5. This antibody does not inhibit ATPase activity. A second epitope for mAb 1218 was also identified. This epitope was between amino acid positions 853 and 946 according to Western blot analysis of tryptic fragments. A synthetic peptide (888-907) containing this second epitope displaced mAb 1218 from vesicles adsorbed to the surface of ELISA wells (Bayle et al., 1992).

Monoclonal antibody 146 was generated against intact parietal cells and subsequently purified and shown to react with rat H^+-K^+-ATPase. In cells it reacts on the outside surface of the canaliculus as shown by immunogold electron microscopy (Mercier et al., 1989). Western analysis of rat, rabbit, and hog enzyme gave the surprising result that it was present on the β subunit of rat and rabbit and absent from the β subunit of hog. Disulfide reduction eliminated reactivity of this antibody. Comparing sequences of the different β subunits, there is an arg, pro substitution in the hog for leu, val in the rat between the disulfide at position 161 and 178. This suggests that the β subunit epitope of mAb 146 is contained within this region of the subunit. On the other hand there was also an epitope on the α subunit of all three species recognized by mAb 146 also using Western analysis. This epitope was defined both by tryptic mapping and octamer walking to be between positions 873 and 877 of the hog α subunit. This is on or close to the extra-cytoplasmic face of H7. The finding using Western blotting that there was an epitope both on the α and β subunits was confirmed by expressing the rabbit subunits individually in SF9 cells using baculovirus transfection. The β subunit reacted in the SF9 cells and on Western blots with mAb 146. The α subunit did not react in the cells but did on Western blots as if the epitope in the α subunit was difficult to access in the absence of a denaturing detergent such as SDS (Mercier et al., 1993). These data may indicate tight binding between the α and β subunits in this region of the enzyme, as is discussed further below.

Molecular Biological Analysis. A molecular biological method was developed to analyze not only for the presence of the membrane segments defined as above, but also to explore the nature of membrane insertion (Bamberg and Sachs, 1994). A cDNA encoding a fusion protein of the 102 *N*-terminal amino acids of the rabbit H^+-K^+-ATPase α subunit linked by a variable segment to the 177 most C-terminal amino acids of the rabbit H^+-K^+-ATPase β subunit was transcribed and translated in a rabbit reticulocyte lysate system using labeled methionine in the absence or presence of microsomes. The cDNA for the variable region containing one or more putative membrane spanning segments is synthesized using selected primers in a PCR reaction and ligated into the cDNA construct. Since the β subunit region has 5 consensus N-glycosylation sites, translocation of the C-terminal β subunit part of the fusion protein into the interior of the microsomes can be determined by glycosylation. The presence of glycosylation is evidence for an odd number of transmembrane segments in the variable region preceding the β subunit part. The absence of glycosylation shows either the presence of an even number of membrane spanning segments or absence of membrane insertion. Figure 2 illustrates the principle of the method.

Membrane segments can act either as an insertion signal, when the protein contains a signal anchor sequence or can act as a stop transfer sequence when the protein contains a sequence which prevents further transport of the newly synthesized peptide across the microsomal membrane. To define a signal anchor sequence, that sequence is placed in the variable region. To define a stop transfer sequence the sequence is placed after a known signal anchor sequence, either its natural companion sequence or after the sequence for M1 in the experiments performed on the H^+-K^+-ATPase.

The sequence inserted into the translation vector for the putative H1 segment (using the rabbit sequence beginning at methionine) began at lys^{102} and extended to thr^{136}. When translation was carried out in the presence of microsomes essentially all of the product was glycosylated as determined by an increase in MWt on SDS gel electrophoresis. Direct evidence for this was provided by digestion of the solubilized product with N-glycanase which reduced the MWt to that seen in the absence of microsomes.

The sequence used for the putative H2 membrane spanning segment began at tyr^{142} and extended to ser^{171}. By itself, it not only acted as a signal anchor sequence,

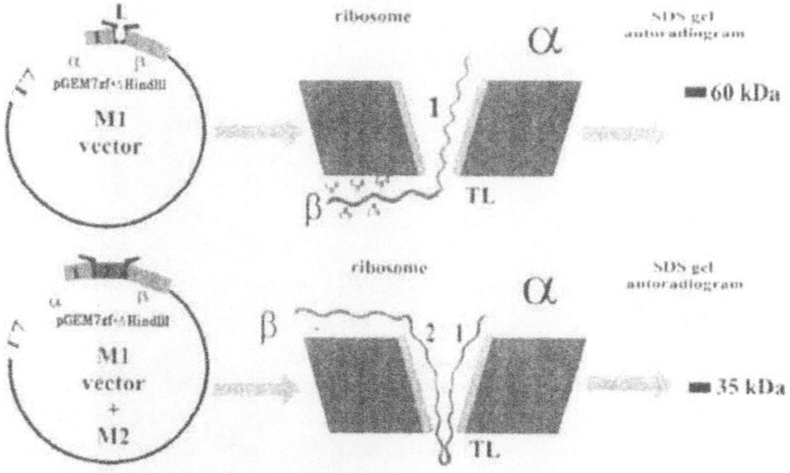

Figure 2. A simplified model illustrating the principle of *in vitro* translation of the M1 vector used to determine membrane spanning segments of the α subunit. Translation is carried out in the presence or absence of microsomes, and the Mwt of the product determined by SDS PAGE and autoradiography. With an odd number of membrane spanning segments, glycosylation alters the Mwt of the product. In the absence of membrane spanning segments, or with an even number, no glycosylation is observed.

but in association with the H1 segment also prevented glycosylation of H1, showing that it could act as a stop transfer sequence as well. These data demonstrate that translation of the H1/H2 sector of the enzyme agrees with the allocation of this membrane segment pair as determined from hydropathy and biochemical analysis. Thus H1 acts as the signal anchor sequence and H2 as the stop transfer sequence in this region of the α subunit. Figure 3 illustrates the SDS gel when translation of Mo alone and Mo containing H1, H2, and H1⁺H2 were carried out. In the absence of a membrane signal sequence there is no glycosylation. In the presence of either H1 or H2 alone there is glycosylation. In the presence of both H1 and H2, glycosylation does not occur.

The sequence for the H3 membrane spanning segment used for insertion into the translation vector began at val^{303} and ended at arg^{330}. When translated, the product was glycosylated, showing that this sequence could act as a signal anchor sequence. When inserted subsequent to the H1 segment, glycosylation was absent. This showed that this sequence could act as both a signal anchor and stop transfer sequence.

A sequence used for investigating the presence of the H4 segment began at arg^{330} and ended at lys^{360}. This sequence was glycosylated on its own and acted as a stop transfer sequence when inserted in conjunction with H3 or even H1. This transla-

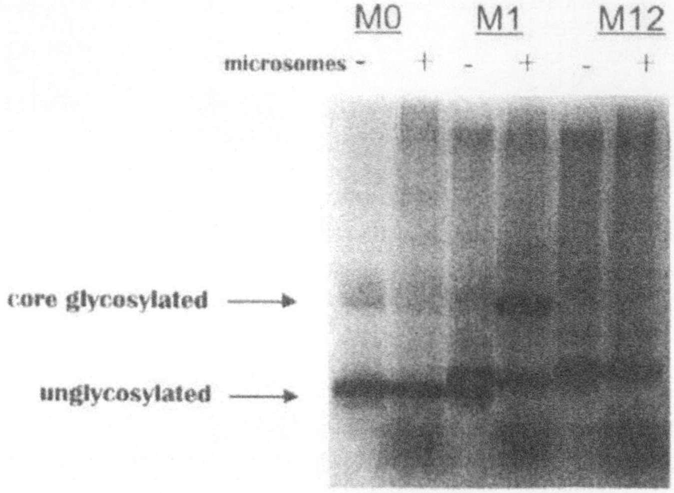

Figure 3. The results of *in vitro* transcription/translation of a DNA sequence containing the 100 first amino acids of the α subunit, a linker sequence and the 200 C-terminal amino acids of the β subunit. The translation is carried out in the absence or presence of microsomes. Inserted into the linker sequence is either no membrane segment (MO), the first (M1), or both together (M1/2). It can be seen that with the M1 membrane segment alone, there is glycosylation of the translation product, showing passage of the two sequence across the microsomal membrane. With a membrane segment pair (M1 + M2), no glycosylation is observed, showing that M2 can act as a stop transfer sequence.

tion data, once again, match the results of hydropathy analysis and biochemical cleavage and labeling data.

The data from translation thus allow the conclusion that it is likely that the first 4 membrane segment pairs are co-inserted during translation and form the membrane anchoring domain prior to translation of the large cytoplasmic loop between H4 and H5 which contains about 430 amino acids.

Various sequences were used to demonstrate the presence of H5 and H6 by this translation methodology. The putative H5 sequences tested began at arg777 or ala788 and ended at leu^{811}, ile^{821}, or thr^{825}. There was some glycosylation evident with the sequence ending at ile^{821}, and this sequence reduced but did not abolish glycosylation when inserted subsequent to H1. This can be considered tentative evidence for an H5 membrane spanning segment, but the data are by no means as clear as those for the first four membrane sequences of the enzyme. None of the other presumed H5 sequences acted as either signal anchor or stop transfer sequences.

The H6 sequences tested began at gly^{814} and ended either at glu^{836} or arg^{852}. These sequences did not act as either signal anchor or as stop transfer sequences. Hence the translation system where co-insertion of H5 or H6 was investigated did not correlate with the trypsinolysis data nor with the labeling observed with the benzimidazoles. Hydropathy analysis predicts only a single transmembrane sequence in the region encompassing H5 and H6.

Various sequences were also tried for translation of the supposed H7 spanning segments. The longest began at arg^{846} and ended at arg^{898}. These acted neither as signal anchor sequences on their own nor as stop transfer sequences when associated with the H1 segment.

The H8 sequence selected for testing began at tyr^{924} and ended at ile^{947}. This sequence did not act as a signal anchor sequence but acted as a stop transfer sequence when translated in association with H1. Thus of the membrane sequences H5 through H8, the only one conforming to its predicted role as a membrane insertion signal on its own is H8. The biochemical data demonstrating the existence of two pairs of membrane segments, H5 and H6 and H7, and H8, are strong. Thus H5, H6, H7 and H8 were all defined both by tryptic cleavage and by labeling by 3 substituted benzimidazoles which act as extra-cytoplasmic cysteine reagents. The negative *in vitro* translation data found in this region suggest that this region of the enzyme is post-translationally inserted and this may depend on preceding sequence or other factors.

The effect of preceding sequence was tested by generating longer insertion sequences beginning at H1 and ending with the different C terminal sequences for H5 or H7. Although glycosylation would be anticipated from an odd number of membrane spanning segments, this was not observed. Hence the signal for membrane insertion of this region might be subsequent to H8.

Although no biochemical data have been obtained for the presence of a ninth or tenth membrane spanning sequence, hydropathy suggests the presence of such sequences. The translation vector containing the sequence indicated by hydropathy beginning at phe^{960} and ending at phe^{990} was glycosylated and prevented glycosylation when inserted subsequent to H1. Hence this sequence can act both as a signal anchor sequence and as a stop transfer sequence.

The sequence representing the tenth hydrophobic domain begins at pro^{994} and ends at arg^{1017}. When this was present alone in the translation vector, no glycosylation was detected. However, when present in association with H1 or the ninth hydrophobic segment, glycosylation was inhibited. This shows that whereas this sequence does not function as a signal anchor sequence it does act as a stop transfer sequence.

The translation data, in contrast to previous data on the H^+-K^+-ATPase, show that there is likely to be an additional pair of membrane spanning sequences H9 and H10 in the α subunit of the H^+-K^+-ATPase.

From this approach, the first 4 membrane segments are present and are co-inserted with translation. The last 2 hydrophobic sequences also appear to be co-inserted. The information within the H5, H6, and H7 sequences is apparently insufficient for these to act on their own as either anchor or stop transfer sequences. The information within the H8 sequence is sufficient for this to act as a stop transfer sequence. However, a stop transfer sequence in the absence of a preceding signal anchor sequence may have no implications for membrane insertion of the stop transfer sequence (Bayle et al., 1995). That H5 through H7 are indeed membrane inserted in the mature enzyme is clear from the biochemical analyses. A reasonable hypothesis that may explain these translation data is that insertion of H5 through H7 is post-translational depending on insertion of H9/10 and that H8 then acts as a stop transfer sequence. The presence of the additional pair of membrane segments predicted by hydropathy is strongly suggested by these translation data. Their absence using fluorescein maleimide as a cysteine labeling reagent may be due to covalent bonding of the 4 cysteines present in H9. These bonds are neither disulfide or thioester in nature since neither reduction nor hydroxylamine treatment elicit F-MI labeling.

The combination of techniques described above, namely, sided proteolysis, epitope mapping, iodination, sided reactivity, and *in vitro* translation provides evidence for a 10 membrane segment model, with a large cytoplasmic loop between H4 and H5 and a large extra-cytoplasmic loop between H7 and H8. Given the evidence presented above, where 10 transmembrane segments have been determined, the transmembrane segments defined above are now named as M1, M2, and so forth.

Effect of Disulfide Reduction on α Subunit Proteolysis. It has been shown that reduction of the disulfides of the β subunit inhibits the activity of the H^+-K^+-ATPase (Chow et al., 1992). In the case of the Na^+-K^+-ATPase, the effect of reducing agents on the ability of the enzyme to hydrolyze ATP and bind ouabain was quantitatively correlated with the reduction of disulfide bonds in the β subunit(Kirley, 1990). When these disulfide bonds are disrupted in the H^+-K^+-ATPase, the pattern obtained following tryptic digestion changes. The normal tryptic cleavage site at the N-terminal end of H5 is at position 776. Following reduction, the cleavage site moves to position 792. This change of tryptic cleavage at the N-terminal end of H5 is similar to that seen with labeling of both cysteines in the H5/H6 region with pantoprazole (Shin et al., 1993).

C. The β Subunit

The primary sequences of the β subunits have been reported for rabbit (Reuben et al., 1990), hog (Toh et al., 1990), rat (Canfield et al., 1990; Shull, 1990; Maeda et al., 1991; Newman and Shull, 1991), mouse (Canfield and Levenson, 1991; Morley et al., 1992) and human (Ma et al., 1991) enzymes. The hydropathy profile of the β subunit appears less ambiguous than the α subunit. There is one membrane spanning region predicted by the hydropathy analysis, which is located at the region between positions 38 and 63 near the N-terminus. The tryptic digestion of the intact gastric H^+-K^+-ATPase produces no visible cleavage of the β subunit on SDS gels. Wheat germ agglutinin (WGA) binding of the β subunit is retained. These data indicate that most of the β subunit is extra-cytoplasmic and glycosylated. When lyophilized hog vesicles are cleaved by trypsin followed by reduction, a small, nonglycosylated peptide fragment is seen on SDS gels with the N-terminal sequence AQPHYS which represents C-terminal region beginning at position 236 (Mercier et al., 1993). This small fragment is not found either after trypsinolysis of intact vesicles nor in the absence of reducing agents. A disulfide bridge must therefore connect this cleaved fragment to the β subunit containing the carbohydrates. The C-terminal end of the disulfide is at position 262. This leaves little room for an additional membrane spanning α helix. Hence it is likely that the β subunit has only one membrane spanning segment.

D. Region of Association of the α and β Subunits

The β subunit of both the Na^+-K^+- and H^+-K^+-ATPase is necessary for targeting the complex from the endoplasmic reticulum to the plasma membrane (Renaud et al., 1991; Jaunin et al., 1992). It also stabilizes a functional form of both the gastric H^+-K^+-and Na^+-K^+-ATPases (Ackermann and Geering, 1990; Geering, 1991).

In the case of the Na^+-K^+-ATPase, the last 161 amino acids of the α subunit are essential for effective association with the β subunit (Lemas et al., 1992). Further more, the last 4 or 5 C-terminal hydrophobic amino acids of the Na^+ pump β subunit are essential for interaction with the α subunit whereas the last few hydrophilic amino acids are not (Beggah et al., 1993). Expression of sodium pump α subunit along with the β subunit of either the sodium or proton pump in *Xenopus* oocytes has shown that the β subunit of the gastric proton pump can act as a surrogate for the β subunit of the sodium pump as far as membrane targeting and $^{86}Rb^+$ uptake, thus suggesting some homology in the associative domains of the β subunits of the two pumps (Horisberger et al., 1991a). The H^+-K^+-ATPase α subunit requires its β subunit for efficient cell-surface expression and the C-terminal half of the α subunit was shown to assemble with the β subunit (Gottard et al., 1993).

In order to specify the region of the α subunit associated with the β subunit, the tryptic digest was solubilized using nonionic detergents such as NP-40 or $C_{12}E_8$. These detergents allow the holoenzyme to retain ATPase activity. The soluble enzyme was then adsorbed to a WGA affinity column. Following elution of the peptides not associated with the β subunit binding to the WGA column, elution of the β subunit with 0.1N acetic acid also eluted almost quantitatively the M7/loop/M8 sector of the α subunit. These data show that this region of the α subunit is tightly associated with the β subunit such that nonionic detergents are unable to dissociate it from the β subunit (Shin and Sachs, 1994).

If tryptic digestion is carried out in the presence of K^+, a fragment of 19-21 kDa is produced which contains the M7 segment and continues to the C terminal region of the enzyme. When this digest is solubilized and passed over the WGA column as outlined above, in addition to the 19-21 kDa fragment, a fragment representing the M5/loop/M6 sector is now also retained by the β subunit. Hence provided there is no hydrolysis between M8 and M9, an additional interaction is present between the α and β subunits or between the M5,6 and M9,10 regions of the α subunit. The antibody mAb 146-14 also recognizes the region of the α subunit at the extracytoplasmic face of the M7 segment as well as the β subunit, a finding consistent with the association found by column chromatography.

E. Two-Dimensional Structure

Reconstruction of two-dimensional crystals of the Ca^{2+} ATPase showed that this P-type enzyme consisted of three distinct segments fitting into a box of 120 Å (height) × 50 Å (perpendicular to the dimer ribbons) × 85 Å (along the dimer ribbons). The enzyme has a highly asymmetric mass distribution across the lipid bilayer. The cytoplasmic region comprises ~70% of the total mass, whereas the luminal region has only ~5%. The cytoplasmic domain was shown to have a complex structure similar to the shape of the head and neck of a bird. The "head" is responsible for forming dimer ribbons and contains the ATP binding domain. The "neck" represents the stalk domain about 25 Å long consisting of two segments. The transmembrane part consists of three segments defined as A, B and C segments: The largest segment (A segment) consists of two parts, one vertical (A2) and the other inclined (A1); the B segment is also tilted and connected to A2 and a third membrane segment (C segment) (Toyoshima et al., 1993).

The two-dimensional structure crystals of the H^+-K^+-ATPase formed in an imidazole buffer containing VO_3^- and Mg^{2+} ions was resolved at about 25 Å (Rabon et al., 1986; Mohraz et al., 1990; Hebert et al., 1992). The average cell edge of the H^+-K^+-ATPase was 115 Å, containing four asymmetric protein units of 50 Å × 30 Å (Hebert et al., 1992), whereas the unit cell dimension of the $Co(NH_3)_4ATP$-induced crystals of Na^+-K^+-ATPase was 141 Å (Skriver et al., 1989; Skriver et al., 1992). This suggests a more compact packing of the H^+-K^+-ATPase than the Ca^{2+} ATPase or the Na^+-K^+-ATPase. More recently, two-dimensional crystals of the Na^+-K^+-ATPase were reported to be best formed at pH 4.8 in sodium citrate buffer and to represent an unique lattice (a = 108.7 , b = 66.2, r = 104.2) by electron cryomicroscopy. There are two high-contrast parts in one unit cell (Tahara et al., 1993). The trypsinized Na^+-K^+-ATPase membranes were analyzed by electron microscopy (Ning et al., 1993). Both surface particles observed by negative staining and the protruding cytoplasmic portion of the α subunit were removed but the general membrane structure was preserved. Intramembranal particles defined by freeze-fracture were preserved after trypsinolysis, showing that the remaining membrane protein fragments retained their native structure within the lipid bilayer after proteolysis.

F. A Model of the H^+-K^+-ATPase

The above data show the presence of probably 10 membrane spanning segments of the catalytic subunit of the H^+-K^+-ATPase, consistent with those found for the

Mg^{2+}-ATPase of *S. typhimurium* using a fusion protein approach (Smith et al., 1993). Furthermore, the boundary amino acids are reasonably well defined by a combination of these techniques with alignment with other P type ATPases and molecular modeling. A model is presented in Figure 4. Here both subunits are represented. The α subunit is shown with ten membrane spanning segments, the β with a single membrane spanning segment. The largest extra-cytoplasmic loop of the a subunit is between membrane segments 7 and 8.

The figure also shows the presence of hydrophilic residues in the membrane domain. It can be seen that these are clustered in general in M4 through M8. It has been suggested that these residues are important in transmembrane ion transport or intramembranal ion binding. Mutational studies have been in general confirmatory of this hypothesis. However, it must be pointed out that ion binding sites in model systems such as valinomycin consist largely of peptide carbonyl oxygens, and there may well be a discrimination between ion binding sites and ion translocation pathways. More detailed information on the tertiary structure of the membrane domain is needed.

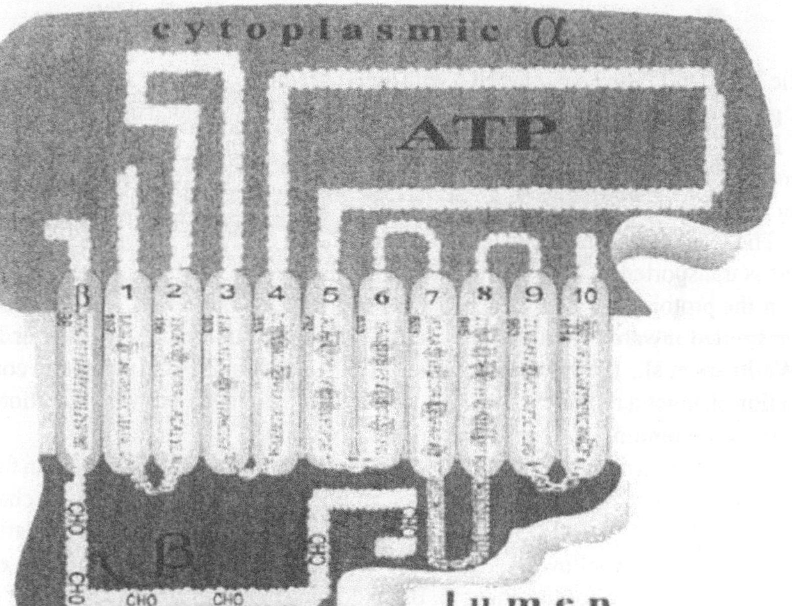

Figure 4. A two-dimensional representation of the H,K ATPase heterodimer. The α subunit is shown as a ten membrane spanning segment model, based on biochemical and *in vitro* translation methods. The β subunit is shown with a single membrane spanning segment.

G. Regions of Association of the α Subunit

There has been much suggestive evidence that the α-β heterodimeric H^+-K^+-ATPase exists as an oligomer. Such evidence includes target size (Rabon et al., 1988) and unit cell size of the enzyme in two dimensional crystals (Hebért et al., 1992). Recently, using Blue native gel separation, and cross linking with Cu^{2+}- phenanthroline, it was shown that the enzyme did indeed exist as an $(\alpha\beta)_2$ heterodimeric dimer. Membrane-bound H,K ATPase reacted with Cu-phenanthroline to provide an α-α dimer. No evidence was obtained for β-β dimerization. ATP prevents this Cu-phenanthroline-induced α-α dimerization. It was possible, by blocking free SH groups and subsequent reduction and labeling with fluorescein maleimide to demonstrate that the site of Cu^{2+}- oxidative cross-linking was either at cys^{565} or cys^{616} (Shin and Sachs, 1996). Hence this region of the α subunit is in close contact with its neighboring α subunit. Expression studies of the H^+-K^+-ATPase in insect SF9 cells and immunoprecipitation reached the conclusion that a similar region of the a subunit of this enzyme was also in close contact (Koster et al., 1995).

IV. CONFORMATIONS OF THE H^+-K^+-ATPase

The H^+-K^+-ATPase generates HCl with an H^+ gradient of 4×10^6 and a K^+ gradient of greater than ten-fold. This transport is achieved by the electroneutral exchange of H^+ for K^+, which is dependent on conformation changes in the protein. The alteration of enzyme conformation changes the affinity and sidedness of the ion binding sites during the cycle of phosphorylation and dephosphorylation.

The ions transported from the cytoplasmic side are H^+ or Na^+ at high pH. Since Na^+ is transported as a surrogate for H^+, it is likely that the hydronium ion, rather than the proton *per se* is the species transported (Polvani et al., 1989). The ions transported inwards from the outside face of the pump are Tl^+, K^+, Rb^+ or NH_4^+ (Wallmark et al., 1980; Lorentzon et al., 1988). Presumably the change in conformation changes a relatively small ion binding domain in the outward direction into a larger ion binding domain in the inward direction.

The E_1 conformation of the H^+-K^+-ATPase binds the hydronium ion from the cytoplasmic side at high affinity. With phosphorylation, the conformation changes from $E_1P{\bullet}H_3O^+$ to $E_2P{\bullet}H_3O^+$ form, which has high affinity for K^+ and low affinity for H_3O^+, thereby allowing release of H_3O^+ and binding of K^+ on the extracytoplasmic surface of the enzyme. Breakdown of the E_2P form requires K^+ or its congeners on the outside face of the enzyme. With dephosphorylation, the E_1K^+ conformation is produced with a low affinity for K^+, releasing K^+ to the cytoplasmic side, 2nd allowing rebinding of H_3O^+ (Wallmark et al., 1980).

The steps of phosphorylation of the H^+-K^+-ATPase were studied by measuring the inorganic phosphate, $P^{18}O_4$ and $P^{16}O_4$ distribution as a function of time at different H^+, K^+, and $[^{18}O]Pi$ concentrations (Faller and Diaz, 1989). The formation of $E{\bullet}Pi$ complex that exchanges ^{18}O with HOH was slower at pH 5.5 than at pH 8 and is not diffusion-controlled, suggesting unimolecular chemical transformation involving an additional intermediate in the phosphorylation mechanism such as, perhaps, a protein conformational change. From competitive binding of ATP and 2',3'-O-(2,4,6-trinitrophenylcyclohexadienylidine)adenosine 5'-phosphate (TNP-ATP), two classes of nucleotide binding sites were suggested (Faller, 1989). TNP-ATP is not a substrate for the H^+-K^+-ATPase. However, TNP-ATP prevents phosphorylation by ATP and inhibits the K^+-stimulated pNPPase and ATPase activities. The number of TNP-ATP binding sites was twice the stoichiometry of phosphoenzyme formation.

Fluorescein isothiocyanate (FITC) binds to the H^+-K^+-ATPase at pH 9.0, inhibiting ATPase activity but not pNPPase activity (Jackson et al., 1983). Fluorescence of the FITC-labeled enzyme, representing the E_1 conformation, was quenched by K^+, Rb^+, and Tl^+ (Jackson et al., 1983; Markus et al., 1989). The quenching of the fluorescence by KCl reflects the formation of E_2K^+. FITC binds at lys^{516} in the hog enzyme sequence (Farley and Faller, 1985). This FITC binding site apparently becomes less hydrophobic when KCl binds to form the E_2•K conformation.

The FITC labeled Na^+-K^+-ATPase has quite similar properties (Karlish, 1980; Farley et al., 1984; Smirnova and Faller, 1993). Two K^+ ions are required to cause the conformational change from E1 to E2 (Smirova and Faller, 1993). The binding site of FITC was at lys^{501}. However, several additional lysines at positions 480 and 766 were shown to react with FITC during inactivation of the Na^+-K^+-ATPase. These lysines were also protected from labeling in the presence of ATP (Xu, 1989).

The fluorescent 1-(2-methylphenyl)-4-methylamino-6-methyl-2,3-dihydropyrrolo [3,2-c]quinoline (MDPQ) was shown to inhibit the H^+-K^+-ATPase and the K^+ phosphatase competitively with K^+(Rabon et al., 1991). MDPQ fluorescence is quenched by the imidazopyridine, SCH 28080. The imidazopyridine Me-DAZIP binds to the M1/loop/M2 sector of the β subunit (Munson et al., 1991). MDPQ binding to the extra-cytoplasmic surface of the pump enhances its fluorescence, suggesting that inhibitor binding occurs to a relatively hydrophobic region of the protein. The fluorescence was quenched by K^+, independently of Mg^{2+}. The binding of Mg-ATP increased the fluorescence due to the formation of an E_2P•[I] complex (Rabon et al., 1991).

The fluorescent changes with FITC and MDPQ may reflect relative motion of the cytoplasmic domain and the connecting loop between M1 and M2 with respect to the hydrophobic domain of the enzyme. In the E1 form, the FITC region is relatively closer to the membrane and the extra-cytoplasmic loop relatively hydrophilic. With the formation of the E_2•K^+ form the FITC region is more distant from the membrane, whereas with formation of the E_2P form the MDPQ binding region between M1 and M2 moves toward the membrane. These postulated conformational changes are therefore reciprocal in the two major conformers of the enzyme.

Two irreversible inhibitors that form cysteine reactive sulfenamides in the acid space generated by the pump, omeprazole and E3810, appear to inhibit the enzyme in different conformations (Morii and Takeguchi, 1993). Both omeprazole and E3810, 2-{[4-(3-methoxypropoxy)-3-methylpyridin-2-yl]methylsulfinyl}-1H-

-benzimidazole, are acid activated in luminal surface to form active sulfenamide derivatives, which can bind cysteines within the H^+-K^+-ATPase.

The omeprazole bound enzyme has a lower FITC fluorescence, perhaps due to a E_2-like conformation. Both ATPase activity and steady state phosphorylation was inhibited. The E3810 bound enzyme showed a high FITC fluorescence more like a E_1 conformation. The fluorescence of the E3810 bound enzyme was quenched by K^+ in contrast to the omeprazole derivatized FITC labeled enzyme. It is not known whether these effects are due to differences in structure of the inhibitors or to differences in location of binding site or both.

Sodium ion is able to substitute for protons in the H^+-K^+-ATPase. Na^+ influx was observed in cytoplasmic side out vesicles at pH 8.5 (Polvani et al., 1989). This influx was inhibited by SCH 28080, required K^+ and was consequently due to the ability of Na^+ to act as a surrogate for H^+. When the fluorescence of FITC-labeled H^+-K^+-ATPase was quenched by K^+, Na^+ ions reversed the K^+-induced quench of the fluorescence (Rabon et al., 1990b).

This demonstrates that Na^+ ion stabilizes the E_1 form by forming the E_1•Na conformation and K^+ ion stabilizes the E_2 form to form the E_2•K conformation. With these ions it was possible to show that the rate of the E_2 to E_1 conformation change in the H^+-K^+-ATPase was much faster than the equivalent step of the Na^+-K^+-ATPase, which accounted for the difficulty of demonstrating Rb occlusion in the H^+-K^+-ATPase as compared to the Na^+-K^+-ATPase (Rabon et al., 1990b; Rabon et al., 1993).

The effect of trypsin on the gastric H^+-K^+-ATPase provides evidence for conformational changes as a function of ligand binding (Saccomani et al., 1979; Helmich-de Jong et al., 1987). When essentially complete ATPase inhibition was observed, 34% of the α subunit remained in the membrane. The tryptic digestion of the H^+-K^+-ATPase in the presence of ATP revealed the appearance of a 78 kDa and a 30 kDa fragment, while the digestion in the absence of ATP produced 87 kDa and 47 kDa peptide fragments. The difference of tryptic pattern can be interpreted in terms of two forms of phosphorylated protein (Saccomani et al., 1979). The conformation of the H^+-K^+-ATPase was carefully studied by limited proteolytic digestion(Helmich-de Jong et al., 1987). The α subunit in the presence of K^+ (the E_2•K conformation) was cleaved into two fragments of 56 and 42 kDa. In the presence of ATP (representing the E_1•P conformation) tryptic cleavage produced two fragments of 67 and 35 kDa. The 42 and 67 kDa fragments were phosphorylated (Helmich-de Jong et al., 1987). Also, it was observed that only K^+ of the ionic ligands provided significant protection against extensive tryptic hydrolysis (Shin and Sachs, 1994). Neither ATP nor ADP affected the tryptic pattern at high trypsin/protein ratios. (Shin, J. M., unpublished observations).

However, two large fragments of 67 and 33 kDa were found in the presence of ATP, Mg^{2+} and SCH 28080 as a K^+ surrogate; several fragments were produced in the absence of ligands. These data suggest that the E_2K^+ or more particularly the E_2P•[SCH] conformation of the H^+-K^+-ATPase severely limits accessibility of trypsin to most of the lysines and arginines in the β subunit (Munson et al., 1991; Besancon et al., 1993).

Extensive tryptic digestion of the gastric H^+-K^+-ATPase in the presence of KCl provided a C-terminal peptide fragment of 20 kDa beginning at the M7 transmembrane segment, a peptide of 9.4 kDa comprising the M1/loop/M2 sector beginning at asp^{84}, and another peptide of 9.4 kDa containing the M5/loop/M6 sector beginning at asn^{753} (Shin and Sachs, 1994). The C-terminal 20 kDa peptide fragment was suggested to be capable of Rb^+ occlusion (Rabon et al., 1993).

In the case of the Na^+-K^+-ATPase, cation occlusion from the cytoplasmic surface was suggested to occur in two steps (Or et al., 1993). In an initial recognition step, transported cations interact with carboxyl groups. The second step is selective for transported cations and involves occlusion of cations which involves a conformational change forming a more compact structure.

Extensive trypsinolysis of the H^+-K^+-ATPase results in peptide fragments of 11 kDa, 7.5 kDa, and 6.5 kDa, representing four pairs of membrane spanning segments corresponding to the M1/loop/M2, M3/loop/M4 and M5/loop/M6 with M7/loop/M8 fragments. The band at 11 kDa contains a single peptide beginning at leu^{853} which is derived from the M7/loop/M8 region. The band at 7.5 kDa also contains a single peptide beginning at thr^{291} which is derived from the M3/loop/M4 region. The band at 6.5 kDa consists of two peptides. One is a peptide beginning at gln^{104}, derived from the M1/loop/M2 segments and the other is a peptide beginning at leu^{776}, hence containing the M5/loop/M6 regions (Besancon et al., 1993; Shin et al., 1993).

When these digests in the absence and presence of KCl are compared, some regions near the membrane can be seen to be K^+ protected. The region between gly^{93} and glu^{104} near the M1 segment, the region between asn^{753} and leu^{776} near the cytoplasmic side of the M5 segment, and the region after the M8 segment, especially the region between ile^{945} and ile^{963} containing 5 arginines and 1 lysine, are protected from the trypsin digestion in the $E_2 \bullet K$ conformation (Shin and Sachs, 1994). Further, there must be protection prior to M3 and for some distance subsequent to M4, since no fragment containing these segments was found at a MWt less than 20 kDa.

V. INHIBITORS OF THE H^+-K^+-ATPase

The H^+-K^+-ATPase in the parietal cell secretes acid into the secretory canaliculus generating a pH of < 1.0 in the lumen of this structure. The acidity of this space is more than a thousandfold greater than anywhere else in the body, and allows accumulation of weak bases. Weak bases of a pK_a less than 4.0 would be selectively accumulated in this acidic space.

Since the extra-cytoplasmic domain of the H^+-K^+-ATPase is composed of relatively few amino acids present in a highly acidic compartment, weak bases target to this domain. These bases could for example act as competitive or non-competitive inhibitors of the enzyme or, following chemical conversion, as covalent inhibitors of enzyme function.

Since a substituted benzimidazole was first reported to inhibit the H^+-K^+-ATPase (Fellenius et al., 1981), many inhibitors of the H^+-K^+-ATPase have been synthesized. Most inhibitors can be classified into two groups: reversible and covalent. The covalent inhibitors all belong to the substituted benzimidazole family (Nagaya et al., 1989; Fujisaki et al., 1991; Sih et al., 1991; Beil et al., 1992; Kohl et al., 1992; Weidmann et al., 1992; Arakawa et al., 1993). Reversible inhibitors contain protonatable nitrogens and have a variety of structures. One type is represented by the imidazo-pyridine derivatives (Kaminski et al., 1991), others are piperidinopyridines (Hoiki et al., 1990), substituted 4-phenylaminoquinolines (Ife et al., 1992), pyrrolo[3,2-c]quinolines (Leach et al., 1992), guanidino-thiazoles (LaMattina et al., 1990), and scopadulcic acid (Asano et al., 1990). Some natural products such as cassigarol A (Murakami et al., 1992) and naphthoquinone (Danzig et al., 1991) also showed inhibitory activity. Many of these reversible inhibitors show K^+ competitive characteristics, in contrast to the benzimidazole type.

A. Substituted Benzimidazoles

The first compound of this class with inhibitory activity on the enzyme and on acid secretion was the 2-(pyridylmethyl)sulfinylbenzimidazole, timoprazole (Fellenius et al., 1981), and the first pump inhibitor used clinically was 2-[[3,5-dimethyl-4-methoxypyridin-2-yl]methylsulfinyl]-5-methoxy-1H-benzimidazole, omeprazole (Wallmark et al., 1984).

Omeprazole is an acid-activated prodrug (Lindberg et al., 1986). Omeprazole can be accumulated in the acidic space and easily converted to a reactive cationic sulfenamide species, which binds to SH groups as shown in Figure 5 (Keeling et al., 1985; Im et al., 1985; Lindberg et al., 1986; Keeling et al., 1987; Lorentzon et al., 1987). Omeprazole has a stoichiometry of 2 mol inhibitor bound per mol phosphoenzyme under acid transporting conditions and bound only to the α subunit (Lorentzon et al., 1987; Keeling et al., 1987).

Substituted benzimidazole inhibitors show slightly different effects depending on the inhibitor structure. The omeprazole-bound enzyme is in the E_2 form. Another inhibitor E3810, 2-[[4-(3-methoxypropoxy)-3-methylpyridin-2-yl]methylsulfinyl]-1H-benzimidazole, produced the E_1 form of the enzyme after binding. It is claimed that the K^+ dependent dephosphorylation from the phosphoenzyme was inhibited in the E3810-bound enzyme but not in the omeprazole-bound enzyme, whereas phosphoenzyme formation in the absence of K^+ was inhibited in both the E3810- and omeprazole-bound enzymes (Morii et al., 1993).

These inhibitors also have different binding sites. Omeprazole binds to cysteines in the extra-cytoplasmic regions of M5/M6 (cys-813 or 822) and M7/M8 (cys-892) (Besancon et al., 1993). Pantoprazole binds only to the cysteines in M5/M6 (Shin et al., 1993) and lansoprazole binds to cysteines in M3/M4(cys-321), M5/M6, and M7/M8 (Sachs et al., 1993). These data suggest that, of the 28 cys-

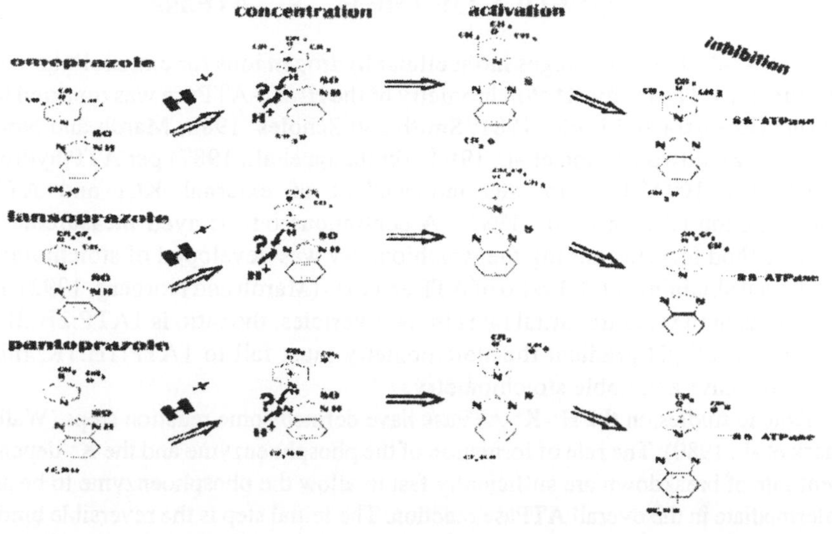

Figure 5. The mechanism of action of the benzimidazole, omeprazole, showing protonation and hence accumulation in acidic spaces with a pH <4.0 (i.e., the pK_a of omeprazole). This is followed by an acid catalyzed conversion to a tetracyclic cationic sulfenamide which then reacts with some of the extra-cytoplasmic cysteines of the α subunit of the H,K ATPase.

teines in the α subunit, only the cysteines present in the M5/M6 region are important for inhibition of acid secretion by the substituted benzimidazoles.

B. Substituted Imidazo[1,2α] Pyridines

Imidazo[1,2α] pyridine derivatives were shown to inhibit gastric secretion(Keeling et al., 1989). SCH 28080, 3-cyanomethyl-2-methyl-8-(phenylmethoxy) imidazo[1,2α]pyridine, inhibited the H^+-K^+-ATPase competitively with K^+ (Wallmark et al., 1987). SCH 28080 binds to free enzyme extra-cytoplasmically in the absence of substrate to form E_2(SCH 28080) complexes. SCH 28080 inhibits ATPase activity with high affinity in the absence of K^+. SCH 28080 has no effect on spontaneous dephosphorylation but inhibits K^+-stimulated dephosphorylation, presumably by forming a E_2-P•[I] complex. Hence SCH 28080 inhibits K^+-stimulated ATPase activity by competing with K^+ for binding E_2P (Mendelein et al., 1990). Steady state phosphorylation is also reduced by SCH 28080, showing that this compound also binds to the free enzyme. Using a photoaffinity reagent, 8-[(4-azidophenyl)methoxy]-1-tritiomethyl-2,3-dimethylimidazo[1,2α]pyridinium iodide, the binding site of this class of K^+ competitive inhibitor was identified to be in or close to the loop between the M1 and M2 segments (Munson et al., 1991).

VI. KINETICS OF THE H^+-K^+-ATPase

The H^+-K^+-ATPase exchanges intracellular hydrogen ions for extracellular potassium ions. The H^+ for K^+ stoichiometry of the H^+-K^+-ATPase was reported to be one (Reenstra and Forte, 1981; Smith and Scholes, 1982; Mardh and Norberg, 1992) or two (Rabon et al., 1982; Skrabanja et al., 1987) per ATP hydrolyzed. The H^+/ATP ratio was independent of external KCl and ATP concentration (Rabon et al., 1982). A continuous but delayed measurement flow method for determining the stoichiometry was developed of stoichiometry, hence obtaining a 1:1:1 ratio of ATP and ions (Mardh and Norberg, 1992). If care is taken to measure initial rates in tight vesicles, the ratio is 1ATP:2H:2K. Since at a full pH gradient the stoichiometry must fall to 1ATP:1H:1K, this pump displays a variable stoichiometry.

Kinetic studies on the H^+-K^+-ATPase have defined some reaction steps (Wallmark et al., 1980). The rate of formation of the phosphoenzyme and the K^+-dependent rate of breakdown are sufficiently fast to allow the phosphoenzyme to be an intermediate in the overall ATPase reaction. The initial step is the reversible binding of ATP to the enzyme in the absence of added K^+ ion, followed by a Mg^{2+} (and proton) dependent transfer of the terminal phosphate of ATP to the catalytic subunit (E_1-P•H^+). The Mg^{2+} remains occluded until dephosphorylation (Rabon et al., 1991). The addition of K^+ to the enzyme-bound acyl phosphate results in a two-step dephosphorylation. The faster initial step is dependent on the concentration of K^+, whereas the slower step is not affected by K^+ concentration. This showed the biphasic effect of K^+ on overall ATPase activity. The second phase of EP breakdown is accelerated in the presence of K^+ but, at K^+ concentration exceeding 500 μM, the ratio becomes independent of K^+ concentration. This shows that two forms of EP exist. The first form E_1P is K^+ insensitive and converts spontaneously in the rate-limiting step to E_2P, the K^+ sensitive form. ATP binding to the H^+-K^+-ATPase occurs in both the E_1 and the E_2 state, but with a lower affinity in the E_2 state (2,000 times lower compared to E_1) (Brzezinski et al., 1988).

H^+ and K^+ interact competitively on the cytoplasmic surface of intact vesicles. The effects of H^+ and K^+ on formation and breakdown of phosphoenzyme were determined using transient kinetics (Stewart et al., 1981). Increasing hydrogen ion concentrations on the ATP-binding face of the vesicles accelerate phosphorylation, whereas increasing potassium ion concentrations inhibit phosphorylation. Increasing hydrogen ion concentration reduces this K^+ inhibition of the phosphorylation rate. Decreasing hydrogen ion concentration accelerates dephosphorylation in the absence of K^+, and K^+ on the luminal surface accelerates dephosphorylation. Increasing K^+ concentrations at constant ATP decreases the rate of phosphorylation, and increasing ATP concentrations at constant K^+ concentration accelerates ATPase activity and increases the steady state phosphoenzyme level (Lorentzon et al., 1988). Therefore, inhibition by cations is due to cation stabilization of a dephospho form at a cytosolically accessible cation-binding site.

In order to determine the role of divalent cations in the reaction mechanism of the H,K ATPase, calcium was substituted for magnesium, which is necessary for phosphorylation (Mendlein and Sachs, 1989). Calcium ion inhibits K^+ stimulation of the H^+-K^+-ATPase by binding at a cytoplasmic divalent cation site. The Ca•EP dephosphorylates 10-20 times more slowly than the Mg•EP in the presence of 10 mM KCl with either 8 mM CDTA or 1 mM ATP. The inability of the Ca•EP to dephosphorylate in the presence of K^+, compared to the Mg•EP, demonstrates that the type of divalent cation which occupies the catalytic divalent cation site required for phosphorylation is important for the conformational transition to a K^+ sensitive phosphoenzyme. Calcium is tightly bound to the divalent cation site of the phosphoenzyme and the occupation of this site by calcium causes slower phosphoenzyme kinetics. Since the presence of CDTA or EGTA does not change the dephosphorylation kinetics of the EP•Ca form of the enzyme, it is concluded that the divalent cation remains occluded in the enzyme until dephosphorylation occurs.

VII. RELATION WITH OTHER P-TYPE ENZYMES

One classification of a division of P-type ATPases into five distinct types was based on ion specificity, biological occurrence, and sequence (Green, 1992). Alignments within and between classes show conserved regions of phosphorylation, nucleotide binding domains, and hinge regions as well as remarkable conservation of hydropathy profiles. However, only a few amino acids are conserved within the membrane domains themselves.

The topology of the Mg^{2+} transport ATPase encoded by the MgtB locus of *Salmonella typhimurium* was particularly suited for applying molecular biological methods (Smith et al., 1993). The membrane topology of MgtB was analyzed by measuring the activity of 35 sites of fusion between MgtB and the reporter enzymes BlaM and LacZ. The lactamase confers penicillin resistance when extracytoplasmic, and the LacZ is functional only when cytoplasmic. The fusion protein reporter data showed that MgtB contains 10 transmembrane segments. An exception was when the Blam protein was fused to pro^{766}. This clone was ampicillin resistant although the 10 segment model predicts a cytoplasmic location for this site.

In the case of SR Ca^{2+}-ATPase, an antipeptide polyclonal antibody against the region between 877 and 888 of SR Ca^{2+}-ATPase (Matthews et al., 1990) and a monoclonal antibody A20 against the region between 870-890 (Clarke et al., 1990b) bound to SR Ca^{2+}-ATPase after solubilization but did not bind to intact vesicles, suggesting that this epitope region is located in the lumen between M7 and M8. Such data are consistent with a 10 membrane segment model. Trypsinolysis of the Ca^{2+}-ATPase of SR resulted in the production of 4 membrane segment pairs and a single transmembrane segment corresponding to H9 (Shin et al., 1994). This latter was due to cleavage between the last two transmembrane segment pairs. *In vitro* translation was able to show membrane insertion properties for the first 4 and last 4 transmembrane segments (Bayle et al., 1995). In combination, these data argue for a 10 transmembrane segment model for this enzyme.

In the case of the Na^+-K^+-ATPase, from the trypsinolysis of the right side-out Na^+-K^+-ATPase, asn^{831} was identified as being at the cytoplasmic surface, and this residue is present at the N-terminal end of the M7 segment (Karlish et al., 1993). This result also provides strong evidence for a 10 transmembrane segment model.

When the cDNA encoding the α subunit of the Na^+-K^+-ATPase was linked to the β subunit via a linker of 17 amino acids and was expressed in a variety of mammalian cell lines as a single protein located primarily on the surface membrane. The β subunit domain was exposed to the extracellular medium. This α-β fused protein functioned as a normal heterodimeric Na pump. Hence the C terminal amino acid of the β subunit and the N terminal amino acid of the β subunit are both cytoplasmic (Emerick and Fambrough, 1993).

As already discussed, in the case of the H^+-K^+-ATPase, the first 8 transmembrane segments were identified by biochemical methods. The likely existence of M9 and M10 was demonstrated by *in vitro* translation. Thus to date there is direct evidence for 10 membrane spanning segments in the Mg^{2+}-ATPase of salmonella and for α subunit of the H^+-K^+-ATPase of gastric mucosa as well as for the Na^+-K^+- and Ca^{2+}-ATPases.

The β subunits of both the H^+-K^+- and Na^+-K^+-ATPases are glycoproteins (Reuben et al., 1990; Toh et al., 1990; Horisberger et al., 1991b). The β subunits of both bind to WGA or tomato lectin (Okamoto et al., 1990; Callaghan et al., 1990; Treuheit et al., 1993). WGA binding demonstrates the existence of N-acetylglucosamine and tomato lectin binding shows the presence of polygalactosamine on the β subunit. The carbohydrates of the β subunit were identified in the case of the Na^+-K^+-ATPase. N-Glycosylation sites are asn at positions 157, 192, and 264. The glycans represent a sequence from the site of asn as follows: two N-acetylglucosamine, mannose, N-acetylglucosamine, galactose, and additional glucosamine and galactose depending on the position (Treuheit et al., 1993). It appears that all potential glycosylation sites of the β subunit of the H^+-K^+-ATPase are occupied (Reuben et al., 1990).

When expression of the cDNAs encoding β subunits of the Na^+-K^+-ATPase that lack some or all of the cytoplasmic N-terminal domain, or that contain deletions in the transmembrane domain in cultured mouse L cells was analysed, the β subunit lacking the cytoplasmic domain is found to assemble with the α subunit of the Na^+-K^+-ATPase, and the resulting hybrid sodium pumps are transported to the plasma membrane. All β subunit mutants capable of membrane insertion were able to assemble with α subunit (Renaud et al., 1991). This shows that the extracellular domain of the β subunit is tightly bound to the α subunit.

Chimeric cDNAs encoding regions of the Na^+-K^+-ATPase α subunit and SR Ca^{2+}-ATPase were expressed with the avian Na^+-K^+-ATPase β subunit cDNA in COS-1 cells to determine which region of the α subunit is required for assembly with the β subunit (Lemas et al., 1992). A chimera, in which 161 amino acids of the Na^+-K^+-ATPase C-terminus replaced the corresponding amino acids of SR Ca^{2+}-ATPase C-terminus, assembled with the β subunit. These data suggest that the C-terminal 161 amino acid region of the Na^+-K^+-ATPase α subunit is critical for subunit assembly with the β subunit. More recently, 26 amino acids have been identified this way (Lemas et al., 1994).

As discussed above, trypsinolysis data and epitope mapping of the H^+-K^+-ATPase provided additional evidence for association of the β subunit with the M7/M8 sector of the α subunit (Shin and Sachs, 1994).Generally similar results were obtained with tryptic digestion in the presence of Rb^+ of the Na^+-K^+-ATPase followed by WGA affinity chromatography. The 10 kDa-peptide fragment, containing the M5/loop/M6 sector, beginning at N-terminal sequence, QAADMI, was retained along with the 20 kDa-peptide fragment consisting of the C-terminus of the α subunit beginning at M7 (Shin, J. M., unpublished data).

When trypsinolysis was carried out at pH 8.2 in the presence of RbCl, the Na^+-K^+-ATPase yielded a 10.8 kDa-peptide fragment containing the M1/loop/M2 sector which begins at asp^{73}, a 10 kDa-peptide fragment containing the M5/loop/M6 sector which begins at gln^{742}, and the C-terminal 19-kDa fragment beginning at the M7 segment. A similar result was obtained after trypsin digestion of the H^+-K^+-ATPase in the presence of KCl. The tryptic digest at pH 8.2 in the presence of KCl provided C-terminal 19-20 kDa fragments and a 9.4 kDa-peptide fragment which comprises two N-terminal sequences of peptides. One is a peptide beginning at asp^{84}, which comprises M1/loop/M2 sector, and the other is a peptide beginning at asn^{753}, which comprises the M5/loop/M6 sector (Shin and Sachs, 1994).

Tryptic digestion of the Na^+-K^+-ATPase in the presence of RbCl provided a stable 19-kDa membrane fragment of the C-terminus beginning at the M7 sector, which can occlude cations (Karlish et al., 1990; Capasso et al., 1992). The C-terminal 20 kDa peptide of the H^+-K^+-ATPase obtained from the trypsin digestion in the presence of KCl was also shown to be able to occlude Rb^+ (Rabon et al., 1993). Site-directed mutagenesis of both glutamic acid residues 955 and 956 of the rat Na^+-K^+-ATPase α1 subunit gave in decrease of Na^+ affinity with a small effect on K^+ affinity, whereas single substitution at glu^{955} or glu^{956} had only slight effects on cation stimulation of Na^+-K^+-ATPase. DCCD binding to glu^{955} inhibited Rb^+ occlusion. These data are equivocal in determining the role of either glu^{955} and glu^{956} in cation binding (Goldshleger et al., 1992; van Huysse et al., 1993). These glutamic acid residues are predicted to be at the cytoplasmic end of M9.

When the tryptic fragments of both the hog gastric H^+-K^+- and the pig kidney Na^+-K^+-ATPase were compared, it was found that there is strong homology of N-terminal sequences of the region located in the cytoplasmic domain before the transmembrane segments. For example, N-terminal sequences located in the cytoplasmic region near the M1 segment are DGPNALRPPRGTPEYVKFAR for the H^+-K^+-ATPase and DGPNALTPPPTTPEWVKFCR for the Na^+-K^+-ATPase. N-terminal sequences near M5 segments show even higher homology between the

Na^+-K^+- and the H^+-K^+-ATPase, namely NAADMILLDDNFASIVTGVEQGR-LIFD for the H^+-K^+-ATPase and QAADMILLDDNFASIVTGVEEGRLIFD for the Na^+-K^+-ATPase. These regions may be involved in cation transport into the membrane domain.

Trypsinolysis of intact SR vesicles from fast twitch rabbit muscle also showed a similar pattern to that found for the H^+-K^+-ATPase. It was possible to identify regions corresponding to the first 8 membrane segment pairs. M9 was also clearly identified, trypsinolysis having occurred between M9 and M10. On the assumption that there was partial access of trypsin to this site, this is also taken as direct evidence for 8 and perhaps 10 membrane spanning pairs for the Ca^{2+}-ATPase of SR. (Shin et al., 1994).

Part of the cation binding sites in the luminal surface of both H^+-K^+-and Na^+-K^+-ATPase are likely to be in the connecting loop between M1 and M2. In the case of the Na^+-K^+-ATPase, the double mutant that contained asp^{111} and arg^{122} showed the most resistance against ouabain inhibition (partially K^+ competitive), suggesting that these positions of the M1-M2 extracellular domain are important for high affinity ouabain binding (Price and Lingrel, 1988; Price et al., 1990). In addition to the identified H1/H2 ectodomain of Na^+-K^+-ATPase, the H3/H4 ectodomain also participates in the ouabain binding site since a mutation in tyr^{317} also confers ouabain resistance (Canessa et al., 1993).

Mutant cDNAs of Ca^{2+}-ATPase were transfected into COS-1 cells, and ATP-dependent Ca^{2+} transport or partial reactions of the expressed Ca^{2+}-ATPase were measured (Clarke et al., 1990a). Possible Ca^{2+} binding sites of the SR Ca^{2+}-ATPase were shown to be associated with glu^{309} in the M4 segment, glu^{771} in the M5 segment, asn^{796}, thr^{799}, and asp^{800} in the M6 segment, and glu^{908} in the M8 segment. Mutants that replaced pro^{312} with ala or gly were defective in the E_1P to E_2P transition, suggesting the M4 segment is important for conformational changes.

In HeLa cells expressing mRNA and protein of rat Na^+-K^+-ATPase $\alpha 1$ mutants that substituted both asp^{712} and asp^{716} to asn cannot survive in 0.1 μM ouabain. The Na^+-K^+-ATPase mutants which represent asn substitutions for asp^{712} or asp^{716} show inhibition of ATP hydrolysis. However, only asn substitution of asp^{716} inhibits phosphorylation (Lane et al., 1993). Probably, these asp residues are close to asp^{369} which is the phosphorylation site (Jorgensen and Andersen, 1988). The region of SR Ca^{2+}-ATPase associated with ATP binding can be shown by the identification of the ATP binding site, the ADP binding site, and the phosphorylation site. Asp^{351} is the phosphorylation site in this enzyme (Allen and Green, 1976). Lys^{492} and lys^{684} reacted with the terminal part of an ATP analog, AP_3PL(Yamamoto et al., 1989). Lys^{492} reacted with TNP-8N_3-ATP, a photo-reactive ATP analog (McIntosh et al., 1992). Thr^{532} and thr^{533} were labeled by 8-N^3-ADP (Lacapere et al., 1993). These data suggest that the amino acids of the Ca^{2+}-ATPase, namely, asp^{351}, lys^{492}, thr^{532}, thr^{533}, and lys^{684} are close to the phosphorylation site.

Illumination of sarcoplasmic reticulum vesicles with ultraviolet light in the presence of vanadate produced two photocleaved fragments. The cleavage patterns differed depending on the presence of calcium. The photocleavage in the absence of calcium ion occurs at the V cleavage site near the phosphorylation site, asp^{351}, producing fragments with molecular masses of 87 kDa (V1) and 22 kDa (V2). In the presence of calcium ion, the vanadate catalyzed photocleavage occurs at the VC cleavage site near the FITC binding site, lys^{515}, producing fragments of 71 kDa (VC1) and 38 kDa (VC2). This suggests that, since vanadate is an analog of Pi, that perhaps there is a conformational change in the phosphorylation site in the presence of Ca^{2+} wherein it is closer to the FITC binding region than in the absence of Ca^{2+}. Under similar conditions, the Na^+-K^+-ATPase was completely resistant to photocleavage, this was also the ease with the H^+-K^+-ATPase (Molnar et al., 1991).

The mammalian P type ATPases and the Mg^{2+} ATPase of *S. typhimurium* conform to this 10 membrane segment model. However, the heavy metal P type ATPases of bacteria appear to have only 8 transmembrane segments, based on *in vitro* translation of hydrophobic sequences of a P type ATPase cloned from *Helicobacter pylori* (Melchers et al., 1995). Since its hydropathy profile is similar to many other bacterial P type ATPases, it is likely that there are two topological families of P type ATPases, the 8 and 10 transmembrane segment families.

VIII. ACID SECRETION AND THE ATPase

The H^+-K^+-ATPase is present mainly in the gastric parietal cell. In the resting parietal cell it is present in smooth surfaced cytoplasmic membrane tubules. Upon stimulation of acid secretion, the pump is found on the microvilli of the secretory canaliculus of the parietal cell. This morphological change results in a several fold expansion of the canaliculus (Helander and Hirschowitz, 1972). In addition to this transition, there is activation of a K^+ and Cl^- conductance in the pump membrane which allows K^+ to access the extra-cytoplasmic face of the pump (Wolosin and Forte, 1983). This allows H^+ for K^+ exchange to be catalyzed by the ATPase (Sachs et al., 1976).

The covalent inhibitors of the H^+-K^+-ATPase that have been developed for the treatment of ulcer disease and esophagitis depend on the presence of acid secreted by the pump. They are also acid activated prodrugs that accumulate in the acid space of the parietal cell. Hence their initial site of binding is only in the secretory canaliculus of the functioning parietal cell (Scott et al., 1993). These data show also that the pump present in the cytoplasmic tubules does not generate HCl.

IX. GENE EXPRESSION OF THE H^+-K^+-ATPase

The upstream DNA sequence of the α subunit in rat H^+-K^+-ATPase contains both Ca^{2+} and cAMP responsive elements (Tamura et al., 1992). There are gastric nuclear proteins present that bind selectively to a nucleotide sequence, GATACC, in this region of the gene (Maeda et al., 1991). These proteins have not been detected in other tissues. Another regulatory site has also been postulated (Kaise et al., 1995).

Stimulation of acid secretion by histamine increases the level of mRNA for the α subunit of the pump (Tari et al., 1991). Elevation of serum gastrin, which secondarily stimulates histamine release from the enterochromaffin-like cell in the vicinity of the parietal cell, also stimulates transiently the mRNA levels in the parietal cell (Tari et al., 1993). H2 receptor antagonists block the effect of elevated serum gastrin on mRNA levels (Tari et al., 1994). It seems therefore that activity of the H2 receptor on the parietal cell determines in part gene expression of the ATPase. It might be expected therefore that chronic stimulation of this receptor would upregulate pump levels whereas inhibition of the receptor would down-regulate levels of the ATPase.

However, chronic administration of these H2 receptor antagonists, such as famotidine, results in an increase in pump protein, whereas chronic administration of omeprazole (which must stimulate histamine release) reduces the level of pump protein in the rabbit (Crothers et al., 1993; Scott et al., 1994). Regulation of pump protein turnover downstream of gene expression must account for these observations.

X. THE H^+-K^+-ATPase AND ACID-RELATED DISEASE

Secretion of acid by the gastric mucosa is required for the presence of duodenal and gastric ulcers and for the presence of esophageal reflux disease. It is therefore natural that medical treatment of these diseases has depended in large measure on reduction of the acidity of the stomach. The major change in medical treatment came with the introduction of H2 receptor antagonists (Black et al., 1972). These drugs have become the mainstay of treatment over the last twenty years. More recently, the substituted benzimidazoles, acting to inhibit the acid pump itself, have shown impressive clinical results in a series of comparative trials (Walan et al., 1989). These approaches to treatment have been a major medical success.

In the last decade, it has also been recognized that a second pathogenetic factor is essential in duodenal and gastric ulcer. This is infection of the gastric or duodenal mucosa by *H. pylori* (Marshall, 1983). Eradication of this organism prevents the recurrence of ulcers of these regions of the GI tract (Burget et al., 1990). The most likely contribution of this organism to duodenal and gastric ulcer is destruction of the tight junction. Thus, with an intact tight junction, even with acid present, there is clinically insignificant acid back diffusion. In the absence of acid, even with a disrupted tight junction, there is no acid to back diffuse and damage the epithelial cells. Both factors must be present for ulceration to occur. Future medical therapy of ulcer disease will be reduction of acid secretion and eradication of *H. pylori*. Prevention of ulcer disease will entail eradication of *H. pylori*.

XI. SUMMARY

The gastric H^+-K^+-ATPase is an ion-motive ATPase which is responsible for producing gastric acid. This enzyme catalyzes an electroneutral exchange of extracytoplasmic K^+ for cytoplasmic H^+ across the secretory membrane of the gastric parietal cells. The H^+-K^+-ATPase consists of two subunits, a catalytic α subunit and the β subunit. The α subunit is composed of about 1,034 amino acids, and it has a phosphorylation site and ion-transporting pathway. The β subunit is composed of about 291 amino acids and is heavily glycosylated. The membrane topology of the α subunit has been studied by protease digestion, sided reagents, antibody epitope studies, and molecular biological analysis. All these studies provide evidence for a 10 transmembrane segment model for the α subunit. The β subunit has a single transmembrane segment. Other P type ATPases such as the Na^+-K^+-and Ca^{2+}-ATPases display similar topology. The H^+-K^+-ATPase α subunit has a strong association with the β subunit. The luminal loop between the seventh transmembrane segment (M7) and the eighth transmembrane segment (M8) of the α subunit is associated with the β subunit. The H^+-K^+-ATPase exists as an $(\alpha\beta)_2$ heterodimeric dimer. The cytoplasmic loop between the fourth and the fifth transmembrane segments is a region of close proximity between two α subunits. The H^+-K^+-ATPase transports H^+ ion from the cytoplasmic region to the lumen by conformational change induced by phosphorylation and dephosphorylation. The H^+-K^+-ATPase binds hydronium ion. Following phosphorylation, the conformation changes spontaneously from the $E_1P{\bullet}H_3O^+$ to the $E_2P{\bullet}H_3O^+$ form. After releasing H_3O^+ and binding K^+ on the extra-cytoplasmic surface of the enzyme the $E_2P{\bullet}K^+$ conformation is formed. The $E_2P{\bullet}K^+$ conformation converts rapidly to the E_1K^+ conformation with dephosphorylation. The E_1K^+ conformation releases K^+ to the cytoplasmic side, allowing the rebinding of H_3O^+. Using extrinsic fluorescent conformational probes such as FITC and the arylquinoline, MDPQ, some of the changes associated with either K^+ binding or phosphorylation can be followed. FITC fluorescence (lys^{518}) quenches with addition of K^+ and the fluorescence of MDPQ (TM1-TM2) increases with phosphorylation. These data show reciprocal changes in the cytoplasmic and extracytoplasmic regions.

ACKNOWLEDGMENT

This work was supported by USVA-SMI and NIH grant numbers RO1 DK 40165 and 41301.

REFERENCES

Ackermann, U., & Geering, K. (1990). Mutual dependence of Na,K-ATPase α- and β-subunits for correct posttranslational processing and intracellular transport. FEBS Lett. 269, 105-108.

Allen, G., & Green, N. M. (1976). A 31-residue tryptic peptide from the active site of the [Ca^{2+}]-transporting adenosine triphosphatase of rabbit sarcoplasmic reticulum. FEBS Lett. 63, 188-192.

Arakawa, T., Fukuda, T., Higuchi, K., Koike, K., Satoh, H., & Kobayashi, K. (1993). NC-1300, a proton pump inhibitor, requires gastric acid to exert cytoprotection in rat gastric mucosa. Jap. J. Pharmacol. 61, 299-302.

Asano, S., Mizutani, M., Hayashi, T., Morita, N., & Takeguchi, N. (1990). Reversible inhibitions of gastric H,K-ATPase by scopadulcic acid and diacetyl scoadol. New biochemical tools of H,K-ATPase. J. Biol. Chem. 265, 22167-22173.

Baldwin, J., Henderson, R., Beckman, E., & Zemlin, F. (1988). Images of purple membrane at 2.8 Å resolution obtained by cryo-electron microscopy. J. Mol. Biol. 202, 585-591.

Bamberg, K., & Sachs, G. (1994). Topological analysis of the H,K-ATPase using *in vitro* translation. J. Biol. Chem. 269, 16909-16919.

Bamberg, K., Mercier, F., Reuben, M. A., Kobayashi, Y., Munson, K. B., & Sachs, G. (1992). cDNA cloning and membrane topology of the rabbit gastric H/K-ATPase α-subunit. Biochim. Biophys. Acta 1131, 69-77.

Bayle, D., Weeks, D., & Sachs G. (1995). The membrane topology of the rat sarcoplasmic and endoplasmic reticulum calcium ATPases by *in vitro* translation scanning. J. Biol Chem. 270, 25678-25684.

Bayle, D., Robert, J. C., Bamberg, K., Benkouka, F., Cheret, A. M., Lewin, M. J. M., Sachs, G., & Soumarmon, A. (1992). Location of the cytoplasmic epitope for a K^+-competitive antibody of the H,K-ATPase. J. Biol. Chem. 267, 19060-19065.

Beggah, A. T., Beguin, P., Jaunin, P., Peittsch, M. C., & Geering, K. (1993). Hydrophobic C terminal aminoacids in the β subunit are involved in assembly with the α subunit of Na^+-K^+-ATPase. Biochemistry 32, 14117-14124.

Beil, W., Staar, U., & Sewing, K. F. (1992). Pantoprazole: a novel H/K-ATPase inhibitor with an improved pH stability, Eur. J. Pharmacol. 218, 265-271.

Besancon, M., Shin, J. M., Mercier, F., Munson, K., Miller, M., Hersey, S., & Sachs, G. (1993). Membrane topology and omeprazole labeling of the gastric H,K-adenosinetriphosphatase. Biochemistry 32, 2345-2355.

Black, J. W., Duncan, W. A. M., Durant, C. J., Ganellin, C. R., & Parsons, E. M. (1972). Definition and antagonism of histamine H2-receptors. Nature 236, 385-387.

Blostein, R., Zhang, R-P., Gottardi, C. J., & Caplan, M. J. (1993). Functional properties of an H,K-ATPase/Na,K-ATPase chimera. J. Biol. Chem. 268, 10654-10658.

Brzezinski, P., Malmstrom, B. G., Lorentzon, P., & Wallmark, B. (1988). The catalytic mechanism of gastric H/K-ATPase: simulations of pre-steady-state and steady-state kinetic results. Biochim. Biophys. Acta 942, 215-219.

Burget, D.W., Chiverton, S.G., & Hunt, R. H. (1990). Is there an optimal degree of acid suppression for healing of duodenal ulcers? Gastroenterology 99, 345-351.

Callaghan, J. M., Toh, B. H., Pettitt, J. M., Humphris, D. C., & Gleeson, P. A. (1990). Poly-N-acetyllactosamine-specific tomato lectin interacts with gastric parietal cells. Identification of a tomato-lectin binding 60-90 KMr membrane glycoprotein of tubulovesicles. J. Cell Science 95, 563-575.

Callaghan, J. M., Toh, B. H., Simpson, R. J., Baldwin, G. S., & Gleeson, P. A. (1992). Rapid purification of the gastric H/K-ATPase complex by tomato-lectin affinity chromatography. Biochem. J. 283(pt 1), 63-68.

Canessa, C. M., Horisberger, J-D., & Rossier, B. C. (1993). Mutation of a tyrosine in the H3-H4 ectodomain of Na,K-ATPase α subunit confers ouabain resistance. J. Biol. Chem. 268, 17722-17726.

Canfield, V. A., & Levenson, R. (1991). Structural organization and transcription of the mouse gastric H,K-ATPase β subunit gene. Proc. Natl. Acad. Sci.USA 88, 8247-8251.

Canfield, V. A., Okamoto, C. T., Chow, D., Dorfman, J., Gros, P., Forte, J. G., & Levenson, R. (1990). Cloning of the H,K-ATPase β subunit. J. Biol. Chem. 265, 19878-19884.

Capasso,J.M., Hoving,S., Tal,D.M., Goldshleger,R., & Karlish,S.J.D. (1992). Extensive digestion of Na,K-ATPase by specific and nonspecific proteases with preservation of cation occlusion sites, J. Biol. Chem. 267, 1150-1158.

Carafoli, E. (1992). The Ca^{2+} pump of the plasma membrane. J. Biol. Chem. 267, 2115-2118.

Ceska, T. A., & Henderson, R. (1990). Analysis of high-resolution electron diffraction patterns from purple membrane labelled with heavy atoms. J. Mol. Biol. 213, 539-560.

Chow, D. C., Browning, C. M., & Forte, J. G. (1992). Gastric H,K-ATPase activity is inhibited by reduction of disulfide bonds in β-subunit. Amer. J. Physiol. 263, C39-C46.

Clarke, D. M., Loo, T. W., & MacLennan, D. H. (1990a). Functional consequences of alterations to polar amino acids located in the transmembrane domain of the Ca-ATPase of sarcoplasmic reticulum. J. Biol. Chem. 265, 6262-6267.

Clarke, D. M., Loo, T. W., & MacLennan, D. H. (1990b). The epitope for monoclonal antibody A20(amino acids 870-890) is located on the luminal surface of the Ca-ATPase of sarcoplasmic reticulum. J. Biol. Chem. 265, 17405-17408.

Crothers, J. M., Chow, D. C., & Forte, J. G. (1993) Omeprazole decreases H^+,K^+-ATPase Protein and increases permeability of oxyntic secretory membrane in rabbits. Am J Physiol. 265, G231-G241.

Crowson, M. S., & Shull, G. E. (1992). Isolation and characterization of a cDNA encoding the putative distal colon H,K-ATPase. Similarity of deduced amino acid sequence to gastric H,K-ATPase and Na,K-ATPase and mRNA expression in distal colon, kidney, and uterus. J. Biol. Chem. 267, 13740-13748.

Danzig, A. H., Minor, P. L., Garrigus, J. L., Fukuda, D. S., & Mynderse, J. S. (1991). Studies on the mechanism of action of A80915, a semi-naphthoquinone natural product, as an inhibitor of gastric H,K-ATPase. Biochem. Pharmacology 42, 2019-2026.

Emerick, M., & Fambrough, D. M. (1993). Intramolecular fusion of Na pump subunits assures exclusive assembly of the fused α and β subunit domains into a functional enzyme in cells also expressing endogenous Na pump subunits. J. Biol. Chem. 268, 23455-23459.

Faller, L. D. (1989). Competitive binding of ATP and the fluorescent substrate analogue 2',3'-O-(2,4,6-trinitrophenylcyclohexadienylidine) adenosine 5'-phosphate to the gastric H,K ATPase : Evidence for two classes of nucleotide sites. Biochemistry 28, 6771-6778.

Faller, L. D., & Diaz, R. A. (1989). Evidence from ^{18}O exchange measurements for steps involving a weak acid and a slow chemical transformation in the mechanism of phosphorylation of the gastric H^+,K^+-ATPase by inorganic phosphate. Biochemistry 28, 6908-6914.

Farley, R. A., & Faller, L. D. (1985). The amino acid sequence of an active site peptide from the H,K-ATPase of gastric mucosa. J. Biol. Chem. 260, 3899-3901.

Farley, R. A., Tran, C. M., Carilli, C. T., Hawke, D., & Shively, J. E. (1984). The amino acid sequence of a fluorescein-labeled peptide from the active site of Na,K-ATPase. J. Biol. Chem. 259, 9532-9535.

Fellenius, E., Berglindh, T., Sachs, G., Olbe, L., Elander, B., Sjöstrand, S-E., & Wallmark, B. (1981). Substituted benzimidazoles inhibit gastric acid secretion by blocking (H,K)ATPase, Nature 290, 159-161.

Fujisaki, H., Shibata, H., Oketani, K., Murakami, M., Fujimoto, M., Wakabayashi, T., Yamatsu, I., Yamaguchi, M., Sakai, H., & Takeguchi, N. (1991). Inhibition of acid secretion by E3810 and omeprazole, and their reversal by glutathione. Biochem. Pharmacol. 42, 321-328.

Futai, M., Noumi, T., & Maeda, M. (1989). ATP synthase(H^+-ATPase) : Results by combined biochemical and molecular biological approaches. Ann. Rev. Biochem. 58, 111-136.

Geering, K. (1991). The functional role of the β-subunit in the maturation and intracellular transport of Na,K-ATPase. FEBS Lett. 285, 189-193.

Goldshleger, R., Tal, D. M., Moorman, J., Stein, W. D., & Karlish, S. J. D. (1992). Chemical modification of glu-953 of the α chain of Na,K-ATPase associated with inactivation of cation occlusion. Proc. Natl. Acad. Sci. USA 89, 6911-6915.

Gottard, C. J., & Caplan, M. J. (1993). Molecular requirements for the cell-surface expression of multisubunit ion-transporting ATPases. J. Biol. Chem. 268, 14342-14347.

Glynn, I. M., & Karlish, S. J. D. (1990). Occluded cations in active transport. Ann. Rev. Biochem. 59, 171-205.

Green, N. M. (1992). Evolutionary relationships within the family of P-type cation pumps. Ann. N. Y. Acad. Sci. 671, 104-112.

Hall, K., Perez, G., Anderson, D., Gutierrez, C., Munson, K., Hersey, S. J., Kaplan, J. H., & Sachs, G. (1990). Location of the carbohydrates present in the H,K-ATPase vesicles isolated from hog gastric mucosa. Biochemistry 29, 701-706.

Hall, K., Perez, G., Sachs, G., & Rabon, E. (1991). Identification of H^+/K^+-ATPase α,β-heterodimers. Biochim. Biophys. Acta 1077, 173-179.

Hebert, H., Xian, Y., Hacksell, I., & Mardh, S. (1992). Two-dimensional crystals of membrane-bound gastric H,K-ATPase. FEBS Lett. 299, 159-162.

Helander, H. F., & Hirschowitz, B.I.(1972). Quantitative ultrastructural studies on gastric parietal cells. Gastroenterology 63, 951.

Helmich-de Jong, M. L., Vries, van E-D., & de Pont, J. J. H. H. M. (1987). Conformational states of (H^++K^+)-ATPase studied using tryptic digestion as a tool. Biochim. Biophys. Acta 905, 358-370.

Hermolin, J., & Fillingame, R. H. (1989). H^+-ATPase activity of *Escherichia coli* F1F0 is blocked after reaction of dicyclohexylcarbodiimide with a single proteolipid(subunit c) of the F0 complex. J. Biol. Chem. 264, 3896-3903.

Hoiki,Y., Takada,J., Hidaka,Y., Takeshita,H., Hosoi,M., & Yano,M. (1990). A newly synthesized pyridine derivative, (Z)-5-methyl-2-[2-(1-naphtyl)ethenyl]-4-piperidinopyridine hydrochloride(AU-1421), as a reversible gastric pump inhibitor. Arch. Intern. Pharma. Therap. 305, 32-44.

Horisberger, J-D., Jaunin, P., Reuben, M. A., Lasater, L. S., Chow, D. C., Forte, J. G., Sachs, G., Rossier, B. C., & Geering, K. (1991a). The H,K-ATPase β-subunit can act as a surrogate for the β-subunit of Na,K-pumps. J. Biol. Chem. 266, 19131-19134.

Horisberger, J-D., Lemas, V., Kraehenbuhl, J-P., & Rossier, B. C. (1991b). Structure-function relationship of Na,K-ATPase. Ann. Rev. Physiol. 53, 565-584.

van Huysse, J. W., Jewell, E. A., & Lingrel, J. B. (1993). Site-directed mutagenesis of a predicted cation binding site of Na,K-ATPase. Biochemistry 32, 819-826.

Ife, R. J., Brown, T. H., Keeling, D. J., Leach, C. A., Meeson,, M. L., Parsons, M. E., Reavill, D. R., Theobald, C. J., & Wiggall, K. J. (1992). Reversible inhibitors of the gastric H,K-ATPase. 3-substituted-4-(phenylamino)quinolines. J. Med. Chem. 35, 3413-3422.

Im, W. B., Blakeman, D. P., & Sachs, G. (1985). Reversal of antisecretory activity of omeprazole by sulfhydryl compounds in isolated rabbit gastric glands. Biochim.Biophys. Acta 845, 54-59.

Jackson, R. J., Mendlein, J., & Sachs, G. (1983). Interaction of fluorescein isothiocyanate with the H,K-ATPase. Biochim. Biophys. Acta 731, 9-15.

Jaunin, P., Horisberger, J. D., Richter, K., Good, P. J., Rossier, B. C., & Geering, K. (1992). Processing, intracellular transport, and functional expression of endogenous and exogenous α-β3 Na,K-ATPase complexes in xenopus oocytes. J. Biol. Chem. 267, 577-585.

Jorgensen, P. L., & Andersen, J. P. (1988). Structural basis for E1-E2 conformational transitions in Na,K-pump and Ca-pump proteins. J. Membrane Biol. 103, 95-120.

Kaise, M., Muraoka, A., Yamada, J., & Yamada, T. (1995). Epidermal growth factor induces H^+,K^+-ATPase α-subunit gene expression through an element homologous to the 3' half-site of the c-fos serum response element. J. Biol. Chem. 270, 18637-18642.

Kaminski, J. J., Wallmark, B., Briving, C., & Andersson, B. M. (1991). Antiulcer agents.5. Inhibition of gastric H/K-ATPase by substituted imidazo[1,2a]pyridines and related analogues and its implication in modeling the high affinity potassium ion binding site of the gastric proton pump enzyme. J. Med. Chem. 34, 533-541.

Karlish,S.J.D. (1980). Characterization of conformational changes in (Na,K)-ATPase labeled with fluorescein at the active site, J. Bioenerg. Biomembrane 12, 111-136.

Karlish,S.J.D., Goldshleger,R., & Stein,W.D. (1990). A 19-kDa C-terminal tryptic fragment of the α chain of Na/K-ATPase is essential for occlusion and transport of cations, Proc. Natl. Acad. Sci. USA 87, 4566-4570.

Karlish,S.J.D., Goldshleger,R., & Jorgensen,P.L. (1993). Location of asn831 of the α chain of Na/K-ATPase at the cytoplasmic surface. Implication for topological models, J. Biol. Chem. 268, 3471-3478.

Keeling, D. J., Fallowfield, C., Milliner, K. J., Tingley, S. K., Ife, R. J., & Underwood, A. H. (1985). Studies on the mechanism of action of omeprazole. Biochem. Pharmacol. 34, 2967-2973.

Keeling, D. J., Fallowfield, C., & Underwood, A. H. (1987). The selectivity of omeprazole as an H,K-ATPase inhibitor depends upon the means of its activation. Biochem. Pharmacol. 36, 339-344.

Keeling, D. J., Taylor, A. G., & Schudt, C. (1989). The binding of a K^+ competitive ligand, 2-methyl, 8-(phenylmethoxy)imidazo[1,2-a]pyridine 3-acetonitrile, to the gastric H,K-ATPase. J. Biol. Chem. 264, 5545-5551.

Kirley, T. L. (1990). Inactivation of Na,K-ATPase by β-mercaptoethanol. Different sensitivity to reduction of the three β subunit disulfide bonds. J. Biol. Chem. 265, 4227-4232.

Kohl, B., Sturm, E., Senn-Bilfinger, J., Simon, W. A., Kruger, U., Schaefer,., Rainer, G., Figala, V., & Klemm, K. (1992). H,K-ATPase inhibiting 2-[(2-pyridylmethyl)sulfinyl]benzimidazoles. A novel series of dimethoxypyridyl-substituted inhibitors with enhanced selectivity. The selection of pantoprazole as a clinical candidate. J. Med. Chem. 35, 1049-1057.

Koster, J. C., Blanco, G., & Mercer, R. W. (1995). A cytoplasmic region of the Na,K-ATPase α-subunit is necessary for specific α/α association. J. Biol. Chem. 270, 14332-14339.

Kyte, J., & Doolittle, R. F. (1982). A simple method for displaying the hydropathic character of a protein. J. Mol. Biol. 157, 105-132.

Lacapere, J-J., Garin, J., Trinnaman, B., & Green, N. M. (1993). Identification of amino acid residues photolabeled with 8-azidoadenosine 5'-diphosphate in the catalytic site of sarcoplasmic reticulum Ca-ATPase. Biochemistry 32, 3414-3421.

LaMattina, J. L., McCarthy, P. A., Reiter, L. A., Holt, W. F., & Yeh, L. A. (1990). Antiulcer agents. 4-Substituted 2-guanidinothiazoles: reversible, competitive, and selective inhibitors of gastric H,K-ATPase. J. Med. Chem. 33, 543-552.

Lane, L. K., Kirley, T. L., & Ball, W. J. Jr. (1986). Structural studies on H,K-ATPase : Determination of the NH_2-terminal amino acid sequence and immunological cross-reactivity with Na,K-ATPase. Biochem, Biophys. Res. Commun. 138, 185-192.

Lane, L. K., Feldmann, J. M., Flarsheim, C. E., & Rybczynski, C. L. (1993). Expression of rat α1 Na,K-ATPase containing substitutions of "essential" amino acids in the catalytic center. J. Biol. Chem. 268, 17930-17934.

Laubinger, W., Deckers-Hebestreit, G., Altendorf, K., & Dimroth, P. (1990). A hybrid adenosinetriphosphatase composed of F1 of *Escherichia coli* and F0 of Propionigenium modestum is a functional sodium ion pump. Biochemistry 29, 5458-5463.

Leach, C. A., Brown, T. H., Ife, R. J., Keeling, D. J., Laing, S. M., Parsons, M. E., Price, C. A., & Wiggall, K. J. (1992). Reversible inhibitors of the gastric H,K-ATPase. 2. 1-Arylpyrrolo[3,2-c]quinlines: effect of the 4-subsituent. J. Med. Chem. 35, 1845-1852.

Lemas, M. V., Takeyasu, K., & Fambrough, D. M. (1992). The carboxyl-terminal 161 amino acids of the Na,K-ATPase α-subunit are sufficient for assembly with the β-subunit. J. Biol. Chem. 267, 20987-20991.

Lemas, M. V., Hamrick, M., Takeyasu, K., & Fambrough, D. M. (1994). 26 Amino acids of an extracellular domain of the H^+-K^+-ATPase α subunit are sufficient for assembly with the H^+-K^+-ATPase β subunit. J. Biol. Chem. 269, 8255-8259.

Lindberg, P., Nordberg, P., Alminger, T., Brandstrom, A., & Wallmark, B. (1986). The mechanism of action of the gastric acid secretion inhibitor, omeprazole. J. Med. Chem. 29, 1327-1329.

Lorentzon, P., Eklundh, B., Brandstrom, A., & Wallmark, B. (1985). The mechanism for inhibition of gastric H,K-ATPase by omeprazole. Biochim. Biophys. Acta 817, 25-32.

Lorentzon, P., Jackson, R., Wallmark, B., & Sachs, G. (1987). Inhibition of H,K-ATPase by omeprazole in isolated gastric vesicles requires proton transport. Biochim. Biohys. Acta 897, 41-51.

Lorentzon, P., Sachs, G., & Wallmark, B. (1988). Inhibitory effects of cations on the gastric H,K-ATPase. A potential sensitive step in the K^+ limb of the pump cycle. J. Biol. Chem. 263, 10705-10710.

Lytton, J., Westlin, M., Burk, S. E., Shull, G. E., & MacLennan, D. H. (1992). Functional comparisons between isoforms of the sarcoplasmic or endoplasmic reticulum family of calcium pumps. J. Biol. Chem. 267, 14483-14489.

Ma, J. Y., Song, Y. H., Sjostrand, S. E., Rask, L. & Mardh, S. (1991). cDNA cloning of the β subunit of the human gastric H,K-ATPase. Biochem. Biophys. Res. Commun. 180, 39-45.

Maeda, M., Ishizaki, J., & Futai, M. (1988a). cDNA cloning and sequence determination of pig gastric H,K-ATPase. Biochem. Biophys. Res. Commun. 157, 203-209.

Maeda, M., Tagaya, M., & Futai, M. (1988b). Modification of gastric H,K-ATPase with pyridoxal 5'-phosphate. J. Biol. Chem. 263, 3652-3656.

Maeda, M., Oshiman, K-I., Tamura, S., & Futai, M. (1990). Human gastric H,K-ATPase gene. Similarity to Na,K-ATPase genes in exon/intron organization but difference in control region, J. Biol. Chem. 265, 9027-9032.

Maeda, M., Oshiman, K., Tamura, S., Kaya, S., Mahmood, S., Reuben, M. A., Lasater, L. S., Sachs, G., & Futai, M. (1991). The rat H/K-ATPase β subunit gene and recognition of its control region by gastric DNA binding protein. J. Biol. Chem. 266, 21584-21588.

Mardh, S., & Norberg, L. (1992). A continuous flow technique for analysis of the stichiometry of the gastric H,K-ATPase. Acta Physiol. Scand. 146, 259-263.

Markus, S., Priel, Z., & Chipman, D. M.(1989). Interaction of calcium and vanadate with fluorescein isothiocyanate labeled Ca^{2+}-ATPase from sarcoplasmic reticulum: kinetics and equilibria. Biochemistry 28, 793-799.

Marshall, B. (1983). Unidentified curved bacilli on gastric epithelium in active chronic gastritis. Lancet i., 1273-1275.

Matthews, I., Sharma, R. P., Lee, A. G., & East, J. M. (1990). Transmembranous organization of (Ca-Mg)-ATPase from sarcoplasmic reticulum. Evidence for luminal location of residues 877-888. J. Biol. Chem. 265, 18737-18740.

McIntosh, D. B., Woolley, D. G., & Berman, M. C. (1992). 2',3'-O-(2,4,6-Trinitrophenyl)-8-azido-AMP and -ATP photolabel lys-492 at the active site of sarcoplasmic reticulum Ca-ATPase. J. Biol. Chem. 267, 5301-5307.

Melchers, K., Weitzenegger, T., Buhmann, A., Steinhilber, W., Sachs, G. & Schäfer, K. P. (1995) Cloning and membrane topology of a P type ATPase from *Helicobacter pylori*. J. Biol. Chem. 271, 446-457.

Mendlein, J., & Sachs, G. (1989). The substitution of calcium for magnesium in H,K-ATPase catalytic cycle. J. Biol. Chem. 264, 18512-18519.

Mendlein, J. & Sachs, G. (1990). Interaction of a K^+-competitive inhibitor, a substituted imidazo[1,2a]pyridine, with the phospho- and dephosphoenzyme forms of H,K-ATPase. J. Biol. Chem. 265, 5030-5036.

Mercier, F., Reggio, H., Devilliers, G., Bataille, D., & Mangeat, P. (1989). Membrane-cytoskeleton dynamics in rat parietal cells: mobilization of actin and spectrin upon stimulation of gastric acid secretion. J. Cell Biol. 108(2), 441-453.

Mercier, F., Bayle, D., Besancon, M., Joys, T., Shin, J. M., Lewin, M. J. M., Prinz, C., Reuben, A. M., Soumarmon, A., Wong, H., Walsh, J. H., & Sachs, G. (1993). Antibody epitope mapping of the gastric H/K-ATPase. Biochim. Biophys. Acta 1149, 151-165.

Mohraz, M., Sathe, S., & Smith, P. R. (1990). Proc. XIIth Intl. Congr. Electr. Microscopy. (Peachley, L. D. & Williams, D. B. eds.) Vol 1, San Francisco Press Inc. San Francisco, pp. 94-95.

Molnar, E., Varga, S., & Martonosi, A. (1991). Difference in the susceptibility of various cation transport ATPases to vanadate-catalyzed photocleavage. Biochim. Biophys. Acta 1068, 17-26.

Morii,M., & Takeguchi,N. (1993). Different biochemical modes of action of two irreversible H^+,K^+-ATPase inhibitors, Omeprazole and E3810, J. Biol. Chem. 268, 21553-21559.

Morley, G. P., Callaghan, J. M., Rose, J. B., Toh, B. H., Gleeson, P. A., & van Driel, I. R. (1992). The mouse gastric H,K-ATPase β subunit. Gene structure and co-ordinate expression with the α subunit during ontogeny. J. Biol. Chem. 267, 1165-1174.

Munson, K. B. & Sachs, G. (1988). Inactivation of H,K-ATPase by a K^+-competitive photoaffinity inhibitor. Biochemistry 27, 3932-3938.

Munson, K. B., Gutierrez, C., Balaji, V. N., Ramnarayan, K., & Sachs, G. (1991). Identification of an extra-cytoplasmic region of H,K-ATPase labeled by a K^+-competitive photoaffinity inhibitor. J. Biol. Chem. 266, 18976-18988.

Murakami,S., Arai,I., Muramatsu,M., Otomo,S., Baba,K., Kido,T.,& Kozawa,M. (1992). Effect of stilbene derivatives on gastric H,K-ATPase, Biochem. Pharmacology 44, 33-37.

Nagaya,H., Satoh,H., Kubo,K., & Maki,Y. (1989). Possible mechanism for the inhibition of gastric H,K-adenosinetriphosphatase by the proton pump inhibitor AG-1749. J. Pharmacol. Exptl. Therapeutics 248, 799-805.

Nelson,N. (1991). Structure and pharmacology of the proton-ATPases. TIPS 12, 71-75.

Newman, P. R., Greeb, J., Keeton, T. P., Reyes, A. A., & Shull, G. E. (1990). Structure of the human gastric H,K-ATPase gene and comparison of the 5'-flanking sequences of the human and rat genes. DNA and Cell Biol. 9, 749-762.

Newman, P. R., & Shull, G. E. (1991). Rat gastric H,K-ATPase β-subunit gene: intron/exon organization, identification of multiple transcription initiation sites, and analysis of the 5'-flanking region. Genomics 11, 252-262.

Ning, G., Maunsbach, A. B., & Esmann, M. (1993). Ultrastructure of membrane-bound Na,K-ATPase after extensive tryptic digestion, FEBS Lett. 330, 19-22.

Okamoto, C. T., Karpilow, J. M., Smolka, A., & Forte, J. G. (1990). Isolation and characterization of gastric microsomal glycoproteins. Evidence for a glycosylated β-subunit of the H/K-ATPase. Biochim. Biophys. Acta 1037, 360-372.

Or, E., David, P., Shainskaya, A., Tal, D. M., & Karlish, S. J. D. (1993). Effects of competitive sodium-like antagonists on Na,K-ATPase suggest that cation occlusion from the cytoplasmic surface occurs in two steps. J. Biol. Chem. 268, 16929-16937.

Oshiman, K., Motojima, K., Mahmood, S., Shimada, A., Tamura, S., Maeda, M., & Futai, M. (1991). Control region and gastric specific transcription of the rat H,K-ATPase β gene. FEBS Lett. 281, 250-254.

Polvani, C., Sachs, G., & Blostein, R. (1989). Sodium ions as substitutes for protons in the gastric H,K-ATPase. J. Biol. Chem. 264, 17854-17859.

Price, E. M., & Lingrel, J. B. (1988). Structure-function relationship in the Na,K-ATPase α subunit: Site-directed mutagenesis of glutamine-111 to arginine and asparagine-122 to aspartic acid generates a ouabain-resistant enzyme. Biochemistry 27, 8400-8408.

Price, E. M., Rice, D. A., & Lingrel, J. B. (1990). Structure-function studies of Na,K-ATPase. Site-directed mutagenesis of the border residues from the H1-H2 extracellular domain of the α subunit. J. Biol. Chem. 265, 6638-6641.

Rabon, E., & Reuben, M. A. (1990). The mechanism and structure of the gastric H,K-ATPase. Ann. Rev. Physiol. 52, 321-344.

Rabon, E. C., McFall, T. L., & Sachs, G. (1982). The gastric H,K-ATPase: H^+/ATP stoichiometry, J. Biol. Chem. 257, 6296-6299.

Rabon, E., Wilke, M., Sachs, G., & Zamphigi, G. (1986). Crystallization of the gastric H,K-ATPase, J. Biol. Chem. 261, 1434-1439.

Rabon, E. C., Gunther, R. D., Bassilian, S., & Kempner, E. S. (1988). Radiation inactivation analysis of oligomeric structure of the H,K-ATPase. J. Biol. Chem. 263, 16189-16194.

Rabon, E. C., Bassilian, S., & Jakobsen, L. J. (1990a). Glutaraldehyde crosslinking analysis of the $C_{12}E_8$ solubilized H,K-ATPase. Biochim. Biophys. Acta 1039, 277-289.

Rabon, E. C., Bassilian, S., Sachs, G., & Karlish, S. J. D. (1990b). Conformational transitions of the H,K-ATPase studied with sodium ions as surrogates for protons. J. Biol. Chem. 265, 19594-19599.

Rabon, E., Sachs, G., Bassilian, S., Leach, C., & Keeling, D. (1991). K^+-Competitive fluorescent inhibitor of the H,K-ATPase. J. Biol. Chem. 266, 12395-12401.

Rabon, E. C., Smillie, K., Seru, V., & Rabon, R. (1993). Rubidium occlusion within tryptic peptides of the H,K-ATPase. J. Biol. Chem. 268, 8012-8018.

Reenstra, W. & Forte, J. (1981). H^+/K^+ ATP stoichiometry for gastric H,K-ATPase. J. Membrane Biol. 61, 55-60.

Reuben, M. A., Lasater, L. S., & Sachs, G. (1990). Characterization of a β subunit of the gastric H/K-ATPase. Proc. Natl. Acad. Sci. USA 87, 6767-6771.

Renaud, K. J., Inman, E. M., & Fambrough, D, M. (1991). Cytoplasmic and transmembrane domain deletions of Na,K-ATPase β-subunit. Effects on subunit assembly and intracellular transport. J. Biol. Chem. 266, 20491-20497.

Saccomani, G., Dailey, D. W., & Sachs, G. (1979). The action of trypsin on the (H^++K^+)-ATPase. J. Biol. Chem. 254, 2821-2827.

Sachs, G., Chang, H. H., Rabon, E., Schackman, R., Lewin, M., & Saccomani, G. (1976). A nonelectrogenic H^+ pump in plasma membranes of hog stomach. J. Biol. Chem. 251, 7690-7698.

Sachs, G., Shin, J.M., Besancon, M., & Prinz, C. (1993). The continuing development of gastric acid pump inhibitors. Alimentary Pharmacol. Therap. 7, 4-12.

Scott, D., Munson, K., Modyanov, N., & Sachs, G. (1992). Determination of the sidedness of the C-terminal region of the gastric H,K-ATPase α subunit. Biochim. Biophys. Acta 1112, 246-250.

Scott, D. R., Helander, H. F., Hersey, S. J., & Sachs, G. (1993). The site of acid secretion in the mammalian parietal cell. Biochim. Biophys. Acta 1146, 73-80.

Scott, D., Besancon, M., Sachs, G., & Helander, H. F. (1994). Effects of anti-secretory agents on parietal cell structure and H/K ATPase levels in rabbit gastric mucosa *in vivo*. Am. J Dig. Dis. 39, 2118-2126.

Shin, J. M., & Sachs, G. (1994). Identification of a region of the H,K-ATPase α subunit associated with the β subunit. J. Biol. Chem. 269, 8642-8646.

Shin, J. M. & Sachs, G. (1996). Dimerization of the gastric H,K ATPase. J. Biol. Chem. 271, 1904-1908.

Shin, J.M., Besancon, M., Simon, A., & Sachs, G. (1993). The site of action of pantoprazole in the gastric H/K-ATPase. Biochim. Biophys. Acta 1148(2), 223-233.

Shin, J. M., Kajimura, M., Arguello, J. M., Kaplan, J. H., & Sachs, G. (1994) Biochemical identification of transmembrane segments of the Ca-ATPase of sarcoplasmic reticulum. J. Biol. Chem. 269, 22533-22537.

Shull, G.E. (1990). cDNA cloning of the β-subunit of the rat gastric H,K-ATPase. J. Biol. Chem. 265, 12123-12126.

Shull, G. E. & Lingrel, J. B. (1986). Molecular cloning of the rat stomach H,K-ATPase. J. Biol. Chem. 261, 16788-16791.

Sih, J. C., Im, W. B., Robert, A., Graber, D. R., & Blakeman, D. P. (1991). Studies on H,K-ATPase inhibitors of gastric acid secretion. Prodrugs of 2-[(2-pyridylmethyl)sulfinyl]benzimidazole proton-pump inhibitors. J. Med. Chem. 34, 1049-1062.

Skrabanja, A.T.P., van der Hijden, H.T.W.M., & de Pont, J.J.H.H.M. (1987). Transport ratios of reconstituted H,K-ATPase. Biochim. Biophys. Acta 903, 434-440.

Skriver, E., Maunsbach, A. B., Herbert, H., Scheiner-Bobis, G., & Schoner, W. (1989). Two-dimensional crystalline arrays of Na,K-ATPase with new subunit interactions induced by cobalt-tetrammine-ATP. J. Ultrastructure Molecular Structure Research 102, 189-195.

Skriver, E., Kaveus, U., Hebert, H., & Maunsbach, A. B. (1992). Three-dimensional structure of Na,K-ATPase determined from membrane crystals induced by cobalt-tetramine-ATP. J. Structural Biology 108, 176-185.

Smith, D. L., Tao, T., & Maguire, M. E. (1993). Membrane topology of a P-type ATPases. The MgtB magnesium transport protein of salmonella typhimurium. J. Biol. Chem. 268, 22469-22479.

Smith, G. & Scholes, P. (1982). The H^+/K^+ ATP stoichiometry of the H,K-ATPase of dog gastric microsomes. Biochim. Biophys. Acta. 688, 803-807.

Smirnova, I. & Faller, L. D. (1993). Mechanism of K^+ interaction with fluorescein 5'-isothiocyanate-modified Na,K-ATPase. J. Biol. Chem. 268, 16120-16123.

Stewart, B., Wallmark, B., & Sachs, G. (1981). The interaction of H^+ and K^+ with the partial reactions of gastric H,K-ATPase. J. Biol. Chem. 256, 2682-2690.

Tahara, Y., Ohnishi, S.-i., Fujiyoshi, Y., Kimura, Y., & Hayashi, Y. (1993). A pH induced two-dimensional crystal of membrane-bound Na,K-ATPase of dog kidney. FEBS Lett. 320, 17-22.

Tamura, S., Tagaya, M., Maeda, M., & Futai, M. (1989). Pig gastric H,K-ATPase. Lys-497 conserved in cation transporting ATPases is modified with pyridoxal 5'-phosphate. J. Biol. Chem. 264, 8580-8584.

Tamura, S., Oshiman, K-I., Nishi, T., Mori, M., Maeda, M., & Futai, M. (1992). Sequence motif in control regions of the H,K ATPase α and β subunit genes recognized by gastric specific nuclear proteins. FEBS Lett. 298, 137-141.

Tari, A., Wu, V., Sumii, M., Sachs, G., & Walsh, J. H. (1991). Regulation of rat gastric H,K ATPase α subunit mRNA by omeprazole. Biochim. Biophys. Acta 1129, 49-56.

Tari, A., Yamamoto, G., Sumii, K., Summii, M., Takehara, Y., Haruma, K., Kajiyama, G., Wu, V., Sachs, G., & Walsh, J. H. (1993). The role of the histamine-2 receptor in the expression of rat gastric H^+,K^+-ATPase α subunit. Amer. J. Physiol. 265, G752-G758.

Tari, A., Yamamoto, G., Yonei, Y., Sumii, M., Sumii, K., Haruma, K., Kajiyama, G., Wu, V., Sachs, G., & Walsh, J. H. (1994). Stimulation of H,K ATPase α subunit expression by histamine. Amer. J. Physiol. 266, G444-G450.

Toh, B-H., Gleeson, P. A., Simpson, R. J., Moritz, R. L., Callaghan, J. M., Goldkorn, I., Jones, C. M., Martinelli, T. M., Mu, F-T., Humphris, D. C., Pettitt, J. M., Mori, Y., Masuda, T., Sobieszczuk, P., Weinstock, J., Mantamadiotis, T., & Baldwin, G. S. (1990). The 60- to 90-kDa parietal cell autoantigen associated with autoimmune gastritis is a β subunit of the gastric H,K-ATPase (proton pump). Proc. Natl. Acad. Sci. USA 87, 6418-6422.

Toyoshima, C., Sasabe, H., & Stokes, D. (1993). Three-dimensional cryo-electron microscopy of the calcium ion pump in the sarcoplasmic reticulum. Nature 362, 469-471

Treuheit, M. J., Costello, C. E., & Kirley, T. L. (1993). Structure of the complex glycans found on the β subunit of Na,K ATPase. J. Biol. Chem. 268, 13914-13919.

Walan, A., Bader, J. P., Classen, M., Lamers, B. H. W., Piper, D. W., Rutgersson, K., & Eriksson, S. (1989). Effect of omeprazole and ranitidine on ulcer healing and relapse rates in patients with benign gastric ulcer, N. Engl. J. Med. 32, 69-75.

Walderhaug, M. O., Post, R. L., Saccomani, G., Leonard, R. T., & Briskin, D. P. (1985). Structural relatedness of three ion-transport adenosine triphosphatases around their active sites of phosphorylation. J. Biol. Chem. 260, 3852-3859.

Wallmark, B., Brandstrom, A., & Lassen, H. (1984). Evidence for acid-induced transformaton of omeprazole into an active inhibitor of H,K-ATPase within the parietal cell. Biochim. Biophys. Acta 778, 549-558.

Wallmark, B., Stewart, H. B., Rabon, E.,Saccomani, G., & Sachs, G. (1980). The catalytic cycle of gastric (H^++K^+)-ATPase. J. Biol. Chem. 255, 5313-5319.

Wallmark, B., Briving, C., Fryklund, J., Munson, K., Jackson, R., Mendlein, J., Rabon, E., & Sachs, G. (1987). Inhibition of gastric H,K-ATPase and acid secretion by SCH 28080, a substituted pyridyl[1,2α]imidazole. J. Biol. Chem. 262, 2077-2084.

Weidmann, K., Herling, A. W., Lang, H. J., Scheunemann, K. H., Rippel, R., Nimmesgern, H., Scholl, T., Bickel, M., & Metzger, H. (1992). 2-[(2-Pyridylemethyl)sulfinyl]-1H-thieno[3,4-d]imidazoles. J. Med. Chem. 35, 438-450.

Wolosin, J. M., & Forte, J. G. (1983). Kinetic properties of the KC1 transport at the secreting apical membrane of the oxyntic cell. J. Membr. Biol. 71, 195-207.

Xu, K-y. (1989). Any of several lysines can react with 5'-isothiocyanatofluorescein to inactivate sodium and potassium ion activated adenosine triphosphatase. Biochemistry 28, 5764-5772.

Yamamoto, H., Imamura, Y., Tagaya, M., Fukui, T., & Kawakita, M. (1989). Ca^{2+}-dependent conformational change of the ATP-binding site of CA_{2+}-transporting ATPase of sarcoplasmic reticulum as revealed by an alteration of the target-site specificty of adenosine triphosphopyridoxal, J. Biochem. (Tokyo) 106, 1121-1125.

Yamaguchi, M., & Kanazawa, T. (1984). Protonation of the sarcoplasmic reticulum Ca-ATPase during ATP hydrolysis. J. Biol. Chem 259, 9526-9531.

ments is less certain and numbers ranging from 8 to 12 have been reported (Serrano et al., 1986, Nakamoto et al., 1989, Rao et al., 1991). However, a value of ten appears most consistent with recent data (Lin and Addison, 1995a) and is in accord with detailed studies with other P-type enzymes (Mata et al., 1992; Sachs et al., 1992; Smith et al., 1993).

The H^+-ATPases from yeast and other fungi have a number of distinguishing features which make them attractive as experimental systems. In yeast and Neurospora, where the enzymes represent a substantial fraction of the total plasma membrane protein (~10 - 40 percent), they can be readily purified and studied in reconstituted systems facilitating detailed biochemical (Malpartida and Serrano, 1981b; Perlin et al., 1984; Scarborough, 1980; Villalobo et al., 1981) and biophysical (Scarborough, 1980; Perlin et al., 1984; Hennesey and Scarborough, 1988; Seto-Young and Perlin, 1991) analyses . The H^+-ATPase is highly electrogenic, resulting in membrane potentials in fungi in excess of 200 mV (interior negative)(Slayman et al., 1973), and its electrical properties have been studied in detail by electrophysiological techniques in *Neurospora* (Slayman et al., 1973; Slayman, 1987; Slayman and Zuckier, 1989) and in reconstituted systems (Seto-Young and Perlin, 1991).

A great deal is known about the biosynthesis and assembly of the H^+-ATPase (Aaronson et al., 1988; Chang et al., 1993), as well as regulation of its gene expression (Capieaux et al., 1989, Rao et al., 1993, Viegas et al., 1994). Yet, it is the potential for structure/function information from genetics that has made the yeast H^+-ATPase an attractive P-type enzyme to study. In particular, it is the availability of classical genetic and molecular biological approaches that makes it possible to manipulate the *PMA1* gene, which encodes the plasma membrane H^+-ATPase. While similar information can sometimes be obtained from site-directed mutations in mammalian enzymes expressed transiently in heterologous systems, the fungal system provides a way to select interesting mutations and to analyze them under a variety of physiological conditions (McCusker et al., 1987; Ulaszewski et al., 1987a). For example, mutations that reduce but do not eliminate H^+-ATPase activity or proton transport can be identified by alterations in cellular growth rate and resistance to the antibiotic hygroymycin B. Moreover, some *PMA1* mutations show a wide variety of pleiotropic phenotypes including slow growth, sensitivity to low pH medium, osmotic sensitivity and sensitivity to low temperature (McCusker et al., 1987). The genetic approaches also permit the identification of mutations that cannot be obtained or analyzed in mammalian cells such as dominant negative mutations (Harris et al., 1994) and intragenic suppressors (Portillo et al., 1991; Harris et al., 1991; Na et al., 1993; Eraso and Portillo, 1994; Anand et al., 1995). Although it is the essentiality of the *PMA1* gene that facilitates the application of diverse approaches, it also creates special problems, like the need to develop conditional gene expression systems (Cid et al., 1987, Harris et al., 1991, Nakamoto et al., 1991; Harris et al, 1994) for the expression of mutant gene products.

In the following sections, we illustrate how genetic probing has been a valuable asset in developing mechanistic information about specific functional domains in the yeast H^+-ATPase. We start by discussing dynamic interactions, which provide information about coupling between ATP hydrolysis and proton transport domains, and detail how genetic probing can aid in structure predictions by focusing on transmembrane segments 1 (M1) and 2 (M2). We then turn to the catalytic center and discuss the nature of the phosphate binding domain by analyzing mutations within the first cytoplasmic loop domain linking M2 and M3 and the large central cytoplasmic domain linking M4 and M5 that confer relative vanadate insensitivity on the H^+-ATPase. This section is followed by a brief discussion of mutations mapping within close proximity to the site of phosphorylation (D378) and the interesting consequences of lethal mutations at this site. Finally, the role of the C-terminus as a regulatory domain of catalytic activity is discussed and we conclude with a forward look at opportunities and challenges.

II. STRUCTURAL DYNAMICS AND COUPLING.

All P-type ATPases couple energy from ATP hydrolysis in a cytoplasmic catalytic domain to ion transport in a membrane-embedded region of the protein and long-range interactions are expected to be an essential feature of the coupling mechanism (Inesi et al., 1990). The mechanism of coupling by P-type ATPases is not understood. Yet, clues to the nature of the long range conformational interactions fundamental to the coupling process have come from a variety of experimental approaches. Biochemical approaches in the analysis of the mammalian ATPases have included drug interaction studies (Forbush, 1983; Kirley and Peng, 1991; Munson et al., 1991; Schultheis and Lingrel, 1993; Jewell-Motz and Lingrel, 1993; Schultheis et al., 1993), immunological probing (Arystarkhova et al., 1992), and limited proteolysis (Lutsenko and Kaplan, 1994). Genetic approaches, involving the analysis of mutant proteins, have been carried out with the Na^+-K^+-ATPase (Jewell-Motz and Lingrel, 1993; Lingrel and Kuntzweiler, 1994) and Ca^{2+}-ATPase (Clarke et al., 1990a; Toyofuku et al., 1992 ;Andersen, 1995; Clarke et al., 1989), and with the yeast H^+-ATPase (Harris et al., 1991; Na et al., 1993; Anand et al., 1995).

Therapeutic antagonists directed against P-type enzymes help illustrate the nature of such coupled interactions. The cardiac glycoside ouabain, which is a potent inhibitor of the Na^+-K^+-ATPase, is known to interact with the extracellular face of the enzyme where it blocks ATP hydrolysis within the cytoplasmically-located catalytic center (Forbush, 1983). Similarly, the H^+-K^+-ATPase antagonists omeprazole and pantoprazole show related inhibition behavior (Besancon et al., 1993, Shin et al., 1993). These compounds form cationic sulfenamides at low pH which rapidly react with the thiol groups of cysteines but, due to the nature of gastric physiology, they will only activate and react with the gastric pump from the extracellular surface. Modification of at least two cysteines near or within the bilayer results in inactivation of the enzyme (Besancon et al., 1993, Shin et al., 1993). How modification of extracellular or near extracellular membrane-embedded residues exerts its effects on catalysis is unclear but it certainly requires conformational interactions over long distances. The actual coupling distance is difficult to assess, although, it has been estimated for the Ca^{2+}-ATPase that the distance between sites within the catalytic center and membrane-embedded ion transport region may exceed 40Å (Bigelow and Inesi, 1991).

Transmembrane segments 5 and 6 (M5/M6) and the stalk segment extending from M5 (S5) have been proposed to play important roles in coupling. Changes in the stalk region during catalysis are believed to influence the ion binding properties associated with M5, M6, and perhaps M4. This proposal is supported by the detailed mutagenesis on the Ca^{2+}-ATPase which show a prominent role for M4, M5, and M6 in ion binding (Rice and MacLennan, 1996), the scanning alanine mutagenesis study of Ambesi et al. (1996) on M4 in the yeast H^+-ATPase, and the report of a Y763G mutation in the Ca^{2+}-ATPase, lying at the cytoplasmic interface of M5/S5, which uncouples ATP hydrolysis from Ca^{2+}-transport (Andersen, 1995.) Yet, despite these promising results, mutations that partially uncouple proton transport from ATP hydrolysis in the yeast H^+-ATPase have been reported in stalk segments S2 (I183A) and S3 (H285Q) (Wach et al., 1995; Wang et al., 1996), which raises the possibility that these segments and their associated transmembrane segments may play a role in coupling.

Yeast genetics provide an important tool to study long-range protein-protein interactions. By examining second site suppressor mutations which either partially or fully complement phenotypes produced by a primary site mutation in *PMA1*, it has been possible to show that the N-terminal transmembrane segments, especially M1 and M2, are conformationally linked to the catalytic ATP hydrolysis domain (Harris et al., 1991; Na et al., 1993.)

This approach is particularly powerful because compensation for the first mutation may arise from either a direct interaction of a second site mutation with the primary site residue to relieve steric stress or an unpaired charge, or it may arise from an indirect conformational shift which stabilizes a mutant enzyme structure (local or global.) It was shown that the low pH growth sensitivity of a primary site mutation, S368F, which is near the site of phosphorylation could be suppressed by a second-site mutation E367V or by F289V. These most likely represent examples of direct interactions. By itself, F289V has phenotypes quite similar to those of S368F, though the double mutant is more nearly wild type. However, the majority of the secondary site mutations that relieved the S368F phenotype were in M1 and M2 (I147M, C148S, A135S, I133F) (Harris et al., 1991)(Figure 1A.) Conversely, the hygromycin B resistant and low pH sensitive growth phenotypes induced by a mutation, A135V, which is near the extracytoplasmic face of M1 could be suppressed by secondary site mutations within the catalytic core ATP binding domain (S660C, S660F, F611L)(Na et al., 1993). It is apparent from the topological positions of the primary site and secondary site mutations that conformational complementation is important here. However, it should be noted that in both studies suppressor mutations were also found in other transmembrane segments, notably M3, M4 and M7, so the potential for direct interactions cannot be excluded (Figure 1B.)

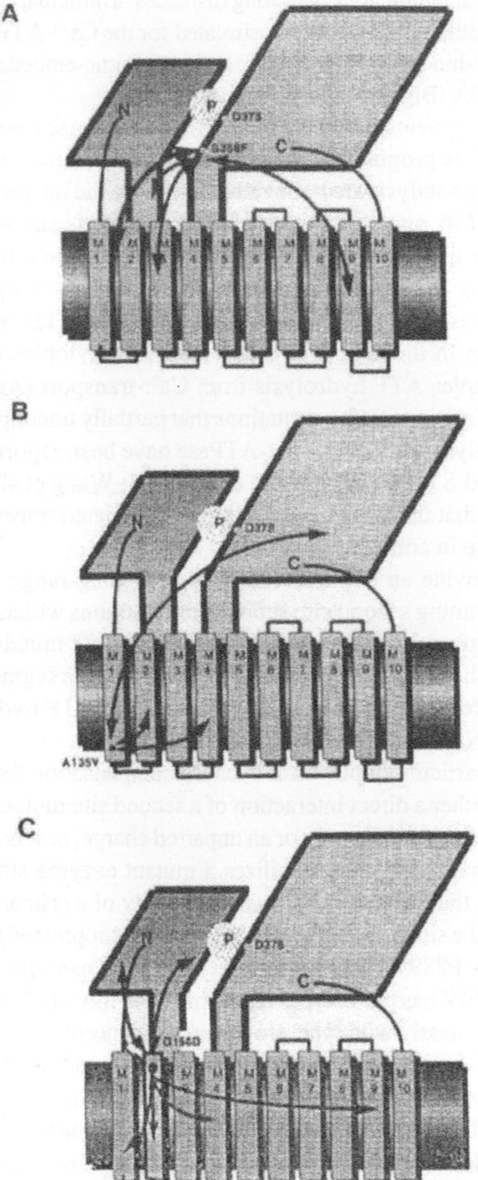

Figure 1. Topological locations of second-site suppressor mutations. Diagram shows the relative location of mutations isolated as second-site intragenic suppressors of phenotypes induced by primary site mutations S368F (panel **A**), A135V (panel **B**) and G158D (panel **C**), as described by Harris et al. (1991), Na et al. (1993), and Anand et al. (1995), respectively.

Diverse studies carried out on higher eukaryotic enzymes support a potential linkage between M1, M2, and the catalytic ATP hydrolysis domain. Genetic modification of residues in M1 and M2 which alters ouabain sensitivity (Lingrel and Kuntzweiler, 1994) was observed to alter catalysis by the enzyme (Price et al., 1990). A monoclonal antibody that recognizes an epitope in the extracellular turn region between transmembrane segments 1 and 2 of the Na^+-K^+-ATPase enhanced ouabain inhibition but also inhibited catalysis (Arystarkhova et al., 1992). Finally, the H^+-K^+-ATPase antagonist SCH28080, which potently blocks ATP hydrolysis, appears to bind within the loop region linking transmembrane segments 1 and 2 (Munson et al., 1991).

III. STRUCTURAL AND FUNCTIONAL CONSIDERATIONS FOR M1 AND M2

A clearer understanding of how a conformationally-sensitive region contained within transmembrane segments 1 and 2 (M1, M2) can interact with the catalytic domain requires a greater understanding of molecular enzyme structure. Unfortunately, the structure of the H^+-ATPase is poorly understood. Secondary structure information has emerged from biophysical studies on the purified preparations of the *Neurospora crassa* and *Saccharomyces cerevisiae* H^+-ATPases (Hennesey and Scarborough, 1988; Park et al., 1992; Goormaghtigh et al., 1994) but it has limited application. The use of cryoelectron microscopy-based image reconstruction techniques has been valuable for the mammalian Ca^{2+}-ATPase (Stokes and Green, 1990; Toyoshima et al., 1993; Stokes et al., 1994) and Neurospora H^+-ATPase (Cyrklaff et al., 1995) in defining a relatively low resolution structure of the enzyme.

In lieu of a high resolution structure, molecular modeling has been used to develop a working model of M1 and M2 that helps account for its role in coupling (Monk et al., 1994). The modeling was possible because a number of well supported constraints could be applied to the modeling process, including topological considerations which limited structural options (Monk et al., 1994). M1 and M2 are predicted to span the bilayer as a helical hairpin structure in which the two α-helical transmembrane segments are joined by a linking loop region to form a partially crossed "V" structure (Figure 2.) This region appears to be an essential structural motif that is common to all members of the P-type ATPase family and a related structure has been predicted for M1 and M2 in the gastric H^+-K^+-ATPase (Munson et al., 1991). The hairpin structure is predicted to have a number of prominent structural features including a short 4-6 amino acid turn linking the α-helices, a tightly packed head region, an N-cap structure stabilizing the turn region and M2, a cluster of 5 interacting phenylalanine residues near the cytoplasmic face of the hairpin structure and a flexible region consisting of G122,P123 that may kink M1. Molecular dynamic simulations of the hairpin structure sug-

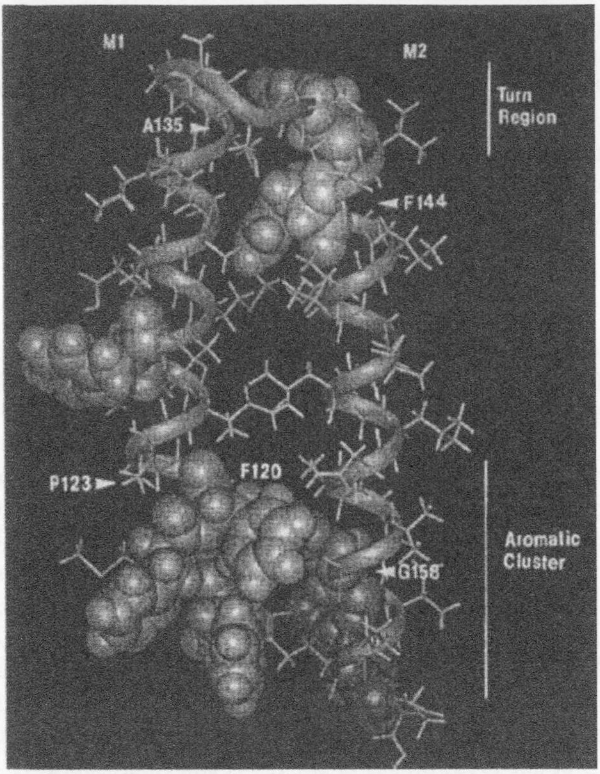

Figure 2. Helical hairpin model for transmembrane segments 1 and 2. A solid ribbon diagram is shown superimposed on the carbon backbone of amino acid residues F116-F163, which are predicted to form a helical hairpin loop structure. The hairpin model was constructed by imposing numerous structural constraints, as described by Monk et al. (1994). Aromatic side chains corresponding to F116, F119, F120, F126, W141, F144, F159, and F163 are shown in CPK format. The positions of amino acid residues inducing prominent phenotypes when mutated are shown by the arrows.

gested that it is highly conformationally active and perturbations in the head region of the hairpin can be propagated throughout the structure, but most prominently down helix 2. This approach provided a possible explanation for how a second site V157F mutation near the cytoplasmic end of M2 could fully complement a destabilizing A135V primary site mutation at the extracytoplasmic face of M1(Monk et al., 1994).

The linking turn between transmembrane segments 1 and 2 appears to have an important role in conformational interactions in the P-type ATPases. A monoclonal antibody to this region in the Na^{+}-K^{+}-ATPase alters the conformation of the enzyme such that ATP hydrolysis is inhibited and cardiac glycoside binding is enhanced (Arystarkhova et al., 1992) and genetic modification of the turn region in the Na^+-K^+-ATPase leads to significantly altered ouabain binding (Price et al., 1989, Price et al., 1990, Lingrel and Kuntzweiler, 1994). The loop region from the various P-type ATPases varies from approximately 4 in the H^+-ATPases to as many as 22 in the Ca^{2+}-ATPase (Monk et al., 1994). Hydrogen bonding and packing interactions are expected to be the most important stabilizing determinants in the hairpin head region. It has been suggested that perturbations in this region from mutation or drug interaction, which disrupt local conformation by altering packing density, diminish the hydrogen bonding potential of the hairpin or by disrupting key hydrogen-bonding pairs would cause main chain displacements (Monk et al., 1994).

The yeast H^+-ATPase contains the shortest turn region, approximately 4-6 amino acids, G137, L138, S139, D140, W141, and V142. Short turns generally depend on the positioning of special residues, usually Gly, Asn, or Pro, which allow a peptide to reverse direction (Chothia, 1984). The P-type ATPases adhere to this convention as tandem proline residues, and Gly and Asn residues frequently occupy border positions (Monk et al., 1994). Genetic modification of the loop region showed that substitution of alanine at G137, L138, and S139 had little effect on phenotype, even though *in vitro* enzyme function was diminished, while the same substitution at D140, W141 and V142 had pronounced effects on phenotype (Seto-Young et al., 1994). Interestingly, extra amino acids could be functionally inserted into the loop structure (insertion between L138 and S139) provided that they were small and uncharged (Seto-Young et al., 1994) suggesting that there is at least limited flexibility of the loop. However, while the insertion of a glycine or alanine between residues 138 and 139 was tolerated, the insertion of a serine yielded a strain that was temperature-sensitive for growth and sensitive to low pH. A threonine at this position was lethal.

An important consequence of the short turn region between transmembrane segments 1 and 2 is that the helical hairpin head region should be tightly packed. Evidence for such tight packing was provided by demonstrating that an A135V mutation near the top of transmembrane segment 1 strongly perturbed the local structure in the region (Na et al., 1993) and the apparent packing in this region was substantially disrupted by substitutions with bulkier residues of neutral charge (Na et al. 1993; Seto-Young et al., 1994). Polar residues were not tolerated at this position. In contrast, F144 at a neighboring position in transmembrane segment 2 could only be altered to the larger Trp and all other residues were lethal including a polar Tyr. Either the retention of a relatively non-polar aromatic character at position 144 is critical or the dramatic reduction of the side chain mass upon substitution with a smaller side group residue produces a cavity in a hydrophobic domain that is strongly destabilizing (Eriksson et al., 1992).

The residues D140 in the loop and D143 near the extracellular border in helix 2 are conserved in all H^+-ATPases (Wach et al., 1992). The conformation of the turn region is predicted to be stabilized by hydrogen bonding interactions involving acidic residues. The carboxyl side group of D140 is predicted to form an H-bond with the amide nitrogen of V142 forming a N-cap and the carboxylic side chain of D143 is proposed to form an H-bond with the backbone amide between S139 and D140 (Monk et al., 1994). Modification of D143 to Ala or more subtle modifications to Glu or Asn result in either dominant or recessive lethal phenotypes indicating the essentiality of this conserved residue (Portillo and Serrano, 1988; Seto-Young et al., 1994).

Another predicted stabilizing feature of the hairpin model is a series of clustered aromatics at the cytoplasmic end, which is also seen in the Na^+-K^+-ATPase α-subunit. Preliminary studies indicate that phenylalanines, F116, F119, F156, and F160 can be functionally replaced with alanine while an F120A mutation is lethal. Both F120 and F159 are highly conserved (Wach et al., 1992) but only F120 appears essential. As might be expected from multiply interacting residues, the substitution of more than one phenylalanine with alanine results in lethality or significantly alters enzyme function (Seto-Young et al., 1996). (In general, it is important to recognize that mutations resulting in lethal phenotypes (dominant or recessive lethality) may produce an enzyme that is catalytically inactive, assembly deficient, or unable to progress to the plasma membrane. Defects in transport to the plasma membrane highlight the complexity of the assembly process and often result in a proliferation of intracellular membranes, as has been observed with heterologously expressed *PMA* genes from plants and yeast, as well as in yeast *PMA1* mutations yielding dominant lethal phenotypes (Supply et al., 1993a, Harris et al., 1994, Villalba et al., 1992). It has been proposed that assembly of the H^+-ATPase occurs in pairs of transmembrane segments, starting with M1, M2, and the numerous internal stabilizing properties of M1, M2 may be necessary to facilitate proper assembly (Lin and Addison, 1995b).

The putative proline kink region in M1 formed by P123 does not appear to be critical for enzyme action, as genetic substitution with alanine produced an active enzyme. P123 is highly conserved amongst the various H^+-ATPases but it is absent from the higher eukaryotic enzymes (Wach et al., 1992). However, glycine, a flexible residue, is found in the Na^+-K^+-ATPase (sheep) α–subunit at equivalent positions to G122 and P123 in the yeast *PMA1*. While a defined kink in the helical structure could be maintained in the absence of proline, as has been demonstrated in the crystal structure of a P61A mutant in the Fis protein (Yuan et al., 1994), there is no direct structural evidence for a kinked structure in M1. However, an extended α-helical structure in M1 appears to be an absolute requirement because a G122P mutation which creates a tandem proline (P122,P123) is lethal (Seto-Young et al., 1996) There are no documented examples of tandem prolines in α-helical structures, as these would be predicted to disrupt the helix (Yun et al., 1991).

Finally, it has been proposed that conformational interactions between M1,M2, and the catalytic ATP binding/phosphorylation domain (cytoplasmic domain bound by M4, M5) may be mediated by the cytoplasmic loop domain bound by M2, M3 (Perlin et al., 1992). In this model, disruptions in M1, M2 due to drug interactions or genetic modifications are propagated to the catalytic core via an interaction with the first cytosolic loop domain linking M2 and M3. This region is known to block catalysis in the active site region by altering the E_1P-E_2P transition (Cid and Serrano, 1988; Andersen et al., 1989; Serrano and Portillo, 1989; Clarke et al., 1990b). Anand et al. (1995) provided some support for this model by demonstrating that a phenotype produced by a G158D mutation near the cytoplasmic end of M2, which partially uncouples the H^+-ATPase (Perlin et al., 1989), could be complemented by second site mutations in the first cytosolic loop domain (D170N, L275S, L275F). As expected, G158 appears to be in a highly conformationally active region since second site suppressor mutations were also found in M2 (G156C, C148W), M1 (V127A), M4 (I332T, V336A), and M9 (F830S) (Figure 1C.) Interestingly, G158 could be replaced with a polar residue, like serine, but hydrophobic residues of increasing size were not well tolerated (Anand et al., 1995). Finally, mutations I183A (Wang et al., 1996) and H285Q (Wach et al., 1995) in the putative stalk region of the first cytoplasmic loop domain which result in partial uncoupling of proton transport from ATP hydrolysis in the yeast H^+-ATPase, support a potential coupling role for this region of the enzyme.

IV. VANADATE SENSITIVITY AND THE SITE OF PHOSPHATE BINDING

A characteristic feature of P-type enzymes is that they form an acyl-phosphate intermediate during catalysis of ATP and they are sensitive to inhibition by vanadate (Bowman and Slayman, 1979). Vanadate is believed to bind at the site from which phosphate is released, and enzyme inhibition results from the formation of a trigonal bipyramid complex that acts as a transition state analog (Pick, 1982). The yeast H^+-ATPase is highly sensitive to vanadate ($K_i \sim 1\ \mu M$) (Perlin et al., 1989; Ulaszewski et al., 1987b; Rao and Slayman, 1993) and it is expected that mutations in the catalytic region which alter the sensitivity of the H^+-ATPase to vanadate would provide important information about the phosphate binding domain. Despite the strong sensitivity of the H^+-ATPase to vanadate, it has not been possible to use this inhibitor as a selection probe, even under conditions where high-affinity phosphate transport systems are induced (Willsky et al., 1984, Willsky et al., 1985). Ulaszewski et al. (1987a, b) first reported a vanadate insensitive (K_i >100-fold that of wild type enzyme) *PMA1* mutant in *S. pombe* that was selected on the basis of growth resistance to Dio-9. This mutation, G268D (Ghislain et al., 1987), and a second mutation conferring vanadate insensitivity, K250T, which was also selected on the basis of Dio-9 resistance (Ghislain et al., 1992) was found to map to the first cytoplasmic loop domain linking M2 and M3. Site-directed mutagenesis of a highly conserved

T231 residue in this region also resulted in a vanadate insensitive enzyme (Serrano and Portillo, 1989).

Vanadate-insensitive *PMA1* mutants have also been identified from growth selections based on resistance to hygromycin B (McCusker et al., 1987, Perlin et al., 1989). The mutations, S368F and P640L mapped near the site of phosphorylation (D378) and within the putative ATP binding domain, respectively. The S368F mutation conferred hygromycin B resistance and sensitivity to low pH and showed >500-fold less sensitivity to vanadate. Interestingly, the level of vanadate insensitivity was directly correlated with the size of the side group substitution at this position. S368A showed normal sensitivity, while Val and Leu showed levels of vanadate sensitivity that were 11- and 228-fold less, respectively, than that of wild type (Harris et al. 1991.) Partial phenotypic suppressors (low pH^r, $HygB^r$ v. low pH^s, $HygB^r$ for *PMA1*-S368F) of S368F were identified which restored by 10-40-fold vanadate sensitivity (Harris et al., 1991). Second site mutations mapped to transmembrane domains (M1, M2, M3 and M7), as well as to the region around S368 (E367V) and the cytoplasmic loop domain linking M2 and M3 (V289F.)(Figure 1A.) Interestingly, the V289F mutation, when examined apart from S368F diminished vanadate sensitivity 11-fold (Harris et al., 1991). Finally, another second-site complementing mutation (L275F) within the first cytoplasmic loop domain, which partially suppressed the phenotype of a G158D mutation in M2, was found to decrease vanadate sensitivity by 75-fold (Anand et al. 1995.)

A potential interaction between the first cytoplasmic loop region and the central cytoplasmic catalytic domain within the vicinity of the site of phosphorylation may be necessary for vanadate binding and may in turn define part of the site of phosphate binding. This notion was supported by the demonstration that vanadate-insensitive *PMA1* mutants show altered interactions with phosphate (Perlin et al., 1989; Goffeau and de Meis, 1990) and that the hydrophobic environment within the β-strand sector influences the binding of phosphate (Goffeau and de Meis, 1990). It is expected that perturbations around the site of phosphorylation, D378, within the large central cytoplasmic domain would alter vanadate sensitivity. Rao and Slayman (1993) demonstrated that *PMA1* mutations K379R, T380S, and T382S within close proximity to D378 induced significant vanadate insensitivity (~10-500-fold) in mutant enzymes. A direct interaction between the central cytoplasmic loop domain containing the site of phosphorylation and the first cytoplasmic loop domain would appear likely. In fact, data supporting this notion has emerged from the application of the dihybrid *GAL4* transcription activation system in yeast in which activation of transcription of the *GAL4* target gene resulted from a direct interaction of the first cytoplasmic loop domain with the large central domain when both domains were co-expressed (C. Padmanabha and C.W. Slayman, unpublished results).

Vanadate interactions at the phosphate binding site may have been altered by a structural defect within the site or by a decrease in the steady-state level of the E2 intermediate known to be essential for vanadate (P_i) binding (Ghislain et al., 1987; Perlin et al., 1988). There is strong genetic evidence to support the notion that perturbations in the first cytoplasmic loop domain due to mutation can alter E_1-E_2 transitions. Portillo and Serrano, (1988) reported that phosphoenzyme levels in the yeast H^+-ATPase during steady-state ATP hydrolysis were enhanced by an E233N mutation or by a T231G mutation (Serrano and Portillo, 1989) suggesting a defect in the hydrolytic step in this enzyme. A comparable mutation to E233N in the Ca^{2+}-ATPase did not support this suggestion, rather, it was found to cause a defect in the E_1P-E_2P transition (Andersen et al., 1989); several other mutations in the β-strand domain also block the E_1P-E_2P transition (Clarke et al., 1990b).

It cannot yet be ascertained whether the mutations described which confer vanadate insensitivity directly or indirectly influence phosphate binding. However, as pointed out by Rao and Slayman, 1993, and as illustrated in Figure 3, the apparent clustering of mutations in the first cytoplasmic loop domain and the central cytoplasmic domain suggests a direct effect on the phosphate binding domain.

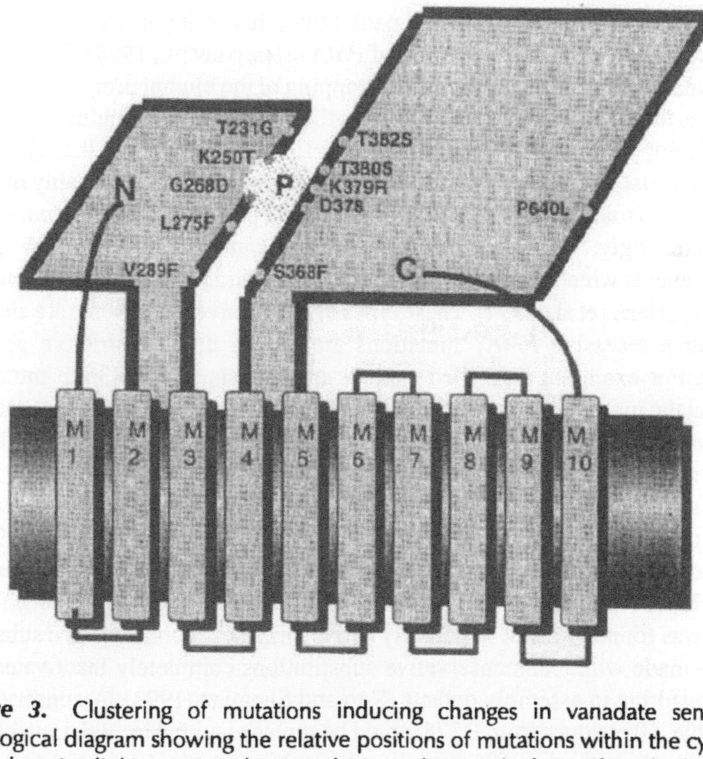

Figure 3. Clustering of mutations inducing changes in vanadate sensitivity. A topological diagram showing the relative positions of mutations within the cytoplasmic loop domains linking M2 and M3 and M4 and M5, which significantly decrease the sensitivity of the H^+-ATPase to vanadate.

V. CATALYTIC SITE PROBING BENEFITS AND SURPRISES

The high degree sequence conservation within the catalytic region of P-type ATPases, and the wealth of information concerning the potential essentiality of specific amino acid residues from chemical modification and proteolysis studies, has prompted numerous mutagenic studies aimed at establishing the importance of these various residues (Maruyama and MacLennan, 1988; Portillo and Serrano, 1988; Rao and Slayman, 1993; Serrano and Portillo, 1989; Lane et al., 1993, Maruyama et al., 1989). In the yeast H^+-ATPase, a great deal of attention has been focused on mutagenesis around the site of phosphorylation, D378. It was initially reported that a D378N mutation produced a viable enzyme (Portillo and Serrano, 1988). However, subsequent studies demonstrated that the D378 residue is essential and that all substitutions at this position were not viable (Rao and Slayman, 1992; Harris et al., 1994), which was consistent with related studies on the higher eukaryotic ATPases (Maruyama and MacLennan, 1988). The apparent discrepancy arose because there is strong selection pressure in yeast to revert lethal amino acid mutations to wild type. This is not surprising since H^+-ATPase function is essential for the cell.

In fact, the D378N mutation is dominant lethal; that is, it prevents cell growth even in the presence of a wild type copy of *PMA1* (Harris et al., 1994). Expression of the dominant lethal mutation causes the trapping of the mutant protein in vesicular structures that are apparently derived from the endoplasmic reticulum (Villalba et al., 1992; Supply et al., 1993a; Harris et al., 1994). Surprisingly, the wild type *PMA1* protein also becomes trapped in these structures; thus cells probably die because they fail to transport *PMA1* (and possibly other proteins) to the plasma membrane. Interestingly, dominant lethal mutations can be eliminated by gene conversion events which restore the wild type amino acid through homologous recombination (Harris et al., 1994). These types of gene conversion events are also recovered when recessive *PMA1* mutations are placed under restrictive growth conditions. For example, a detailed suppressor analysis of the S368F mutation showed that the majority of apparent revertants were in fact the result of ectopic recombination with *PMA2* to repair the catalytic site sequence in *PMA1* with a homologous region from *PMA2* (Harris et al., 1993).

The catalytic region around D378 has been explored by site-directed mutagenesis (Portillo and Serrano, 1988; Rao and Slayman, 1992; Rao and Slayman, 1993), random selection (McCusker et al., 1987; Perlin et al., 1989) and suppression analysis (Harris et al., 1991). Mutagenesis of invariant residues, K379, T380, T382, and T384 was found to result in partially active enzymes if conservative substitutions were made while nonconservative substitutions completely inactivated the enzymes resulting in assembly defects (Rao and Slayman, 1993). A conservative S368 residue, 10 residues from D378, could be replaced with Ala, Val, Leu, or Phe while retaining greater than 60% activity suggesting that a hydroxyl side group at this position is not critical (Harris et al., 1991).

In general, mutations within close sequence proximity to the site of phosphorylation inactivate the enzyme or diminish catalytic activity and induce a decrease in vanadate sensitivity (see previous section.). Other residues, such as K474 which has been implicated in catalysis based on chemical modification studies involving fluorescein 5'-isothiocyanate (FITC) (Serrano, 1988) or conserved acidic residues, D534, D560, D638, in the ATP binding domain have been probed to assess essentiality (Portillo and Serrano, 1988; Serrano and Portillo, 1989). In each case, mutations of K->Q or D->N resulted in significantly diminished catalytic function with the K474Q mutation producing the most defective enzyme. Second site suppressor analysis of a K474R mutation suggested that the FITC site may interact, either directly or indirectly, with the region around the site of phosphorylation (D378) (Maldonado and Portillo, 1995.) The apparent essentiality of K474 in yeast *PMA1* is consistent with chemical modification data (Serrano and Portillo, 1989), although it contrasts with genetic data on the Na^+-K^+-ATPase which indicates that the equivalent residue is not essential (Wang and Farley, 1992).

There have been several attempts to develop a template model for nucleotide binding by P-type ATPases by overlaying the primary sequence onto the coordinates of an existing crystal structure of a nucleotide-binding enzyme (Serrano, 1988; Serrano and Portillo, 1989; Taylor and Green, 1989). However, none of the existing structure models are sufficiently good fits to allow detailed mutagenic probing. Thus, detailed information about the structural architecture of the active site remains uncertain. A promising approach involves the overexpression in bacteria constructs encoding the catalytic domains bound by transmembrane segments 2 and 3 and/or 4 and 5 as fusion proteins which then bind nucleotide analogs (Capieaux et al., 1993). These preparations do not hydrolyze ATP but may still prove valuable in developing a more detailed model of the active site region.

VI. REGULATION OF CATALYTIC ACTIVITY BY THE C-TERMINUS.

When yeast cells are deprived of glucose, the activity of the H^+-ATPase is reduced by more than 10-fold (Serrano, 1983). This effect is retained during *in vitro* membrane isolation, as the enzyme is strongly inactivated. Upon addition of glucose, the H^+-ATPase is rapidly activated which is observed as a restored high level of catalytic turnover. Stimulation of the H^+-ATPase by glucose metabolism results from a combined effect on the K_m, V_{max}, pH optimum and vanadate binding site of the enzyme. The molecular mechanism of this regulation is not fully understood, but it seems that the carboxy terminus may act as an autoinhibitory domain. Deletion of the last 11 amino acids of the enzyme results in constitutive activation of the enzyme (Portillo et al., 1989) and several C-terminal residues (S899, E901, R909, S911, T912) have been implicated from site-directed mutagenesis studies in this process (Portillo et al., 1991, Portillo and Serrano, 1988, Eraso and Portillo, 1994). Regulation of the H^+-A-

TPase by glucose may be mediated by phosphorylation and dephophosphorylation of critical residues on the enzyme in response to nutritional signals (Kolarov et al., 1988, Ulaszewski et al., 1989, Chang and Slayman, 1991). Eraso and Portillo (1994) reported that T912, R909 along with S899 and E901 could define potential phosphorylation sites for calmodulin-dependent protein kinase II and casein kinase II, respectively. (The latter two sites, S899 and E901, were also implicated in the glucose-induced change in K_m (Eraso and Portillo, 1994.) Mutagenesis of these residues supports the notion that both S899 and T912 may be phosphorylated during glucose activation (Portillo et al., 1991, Eraso and Portillo, 1994).

Primary site mutations S911A and T912A, which make the H^+-ATPase inactivatable by glucose, were used to identify second site suppressors (Portillo et al., 1991, Eraso and Portillo, 1994). The suppressor mutations were found in the putative stalk region (A165V, V169I, D170N, A350T, A351T), the ATP binding region (G670S, P669L, G648S, P536L, A547V, A565T, G587N), and the regulatory domain at the C-terminus (S896F, R898K, M907I)(Figure 4.) The location of second site revertants within the stalk region and the ATP binding domain supports the notion that the regulatory domain interacts with these two regions to inhibit activity in the absence of glucose metabolism. A point mutation, K250T, within LOOP1, which constitutively

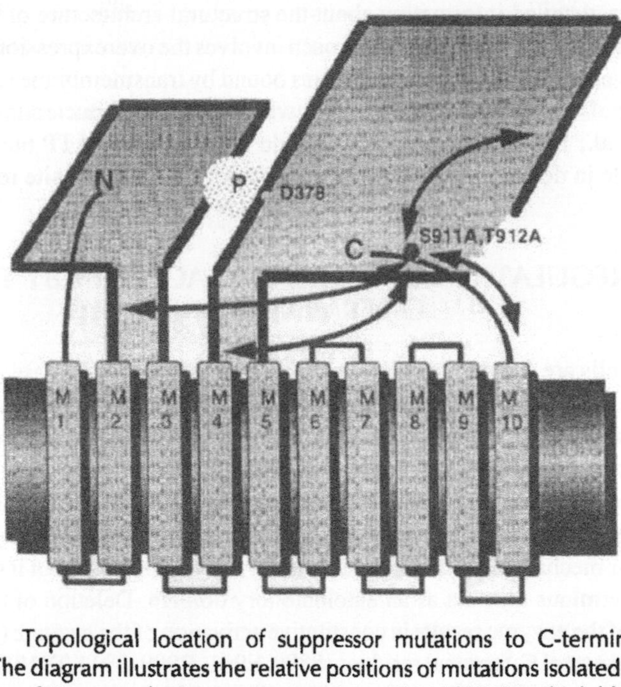

Figure 4. Topological location of suppressor mutations to C-terminal regulatory domain. The diagram illustrates the relative positions of mutations isolated as second site suppressors of C-terminal primary site mutations, S911A/T912A which block growth, as decribed by Eraso and Portillo, (1994).

activates the enzyme in the absence of glucose also supports this notion (Ghislain et al., 1992). Phenotypic suppression resulting in an active enzyme is probably due to compensatory structural alterations within the regulatory domain itself.

A model has been proposed whereby the C-terminus would interact directly or indirectly with at least two distinct functional domains (stalk and ATP binding) to inhibit activity. Glucose metabolism would induce a modification of the H^+-ATPase, perhaps through phosphorylation of the enzyme at S899 or T912 which would release this interaction. A similar model has been proposed for the Ca^{2+}-ATPase from red blood cells (Carafoli, 1992) on the basis of labeling studies whereby the C-terminus inhibits the activity of the enzyme by interacting with the active site, and this inhibition is relieved by phosphorylation or calmodulin binding.

VII. NEW DIRECTIONS, OPPORTUNITIES, AND LIMITATIONS

A great deal has been learned about structure and function of the H^+-ATPase in yeast through the application of diverse genetic approaches. Yet, more detailed information is desperately needed to better understand the role of individual domains or subdomains within the catalytic center and to define their interactions, to elucidate the role of individual transmembrane segments in proton transport, to explore the nature of coupling interactions, and to probe the nature of dimeric interactions. The ability to select primary site mutations that alter function, along with the application of suppressor analyses to define short- and long-range interactions will continue to provide important information about function and dynamics. Even after a high resolution structure is developed, it will be the resolution of structural dynamics that will provide an understanding of coupled catalysis by the H^+-ATPase.

There are technical obstacles along the way but they are not insurmountable. Unlike the higher eukaryotic enzymes, it is difficult to obtain detailed kinetic data for the phosphorylation and dephosphorylation reactions in the H^+-ATPase. This is due to the fact that ATP binds to a loose site (K_m~1 mM) and steady-state phosphorylated intermediate formation rarely exceeds 10% (Smith and Hammes, 1988), most likely because protons cannot be eliminated from the system. However, kinetic information which yields rate constants for partial reactions of ATP hydrolysis has been obtained from classical oxygen exchange measurements (Amory et al., 1982, Amory et al., 1984). It may be valuable to apply this technique to certain classes of mutants perturbed in catalysis and/or vanadate sensitivity. Similarly, it is difficult to accurately resolve proton transport kinetics by the enzyme due to the ubiquitous nature of protons. Yet, if *PMA1* mutants can be expressed in *Neurospora* or if patch clamp techniques can be improved, then it may be possible to extract detailed kinetic information from current-voltage studies by applying reaction-kinetic models (Hansen et al., 1981, Slayman, 1987). In addition, the construction of chimeric enzymes, incorporating transport features (ion specificity) of higher eukaryotic enzymes, may prove valuable as a way to resolve the transport pathway. These and other challenges need to be met to gather as much data as possible from the rich source of genetic information that is available and will continue to flow to this field from yeast studies.

The genetic potential of yeast will continue to be an important driving force in the P-type ATPase field for the elucidation of essential structure and function relationships, and ultimately, for a structurally-based mechanistic description of ATP-driven ion transport.

VIII. SUMMARY

The combination of yeast genetics and biochemistry has facilitated a greater understanding of structure-function relationships in the plasma membrane H^+-ATPase. Because the H^+-ATPase is essential to cell physiology and growth, it has been possible to merge classical genetic screening methods with molecular genetic probing. This approach makes it possible to identify not only individual mutations which influence function but also second site mutations which suppress the effects of a given mutation, thereby providing important information about local and long range structural interactions. In addition, it is possible to construct strains containing only a mutant H^+-ATPase and directly assess the physiological consequences. These diverse approaches have been used to develop notions about conformational coupling involving ATP hydrolysis and proton transport domains, the site for phosphate and vanadate interactions, and the nature of autoregulation via the C-terminal domain.

ACKNOWLEDGMENTS

We wish to acknowledge NIH grants GM 38225 and GM 39739 to D.S.P and J.E.H., respectively.

REFERENCES

Aaronson, L. R., Hager, K. M., Davenport, J. W., Mandala, S. M., Chang, A., Speicher, D. W., & Slayman, C. W. (1988). Biosynthesis of the plasma membrane H^+-ATPase of Neurospora crassa J. Biol. Chem. 263, 14552-14558.

Addison, R., & Scarborough, G. A. (1982). Conformational changes of the *Neurospora* Plasma membrane H^+-ATPase during its catalytic cycle. J. Biol. Chem. 257, 10421-10426.

Ambesi, A., Pan, R.L., & Slayman, C.W. (1996). Alanine-scanning mutagenesis along membrane segment 4 of the yeast plasma membrane H^+-ATPase. J. Biol. Chem. 271, 22999-23005.

Amory, A., & Goffeau, A. (1982). Characterization of the β-Aspartyl phosphate intermediate formed by the H^+-translocating ATPase from the yeast *Schizosaccharomyces pombe*. J. Biol. Chem. 257, 4723-4730.

Amory, A., Goffeau, A., .McIntosh, D. B., & Boyer, P. (1984). Contribution of 18-O technology to the mechanism of the H^+-ATPase from yeast plasma membrane. Curr. Topics Cell. Reg. 24, 471-483.

Amory, A., Goffeau, A., McIntosh, D. B., & Boyer, P. D. (1982). Exchange of oxygen between phosphate and water catalyzed by the plasma membrane ATPase from the yeast *Schizosaccharomyces pombe*. J. Biol. Chem. 257, 12509-12516.

Anand, S., Seto-Young, D., Perlin, D. S., & Haber, J. E. (1995). Mutations of G158 and their second-site revertants in the plasma membrane H^+-ATPase gene (*PMA1*) in *Saccharomyces cerevisiae*. Biochim. Biophys. Acta 1234, 127-132.

Andersen, J. P. (1995). Functional consequences of alterations to amino acids at the M5S5 boundary of the Ca^{2+}-ATPase of sarcoplasmic reticulum. Mutation of Tyr763->Gly uncouples ATP hydrolysis from Ca^{2+} transport. J. Biol. Chem. 270, 908-914.

Andersen, J. P., Vilsen, B., Leberer, E., & MacLennan, D. H. (1989). Functional consequences of mutations in the beta-strand sector of the Ca^{2+}-ATPase of sarcoplasmic reticulum. J. Biol. Chem. 264, 21018-21023.

Arystarkhova, E., Gasparian, M., Modyanov, N. N., & Sweadner, K. J. (1992). Na,K-ATPase extracellular surface probed with a monoclonal antibody that enhances ouabain binding. J. Biol. Chem. 19, 13694-13701.

Besancon, M., Shin, J. M., Mercier, F., Munson, K., Miller, M., Hersey, S., & Sachs, G. (1993). Membrane topology and omeprazole labeling of the gastric H^+,K^+-adenosinetriphosphatase. Biochemistry 32, 2345-2355.

Bigelow, D. J., & Inesi, G. (1991). Frequency-domain fluorescence spectroscopy resolves the location of maleimide-directed spectroscopic probes within the tertiary structure of the Ca-ATPase of sarcoplasmic reticulum. Biochemistry 30, 2113-2125.

Bowman, B. J., Berenski, C. J., & Jung, C. Y. (1987). Size of the plasma membrane H^+-ATPase from *Neurospora crassa* determined by radiation inactivation and comparison with the sarcoplasmic reticulum Ca^{2+}-ATPase. J. Biol. Chem. 15, 8726-8730.

Bowman, B. J., & Slayman, C. W. (1979). The effects of vanadate on the plasma membrane ATPase of *Neurospora crassa*. J. Biol. Chem. 254, 2928-2934.

Capieaux, E., Rapin, C., Thinès, D., Dupont, Y., & Goffeau, A. (1993). Overexpression in *Eschericia coli* and purification of an ATP-binding peptide from the yeast plasma membrane H^+-ATPase. J. Biol. Chem. 268, 21895-21900.

Capieaux, E., Vignais, M. L., Sentenac, A., & Goffeau, A. (1989). The yeast H^+-ATPase gene is controlled by the promoter binding factor TUF. J. Biol. Chem. 264, 7437-7446.

Carafoli, E. (1992). The Ca^{2+} pump of the plasma membrane. J. Biol. Chem. 267, 2115-2118.

Chadwick, C. C., Goormaghtigh, E., & Scarborough, G. A. (1987). A hexameric form of the *Neurospora crassa* plasma membrane H^+-ATPase. Arch. Biochem. Biophys. 252, 348-356.

Chang, A., Rose, M. D., & Slayman, C. W. (1993). Folding and intracellular transport of the yeast plasma membrane H^+-ATPase: Effects of mutations in KAR2 and SEC65. Proc. Natl. Acad. Sci. USA 90, 5808-5812.

Chang, A., & Slayman, C. W. (1991). Maturation of the yeast plasma membrane $[H^+]$ATPase involves phosphorylation during intracellular transport. J. Cell Biol. 115, 289-295.

Chothia, C. (1984). Principles that determine the structure of proteins. Ann. Rev. Biochem. 53, 537-572.

Cid, A., Perona, R., & Serrano, R. (1987). Replacement of the promoter of the yeast plasma membrane ATPase gene by a galactose-dependent promoter and its physiological consequences. Curr. Genet. 12, 105-110.

Cid, A., & Serrano, R. (1988). Mutations of the yeast plasma membrane H^+-ATPase which cause thermosensitivity and altered regulation of the enzyme. J. Biol. Chem. 263, 14134-14139.

Clarke, D. M., Loo, T. W., Inesi, G., & MacLennan, D. H. (1989). Location of high affinity Ca^{2+}-binding sites within the predicted transmembrane domain of the sarcoplasmic reticulum Ca^{2+}-ATPase. Nature 339, 476-478.

Clarke, D. M., Loo, T. W., & MacLennan, D. H. (1990a). Functional consequences of alterations to polar amino acids located in the transmembrane domain of the Ca^{2+}-ATPase of sarcoplasmic reticulum. J. Biol. Chem. 265, 6262-6267.

Clarke, D. M., Loo, T. W., & MacLennan, D. H. (1990b). Functional consequences of mutations of conserved amino acids in the beta-strand domain of the Ca^{2+}-ATPase of sarcoplasmic reticulum. J. Biol. Chem. 265, 14088-10492.

Cyrklaff, M., Auer, M., Kuhlbrandt, W., & Scarborough, G.A. (1995). 2-D structure of the *Neurospora crassa* plasma membrane ATPase as determined by electron cryomicroscopy. EMBO J. 14, 1854-1857.

Dame, J. B., & Scarborough, G. A. (1980). Identification of the hydrolytic moiety of the *Neuropora* plasma membrane H^+-ATPase and demonstration of a phosphoryl-enzyme intermediate in its catalytic mechanism. Biochemistry 19, 2931-2937.

Eraso, P., & Portillo, F. (1994). Molecular mechanism of regulation of yeast plasma membrane H^+-ATPase by glucose. Interaction between domains and identification of new regulatory sites. J. Biol. Chem. 269, 10393-10399.

Eriksson, A. E., Baase, W. A., Zhang, X-J., Heinz, D. W., Blaber, M., Baldwin, E. P., & Mathews, B. W. (1992). The response of a protein structure to cavity-creating mutations and its relationship to the hydrophobic effect. Science 255, 178-183.

Forbush, B., III. (1983). Cardiotonic steroid binding to Na,K-ATPase. Curr. Top. Memb. Transp. 19, 167-201.

Ghislain, M., De Sadeleer, M., & Goffeau, A. (1992). Altered plasma membrane H^+-ATPase from Dio-9-resistant *PMA1-2* mutant of *Schizosaccharomyces pombe*. Eur. J. Biochem. 209, 275-279.

Ghislain, M., & Goffeau, A. (1991). The *PMA1* and *PMA2* H^+-ATPases from *Schizosaccharomyces pombe* are functionally interchangeable. J. Biol. Chem. 266, 18276-18279.

Ghislain, M., Schlesser, A., & Goffeau, A. (1987). Mutation of a conserved glycine residue modifies the vanadate sensitivity of the plasma membrane H^+-ATPase from *Schizosaccharomyces pombe*. J. Biol. Chem. 262, 17549-17555.

Goffeau, A., & de Meis, L. (1990). Effects of phosphate and hydrophobic molecules on two mutations in the beta-strand sector of the H^+-ATPase from the yeast plasma membrane. J. Biol. Chem. 265, 15503-15505.

Goffeau, A., & Slayman, C. W. (1981). The proton-translocating ATPase of the fungal plasma membrane. Biochim. Biophys. Acta. 639, 197-223.

Goormaghtigh, E., Vigneron, L., Scarborough, G. A., & Ruysschaert, J-M. (1994). Tertiary conformational changes of the *Neurospora crassa* plasma membrane H^+-ATPase monitored by hydrogen.deuterium exchange kinetics. J. Biol. Chem. 269, 27409-27413.

Hansen, U.-P., Gradmann, D., Sanders, D., & Slayman, C. L. (1981). Interpretation of current-voltage relationships for "active" ion transport systems: I. Steady-state reaction-kinetic analysis of class-I mechanisms. J. Memb. Biol. 63, 165-190.

Harris, S., Rudnicki, K. S., & Haber, J. E. (1993). Gene conversions and crossing over during homologous and hoemlogous ectopic recombination in *Saccharomyces cerevisiae*. Genetics 135, 5-16.

Harris, S. L., Na, S., Zhu, X., Seto-Young, D., Perlin, D. S., Teem, J. H., & Haber, J. E. (1994). Dominant lethal mutations in the plasma membrane H^+-ATPase gene of *Saccharomyces cerevisiae*. Proc. Natl. Acad. Sci. USA 91, 10531-10535.

Harris, S. L., Perlin, D. S., Seto-Young, D., & Haber, J. E. (1991). Evidence for coupling between membrane and cytoplasmic domains of the yeast plasma membrane H^+-ATPase. An analysis of intragenic revertants of *PMA1-105*. J. Biol. Chem. 266, 24439-24445.

Hennesey, J. P., & Scarborough, G. A. (1988). Secondary structure of the *Neurospora crassa* plasma membrane H^+-ATPase as estimated by circular dichroism. J. Biol. Chem. 263, 3123-3130.

Hennesey, J. P., & Scarborough, G. A. (1990). Direct evidence for the cytoplasmic location of the NH_2- and COOH-terminal ends of the *Neurospora crassa* plasma membrane H^+-ATPase. J. Biol. Chem. 265, 532-537.

Inesi, G., Sumbilla, C., & Kirtley, M. E. (1990). Relationship of molecular structure and function in Ca^{2+}-transport ATPase. Physiol. Rev. 70, 749-760.

Jewell-Motz, E. A., & Lingrel, J. B. (1993). Site-directed mutagenesis of the Na,K-ATPase: Consequences of substitutions of negatively-charged amino acids localized in the transmembrane domains. Biochemistry 32, 13523-13530.

Kirley, T. L., & Peng, M. (1991). Identification of cysteine residues in lamb kidney (Na,K)-ATPase essential for ouabain binding. J. Biol. Chem. 266, 19953-19957.

Kolarov, J., Kulpa, J., Baijot, M., & Goffeau, A. (1988). Characterization of a protein serine kinase from yeast plasma membrane. J. Biol. Chem. 263, 10613-10619.

Lane, L. K., Feldmann, J. M., Flarsheim, C. E., & Rybeczynski, C. L. (1993). Expression of rat α1 Na-K-ATPase containing substitutions of "essential" amino acids in the catalytic center. J. Biol. Chem. 268, 17930-17934.

Lin, J., & Addison, R. (1995a). The membrane topology of the carboxyl-terminal third of the *Neurospora plasma* membrane H^+-ATPase. J. Biol. Chem. 270, 6942-6948.

Lin, J., & Addison, R. (1995b). A novel integration signal that is composed of two transmembrane segments is required to integrate the *Neurospora plasma* membrane H^+-ATPase into microsomes. J. Biol. Chem. 270, 6935-6941.

Lingrel, J. B., & Kuntzweiler. (1994). Na^+-K^+-ATPase. J. Biol. Chem. 269, 19659-19662.

Lutsenko, S., & Kaplan, J. H. (1994). Molecular events in close proximity to the membrane associated with the binding of ligands to the Na,K-ATPase. J. Biol. Chem. 269, 4555-4564.

Maldonado, A.M., & Portillo, F. (1995). Genetic analysis of the fluorescein isothiocyanate binding site of the yeast plasma membrane H^+-ATPase. J. Biol. Chem. 270, 8655-8659.

Malpartida, F., & Serrano, R. (1981a). Phosphorylated intermediate of the ATPase from the plasma membrane of yeast. Eur. J. Biochem. 116, 413-417.

Malpartida, F., & Serrano, R. (1981b). Reconstitution of the proton translocating adenosine triphosphatase of yeast plasma membranes. J. Biol. Chem. 256, 4175-4177.

Mandala, S. M., & Slayman, C. W. (1988). Idetification of tryptic cleavage sites for two conformational states of the *Neurospora plasma* membrane H^+-ATPase. J. Biol. Chem. 263, 15122-15128.

Mandala, S. M., & Slayman, C. W. (1989). The amino and carboxyl termini of the Neurospora plasma membrane H^+-ATPase are cytoplasmically located. J. Biol. Chem. 264, 16276-16282.

Maruyama, K., Clarke, D. M., Fujii, J., Inesi, G., Loo, T. W., & MacLennan, D. H. (1989). Functional consequences of alterations to amino acids located in the catalytic center (isoleucine 348 to threonine 357) and nucleotide-binding domain of the Ca^{2+}-ATPase of sarcoplasmic reticulum. J. Biol. Chem. 264, 13038-13042.

Maruyama, K., & MacLennan, D. H. (1988). Mutation of aspartic acid-351, lysine-352, and lysine-515 alters the Ca^{2+} transport activity of the Ca^{2+}-ATPase expressed in COS-1 cells. Proc. Natl. Acad. Sci. USA 85, 3314-3318.

Mata, A. M., Matthews, I., Tunwell, R. E. A., Sharma, R. P., Lee, A. G., & East, J. M. (1992). Definition of surface-exposed and trans-membranous regions of the (Ca^{2+}-Mg^{2+})-ATPase of sarcoplasmic reticulum using anti-peptide antibodies. Biochem. J. 286, 567-580.

McCusker, J. H., Perlin, D. S., & Haber, J. E. (1987). Pleiotropic plasma membrane ATPase mutations of *Saccharomyces cerevisiae.* Mol. Cell. Biol. 7, 4082-4088.

Monk, B. C., Mason, A.B., Abramochkin, G., Haber, J.E., Seto-Young, D. and Perlin, D.S. (1995). The yeast plasma membrane proton pumping ATPase is a viable antifungal target. I. Effects of the cysteine modifying reagent omeprazole. Biochim. Biophys. Acta 1239, 81-90.

Monk, B. C., Feng, W. C., Marshall, C. J., Seto-Young, D., Na, S., Haber, J. E., & Perlin, D. S. (1994). Modeling a conformationally sensitive region of the membrane sector of the fungal plasma membrane proton pump. J. Bioenerg. Biomembr. 26, 101-115.

Monk, B. C., Montesinos, C., Ferguson, C., Leonard, K., & Serrano, R. (1991). Immunological approaches to the transmembrane topology and conformational changes of the carboxyl terminal regulatory domain of the yeast plasma membrane H^+-ATPase. J. Biol. Chem. 266, 18097-18103.

Munson, K. B., Gutierrez, C., Balaji, V. N., Ramnarayan, K., & Sachs, G. (1991). Identification of an extracytoplasmic region of H^+,K^+-ATPase labeled by a K^+-competitive photaffinity inhibitor. J. Biol. Chem. 266, 18976-18988.

Na, S., Hincapie, M., McCusker, J. H., & Haber, J. E. (1995). *MOP2* (*SLA2*) affects the abundance of the plasma membrane H^+-ATPase of *Saccharomyces cerevisiae.* J. Biol. Chem. 270, 6815-6823.

Na, S., Perlin, D. S., Seto-Young, D., Wang, G., & Haber, J. B. (1993). Characterization of yeast plasma membrane H^+-ATPase mutant *PMA1*-A135V and its revertants. J. Biol. Chem. 268, 11972-11797.

Nakamoto, R., Rao, R., & Slayman, C. W. (1989). Transmembrane segments of the P- type cation-transporting ATPases. Ann. N.Y. Acad. Sci. 574, 165-179.

Nakamoto, R. K., Rao, R., & Slayman, C. W. (1991). Expression of the yeast plasma membrane H^+-ATPase in secretory vesicles. A new strategy for directed mutagenesis. J. Biol. Chem. 266, 7940-7949.

Navarre, C., Catty, P., Leterme, S., Dietrich, F., & Goffeau, A. (1994). Two distinct genes encode small isoproteolipids affecting plasma membrane H^+-ATPase activity of *Saccharomyces cerevisiae*. J. Biol. Chem. 269, 21262-21268.

Odermatt, A., Suter, H., Krapf, R., & Solioz, M. (1993). Primary structure of two P-type ATPase involved in copper homeostasis in *Enterococcus hirae*. 268, 12775-12779.

Pardo, J. M., & Serrano, R. (1989). Structure of a plasma membrane H^+-ATPase gene from the plant *Arabidopsis thaliana*. J. Biol. Chem. 264, 8557-8562.

Park, K., Perczel, A., & Fasman, G. D. (1992). Differential between transmembrane helices and peripheral helices by the deconvolution of circular dichroism spectra of membrane proteins. Protein Sci. 1, 1032-1049.

Pedersen, P. L., & Carafoli, E. (1987). Ion motive ATPases. I. Ubiquity, properties, and significance to cell function. Trend. Biochem. Sci. 12, 146-150.

Perlin, D. S., & Brown, C. L. (1987). Identification of structurally distinct catalytic intermediates of the H^+-ATPase from yeast plasma membranes. J. Biol. Chem. 262, 6788-6794.

Perlin, D. S., Brown, C. L., & Haber, J. E. (1988). Membrane potential defect in hygromycin B-resistant pmal mutants of *Saccharomyces cerevesiae*. J. Biol. Chem. 263, 18118-18122.

Perlin, D. S., Harris, S. L., Monk, B. C., Seto-Young, D., Na, S., Anand, S., & Haber, J. E. (1992). Genetic probing of the yeast plasma membrane H^+-ATPase. Acta Physiol. Scand. 146, 183-192.

Perlin, D. S., Harris, S. L., Seto-Yong, D., & Haber, J. E. (1989). Defective H^+-ATPase of hygromycin B-resistant *PMA1* mutants from *Saccharomyces cerevisiae*. J.Biol. Chem. 264, 21857-21864.

Perlin, D. S., Kasamo, K., Brooker, R. J., & Slayman, C. W. (1984). Electrogenic H^+-translocation by the plasma membrane ATPase of *Neurospora* : Studies on plasma membrane vesicles and reconstituded enzyme. J. Biol. Chem. 259, 7884-7892.

Pick, U. (1982). The interaction of vanadate ions with the Ca-ATPase from sarcoplasmic reticulum. J. Biol. Chem. 257, 6111-6119.

Portillo, F., De Larrinoa, I. F., & Serrano, R. (1989). Deletion analysis of yeast plasma membrane H^+-ATPase and identification of a regulatory domain at the carboxyl terminus. FEBS Lett. 247, 381-385.

Portillo, F., Eraso, P., & Serrano, R. (1991). Analysis of the regulatory domain of yeast plasma membrane H^+-ATPase by directed mutagenesis and intragenic suppression. FEBS Lett. 287, 71-74.

Portillo, F., & Serrano, R. (1988). Dissection of functional domains of the yeast proton-pumping ATPase by directed mutagenesis. EMBO J. 7, 1793-1798.

Price, E. M., Rice, D. A., & Lingrel, J. B. (1989). Site-directed mutagenesis of a conserved, extracellular aspartic acid residue affects the ouabain sensitivity of sheep Na, K-ATPase. J. Biol.Chem. 264, 21902-21906.

Price, E. M., Rice, D. A., & Lingrel, J. B. (1990). Structure-function studies of Na,K-ATPase. Site-directed mutagenesis of the border residues from the H1-H2 extracellular domain of the alpha subunit. J. Biol. Chem. 265, 6638-6641.

Rao, R., Drummond-Barbosa, D., & Slayman, C. W. (1993). Transcriptional regulation by glucose of the yeast *PMA1* gene encoding the plasma membrane H^+-ATPase. Yeast 9, 1075-1084.

Rao, R., & Slayman, C. (1992). Mutagenesis of the yeast plasma membrane H^+-ATPase. Biophys. J. 62, 228-237.

Rao, R., & Slayman, C. W. (1993). Mutagenesis of conserved residues in the phosphorylation domain of the yeast plasma membrane H^+-ATPase. J. Biol. Chem. 268, 6708-6713.

Rao, U. S., Hennessey, J. P., & Scarborough, G. A. (1991). Identification of the membrane-embedded regions of the *Neurosopora crassa* plasma membrane H^+-ATPase. J. Biol. Chem. 266, 14740-14746.

Rice, W.J., & MacLennan, D.H. (1996). Scanning mutagenesis reveals a similar pattern of mutation sensitivity in transmembrane sequences M4, M5, and M6, but not in M8, of the Ca^{2+}-ATPase of Sarcoplasmic reticiulum (SERCA1a). J. Biol. Chem. 271, 31412-31419.

Rudolph, H.K., Antebi, A.,Fink, G.R., Buckley, C.M., Dorman, T.E., LeVitre, J., Davidow, L.S., Mao, J-I. & Moir, D.T. (1989). The yeast secretory pathway is perturbed by mutations in PMRI, a member of Ca^{2+} ATPase family. Cell 58, 133-145.

Sachs, G., Besancon, M., Shin, J. M., Mercier, F., Munson, K., & Hersey, S. (1992). Structural aspects of the gastric H,K-ATPase. J. Bioenerg. Biomembr. 24, 301-308.

Scarborough, G. A. (1980). Proton translocation catalyzed by the electrogenic ATPase in the plasma membrane of *Neurospora*. Biochemistry 19, 2925-2931.

Scarborough, G. A., & Addison, R. (1984). On the subunit composition of the *Neurospora* plasma membrane H^+-ATPase. J. Biol. Chem. 259, 9109-9114.

Scarborough, G. A., Goormaghtigh, E., & Chadwick, C. (1986). Monomers of the *Neurospora* plasma membrane H^+-ATPase catalyze efficient proton translocation. J. Biol. Chem. 261, 7466-7471.

Schlesser, A., Ulaszewski, S., Ghislain, M., & Goffeau, A. (1988). A second transport ATPase gene in *Saccharomyces cerevisiae* . J. Biol. Chem. 263, 19480-19487.

Schultheis, P. J., & Lingrel, J. B. (1993). Substitution of transmembrane residues with hydrogen-bonding potential in the alpha subunit of Na,K-ATPase reveals alterations in ouabain sensitivity. Biochemistry 32, 544-550.

Schultheis, P. J., Wallick, E. T., & Lingrel, J. B. (1993). Kinetic analysis of ouabain binding to native and mutated forms of Na,K-ATPase and identification of a new region involved in cardiac glycoside interactions. J. Biol. Chem. 268, 22686-22694.

Serrano, R. (1983). *In vivo* glucose activation of the yeast plasma membrane ATPase. FEBS Lett. 156, 11-14.

Serrano, R. (1988). Structure and function of proton translocating ATPase in plasma membranes of plants and fungi. Biochim. Biophys. Acta. 947, 1-28.

Serrano, R., Kielland-Brandt, M. C., & Fink, G. R. (1986). Yeast plasma membrane ATPase is essential for growth and has homology with (Na^++K^+),K^+-and Ca^{2+}-ATPase. Nature 319, 689-693.

Serrano, R., & Portillo, F. (1989). Active sites of yeast H^+-ATPase studied by directed mutagenesis. Biochem. Soc. Trans. 17, 973-975.

Seto-Young, D., Hall, M.J., Na, S., Haber, J.E., & Perlin, D.S. (1996). Genetic probing of the first and second transmembrane helices of the plasma membrane H^+-ATPase from *Saccharomyces cerevisiae*. J. Biol. Chem. 271, 581-587.

Seto-Young, D., Na, S., Monk, B. C., Haber, J. E., & Perlin, D. S. (1994). Mutational analysis of the first extracellular loop region of the H^+-ATPase from *Saccharomyces cerevisiae*. J. Biol. Chem. 269, 23988-23995.

Seto-Young, D., & Perlin, D. S. (1991). Effect of membrane voltage on the plasma membrane H^+-ATPase of *Saccharomyces cerevisiae*. J. Biol. Chem. 266, 1383-1389.

Shin, J. M., Besancon, M., Simon, A., & Sachs, G. (1993). The site of action of pantoprazole in the gastric H^+. K^+-ATPase. Biochim. Biophys. Acta 1148, 223-233.

Slayman, C. L. (1987). The plasma membrane ATPase of *Neurospora*: a proton-pumping electroenzyme. J. Bioenerg. Biomembr. 19, 1-20.

Slayman, C. L., Long, W. S., & lu, C. Y.-H. (1973). The relationship between ATP and an electrogenic pump in the plasma membrane of *Neurospora crassa*. J. Membrane Biol. 14, 305-338.

Slayman, C. L., & Zuckier, G. R. (1989). Differential function properties of a P-type ATPase. proton pump. Ann. N.Y. Acad. Sci. 574, 233-245.

Smith, D. L., Tao, T., & Maguire, M. E. (1993). Membrane topology of a P-type ATPase. The *MgtB* magnesium transport protein of *Salmonella typhimurium*. J. Biol. Chem. 268, 22469-22479.

Smith, E. K., & Hammes, G. G. (1988). Studies of the phosphoenzyme intermediate of the yeast plasma membrane proton-translocating ATPase. J. Biol. Chem. 263, 13774-13778.

Snavely, M. D., Miller, C. G., & Maquire, M. E. (1991). The *mgt*B Mg^{2+} transport locus of *Salmonella typhimurium* encodes a P-type ATPase. J. Biol. Chem. 266, 815-823.

Stokes, D. L., & Green, N. M. (1990). Three-dimensional crystals of Ca^{2+}-ATPase from sarcoplasmic reticulum. Symmetry and molecular packing. Biophys. J. 57, 1-14.

Stokes, D. L., Taylor, W. R., & Green, N. M. (1994). Structure, transmembrane topology and helix packing of P-type ion pumps. FEBS Lett. 346, 32-38.

Supply, P., Wach, A., & Goffeau, A. (1993a). Enzymatic properties of the PMA2 plasma membrane-bound H^+-ATPase of *Saccharomyces cerevisiae*. J. Biol. Chem. 268, 19753-19759.

Supply, P., Wach, A., Thinès-Sempoux, D., & Goffeau, A. (1993b). Proliferation of intracellular structures upon overexpression of the PMA2 ATPase in *Saccharomyces cerevisiae*. J. Biol. Chem. 269, 19744-19752.

Taylor, W. R., & Green, N. M. (1989). The homologous secondary structures of the nucleotide-binding sites of six cation-transporting ATPases lead to a probable tertiary fold. Eur. J. Biochem. 179, 241-248.

Toyofuku, T., Kurzydlowski, K., Lytton, J., & MacLennan, D. H. (1992). The nucleotide binding hinge domain plays a crucial role in determining isoform-specific Ca^{2+} dependence of organellar Ca^{2+}-ATPases. J. Biol. Chem. 267, 14490-14496.

Toyoshima, C., Sasabe, H., & Stokes, D. L. (1993). Three-dimensional cryo-electron microscopy of the calcium ion pump in the sarcoplasmic reticulum membrane. Nature 362, 469-471.

Ulaszewski, S., Balzi, E., & Goffeau, A. (1987a). Genetic and molecular mapping of the *PMA1* mutation conferring vanadate resistance to the plasma membrane ATPase from *Saccharomyces cerevisiae*. Mol. Gen. Genet. 207, 38-46.

Ulaszewski, S., Hilger, F., & Goffeau, A. (1989). Cyclic AMP controls the plasma membrane H^+-ATPase activity from *Saccharomyces cerevisiae*. FEBS-Lett. 245, 131-136.

Ulaszewski, S., Van Herck, J. C., Dufour, J. P., Kulpa, J., Nieuwenhuis, B., & Goffeau, A. (1987b). A single mutation confers vanadate resistance to the plasma membrane H^+-ATPase from the yeast *Schizosaccharomyces pombe*. J. Biol. Chem. 262, 223-228.

Vallejo, C. G., & Serrano, R. (1989). Physiology of mutants with reduced expressions of plasma membrane H^+-ATPase. Yeast 5, 307-319.

Viegas, C. A., Supply, P., Capieaux, E., Van Dyck, L., Goffeau, A., & Sá-Correia, I. (1994). Regulation of the expression of the H^+-ATPase genes *PMA1* and *PMA2* during growth and effects of octanoic acid in *Saccharomyces cerevisiae*. Biochim. Biophys. Acta 1217, 74-80.

Villalba, J. M., Palmgren, M. G., Berberian, G. E., Ferguson, C., & Serrano, R. (1992). Functional expression of plant plasma membrane H^+-ATPase in yeast endoplasmic reticulum. J. Biol. Chem. 267, 12341-12349.

Villalobo, A., Boutry, M., & Goffeau, A. (1981). Electrogenic proton translocation coupled to ATP hydrolysis by the plasma membrane Mg^{2+}-dependent ATPase of yeast in reconstituted proteoliposomes. J. Biol. Chem. 256, 12081-12087.

Wach, A., Schlesser, A., & Goffeau, A. (1992). An alignment of 17 deduced protein sequences from plant, fungi, and ciliate H^+-ATPase genes. J. Bioenerg. Biomembr. 24, 309-317.

Wang, K., & Farley, R. A. (1992). Lysine 480 is not an essential residue for ATP binding or hydrolysis by Na,K-ATPase. J. Biol. Chem. 267, 3577-3580.

Wang, G., Tamas, M.J., Hall, M.J., Pascual-Ahuir, A., & Perlin, D.S. (1996). Probing conserved regions of the cytoplasmic LOOP1 segment linking transmembrane segments 2 and 3 of the *Saccharomyces cerevisiae* plasma membrane H^+-ATPase. J. Biol. Chem. 271, 25438-25445.

Willsky, G. R., Leung, J. O., Offermann, P. V. J., Plotnick, E. K., & Dosch, S. F. (1985). Isolation and characterization of vanadate-resistant mutants of *Saccharomyces cerevisiae*. J. Bacteriol. 164, 611-617.

Willsky, G. R., White, D. A., & McCabe, B. C. (1984). Metabolism of added orthovanadate to vanadyl and high-molecular-weight vanadates by *Saccharomyces cerevisiae*. J. Biol. Chem. 259, 13273-13281.

Yuan, H. S., Wang, S. S., Yang, W.-Z., Finkel, S. E., & Johnson, R. C. (1994). The structure of Fis mutant Pro[61]Ala illustrates that the kink within the long α-helix is not due to the presence of the proline residue. J. Biol. Chem. 269, 28947-28954.

Yun, R. H., Anderson, A., & Hermans, J. (1991). Proline in α-helix: stability and conformation studied by dynamic simulation. Proteins: Structure, Function, and Genetics 10, 219-228.

COPPER HOMEOSTASIS BY CPX-TYPE ATPases

THE NEW SUBCLASS OF HEAVY METAL P-TYPE ATPases

Marc Solioz

Advances in Molecular and Cell Biology
Volume 23A, pages 167-203.

ISBN: 0-7623-0287-9

I. INTRODUCTION

The element copper has been known since prehistoric times. By charcoal reduction of its ores, copper was available in the Middle East by about 3500 B.C. Some 500 years later, the advantage of adding tin to produce the harder bronze was discovered; this established the Bronze Age. Copper has continued to this day to be one of man's most important metals. Copper is widely distributed in nature. Its relative abundance in the earth's crust amounts to 68 ppm. Copper is found mainly as the sulfide, oxide, or carbonate, its major ores being copper pyrite (chalcopyrite), $CuFeS_2$; copper glance (chalcocite), Cu_2S; cuprite, Cu_2O; and malachite, $Cu_2CO_3(OH)_2$. Weathered minerals give rise to soluble cupric and cuprous salts, such as carbonates and chlorides (Tylecote, 1992).

During evolution, living organisms developed mechanisms that take advantage of the ability of copper to act as an electron donor and acceptor by the transition between Cu^+ and Cu^{2+} (Karlin, 1993). Over 30 known proteins contain copper. A few examples are lysyl oxidase, tyrosinase, dopamine β-hydroxylase, cytochrome *c* oxidase, ascorbate reductase and superoxide dismutase. Another class of copper proteins, exemplified by plastocyanins and azurines, act as electron carriers. In all these proteins, copper serves as an electron acceptor/donor, by alternating between the redox states Cu(I) and Cu(II). Depending on the coordination of the copper to the protein, the redox potentials of these proteins can vary over the range of less than 200 mV to nearly 800 mV. However, the redox properties of copper can also be deleterious, especially in the presence of molecular oxygen. Highly reactive hydroxyl radicals generated in the reaction $Cu^+ + H_2O = Cu^{2+} + OH^- + OH$ can participate in a number of reactions detrimental to cellular molecules, such as the oxidation of proteins and lipids (Yoshida et al., 1993). The toxic effect of copper is utilized in agriculture for the control of bacterial and fungal diseases (Cha and Cooksey, 1991).

Copper is the third most abundant transition metal in the body, following iron and zinc. Human adults contain around 100 mg of copper, mostly attached to proteins, and require a daily intake of 3 to 5 mg. To cope with changes in the external copper concentration, and yet still maintain internal copper concentrations within narrow limits, appropriate mechanisms for copper absorption, cellular transport, incorporation into protein, storage and excretion are required and must be tightly regulated.

In spite of the essential role of copper in living cells, very little is known about copper homeostasis. Until a few years ago, the notion prevailed that intracellular copper availability in eukaryotic cells is chiefly regulated by metallothioneins. These cysteine-rich proteins of 6-7 kDa can bind, in order of increasing affinity, seven Cd^{2+} or Zn^{2+} ions, or 12 Cu^{+} or Ag^{+} ions (Nielson and Winge, 1985; Kägi and Schaffer, 1988; Presta et al., 1995). While metallothioneins have been studied extensively in eukaryotic cells, they were only recently described in plant cells and bacteria (Huckle et al., 1993; Zhou and Goldsbrough, 1994). Metallothionein expression is induced by increased heavy metal ion concentration. This regulation has been investigated in some detail and it certainly seemed plausible that it is sufficient for heavy metal homeostasis (Culotta et al., 1989; Jungmann et al., 1993; Morby et al., 1993; Samson and Gedamu, 1995). However, with the recent discovery of specific copper transport systems in bacteria, yeast and mammals, this view has been abandoned. It now seems likely that most, if not all cells possess copper transport systems that serve in the uptake, intracellular distribution, and the secretion of copper. While metallothioneins probably fulfill an important role in metal detoxification, their function has to be reevaluated in the context of transport systems involved in copper homeostasis.

Many of the copper transport systems that have now been cloned are P-type ATPases. However, these copper ATPases differ significantly in their structure from the previously known P-type ATPases, such as the Ca^{2+}-ATPases or the $Na^{+}K^{+}$-ATPases and form a distinct subclass, the CPX-type ATPases. CPX-type ATPases are the main subject of this chapter. Since the bacterial cadmium ATPases are close relatives of the copper ATPases and also members of the CPX-type ATPase subclass, they are included in the discussion. For some of the ATPases considered here there is little or no information about their function. While most of them may well be copper ATPases, a role in the transport of other heavy metal ions, notably molybdenum, zinc, nickel, iron or cobalt cannot be excluded. Besides examining CPX-type ATPases *per se*, it will also be instructive to look at the very special problems of copper in biology from a broader viewpoint.

II. CHEMICAL STATE OF COPPER IN BIOLOGY

What makes copper a valuable cellular constituent are its two oxidation states: Cu^{+} [copper(I)] and Cu^{2+} [copper(II)]. While free copper(II) ions are stable in neutral, aqueous solutions exposed to the atmosphere, copper(I) ions can only be maintained at very acidic pH or in complexed form. Highly efficient chelators of copper(I) *in vitro* are o-phenanthroline, bathocuprous disulfonate or 8-hydroxyquinoline (reported formation constants for copper-phenanthroline complexes are 21, irrespective of the ligands on the phenanthroline dipyrimidine ring system (Bell et al., 1991)). Stable copper(I) complexes are also formed by acetonitrile, CN- , or even Tris-buffer (McPhail and Goodman, 1984). Phenanthrolines complex Cu(I) (and Fe(II)) so strongly that they effectively raise the redox potential to a point at which any reducing equivalent can support the reduction of the higher valency metal ion. This reaction can be counteracted by high concentrations (20 mM) of citrate or lactate, which preferentially bind the oxidized form of copper. These physico-chemical aspects have led to difficulties in experimentation with copper in biological systems which may have contributed to the neglect in studying the homeostasis of this biologically important metal.

A. Cytoplasmic Copper

The cytoplasmic condition of cells is reducing, with glutathione (GSH) present in millimolar concentrations acting as the major reducing agent in prokaryotic and eukaryotic cells (Samuni et al., 1981; McLaggan et al., 1990). GSH and other reducing compounds present in cells, such as cysteine or ascorbate, reduce copper(II) to copper(I) and, in addition, avidly bind to it. GSH also protects biological molecules like metallothioneins from oxidation. It is not clear if copper enters the cell only as copper(I), for which there is good evidence, or also as copper(II). But once in the cytoplasm, copper is probably in the reduced Cu^{+}-form and possibly complexed by glutathione. Increased levels of GSH were observed in hepatoma cells treated with copper while inhibition of GSH synthesis with buthionine sulfoximine reduced the incorporation of copper into metallothioneins (Freedman et al., 1989). It was also shown *in vitro* that Cu^{+}-GSH could mediated Cu^{+}-transfer into metal depleted metallothionein or copper-free Cu,Zn-superoxide dismutase (Ciriolo et al., 1990; Ferreira et al., 1993; Ascone et al., 1993). *In vivo* pulse-chase experiments with radioactive Cu^{+} revealed that Cu^{+} could also be transferred in the reverse direction from metallothionein to GSH and then to Cu, Zn-superoxide dismutase (Freedman and Peisach, 1989). All these observations favor the view that GSH plays a key role in copper metabolism. However, inhibition of glutathione synthase in *E. coli* failed to alter the response of cells to copper. So at least in bacteria, GSH may be a dispensable component of copper homeostasis.

Resistance to copper toxicity is also influenced by the level of glutathione-peroxidase. An increased level of this enzyme was observed in copper treated hepatoma cells, allowing efficient accommodation of increased cellular hydrogen peroxide concentrations caused by the oxidation of Cu^{+}-GSH or Cu^{+}-metallothionein (Freedman et al., 1989).

Attempts at defining the role of glutathione and metallothioneins in copper homeostasis are likely to only give a view of the processing of the bulk copper. There must be several mechanisms to insure that copper is distributed to subcellular compartments as needed, maintained in the proper oxidation state, and transferred specifically into copper enzymes during biosynthesis. In addition, regulatory proteins will most likely control the machinery for copper homeostasis at the genetic and functional level. Thus, copper homeostasis may well involve a level of complexity not yet reached by our investigations. For a thorough review of the current understanding of cellular copper transport, the reader is referred to other reviews (Vulpe and Packman, 1995; Linder, 1996).

B. Extracellular Copper

Cells can be exposed to copper(I) or copper(II), depending on the condition of the ambient. There is evidence that specific reductases/oxidases can interconvert copper(I) and copper(II). In the yeast *Saccharomyces cerevisiae*, the *FRE1* locus encodes a Fe(III) reductase required for the uptake of iron, which is transported in the Fe(II) form (Askwith et al., 1994). More recently it was shown that Cu(II) is also a good substrate for the Fre1 reductase. The Fre1 activity, which can be efficiently inhibited by Pt(II), was shown to support 50 to 70% of copper uptake in wild-type cells. In addition, a second, platinum-insensitive reductase relatively specific for Cu(II) appeared to be present in the plasma membrane (Hassett and Kosman, 1995). This suggests that in yeast, copper uptake involves a Cu(I) intermediate (Chang and Fink, 1994). In further support of this concept, it was found that ascorbate, which reduces Cu(II) to Cu(I), suppresses the dependence of copper uptake on plasma membrane reductase activity. The observation that ascorbate also enhances copper transport from ceruloplasmin into human K562 cells suggests that reductive accumulation of copper is a general feature of copper uptake by eukaryotes (Percival and Harris, 1989). Ceruloplasmin is a donor of copper to other tissues (Harris and Percival, 1989). It is a 132 kDa glycoprotein that is synthesized in the liver and contains six atoms of tightly bound copper. Copper is incorporated during the biosynthesis of ceruloplasmin, and copper incorporation and secretion of ceruloplasmin appear to be independent of one another (Vulpe and Packman, 1995). *In vitro*, copper incorporation can only be accomplished in the presence of reducing agents, suggesting that it occurs in the copper(I) form. How copper is taken up and released by ceruloplasmin in the organism is largely unknown.

Reduction of copper(II) to copper(I) for uptake by cells may have thermodynamic reasons. The lower valency states of transition metals are more exchange labile. For example, the water exchange rate for Fe(II) is 3×10^6 s^{-1} compared to 3×10^3 s^{-1} for Fe(III). Copper in the extracellular fluid is normally complexed to organic molecules. Reduction of complexed Cu(II) would strongly facilitate the displacement of the ligand to which the metal ion is bound in the medium, making the metal available for transport across the membrane. In essence, the reduction of copper could increase its bioavailability.

III. COPPER TRANSPORT SYSTEMS

Clearly, copper homeostasis in cells represents a highly complex process, involving different proteins for transport, storage, regulation, transfer, and oxidation and reduction of copper. The current view is that copper transport systems that can pump copper across membranes are vital in both eukaryotic and prokaryotic cells, not only in terms of copper toxicity, but also copper limitation.

The first copper transport systems were discovered more than a decade ago as a result of studies in microorganisms naturally resistant to copper. Some of these bacterial mechanisms, notably in *Escherichia coli* and *Pseudomonas syringae*, have subsequently been investigated in great detail and shown to encompass copper sequestration, export and possibly oxidation or reduction of the copper ions. Since neither of these systems appears to involve P-type copper ATPases, they will not be discussed here. The interested reader is referred to recent reviews (Cervantes and Gutierrez-Corona, 1994; Silver and Ji, 1994; Cooksey, 1994; Silver, 1996). Another copper uptake system, *CTR1*, has been cloned from yeast; as it is not an ATPase, it will not be considered here (Dancis et al., 1994).

An entirely new development in the field of copper transport has been initiated by the first description of two copper ATPases in *Enterococcus hirae* (Odermatt et al., 1992). In an exploding fashion, over two dozen putative copper ATPases have since been described. They all appear to be P-type ATPases, classically represented by the $Na^{+}K^{+}$-ATPases and the Ca^{2+}-ATPases of higher cells. They are presented in detail in other chapters of this book. Some currently known putative copper ATPases and their most important properties are listed in Table 1. Interestingly, the majority of the enzymes listed in this Table have been discovered by serendipity or in the course of sequencing projects and many have therefore not been studied biochemically. Transport has been directly demonstrated only for CadA of *Staphylococcus aureus*, which expels cadmium from the cell, and for CopB of *E. hirae*, which functions in copper extrusion. For other ATPases in this list there is circumstantial evidence for their function, notably for PacS of *Synechococcus*, which appears to export copper, and for the human Menkes and Wilson ATPases, which underlie copper metabolic diseases.

Not included in this list are several mammalian homologues of the human Menkes and Wilson ATPases, because of the great similarity of the gene products . But those included in Table 1 and in the discussion that follows are the cadmium ATPases. They share many properties with the copper ATPases and can be considered the archetype of heavy metal ATPases.

A. P-Type ATPases

P-type ATPases (earlier called E_1E_2-type ATPases) get their name from the fact that they form a phosphorylated intermediate in the course of the reaction cycle (Pedersen and Carafoli, 1987). The residue that has been demonstrated to be phosphorylated is the aspartic acid in the conserved sequence DKTGT (given in the one-letter amino acid code, used throughout this chapter). This sequence, also called the aspartyl kinase domain, is present in all of the more than 100 P-type ATPases sequenced today. Other motifs common to P-type ATPases are the so-called phosphatase domain of consensus sequence TGES, and the ATP binding domains of consensus GCGINDAP, discussed elsewhere in this book. However, the (putative) copper ATPases as well as the cadmium ATPases exhibit several striking features not found in any of the other P-type ATPases: they have putative heavy metal binding sites in the polar N-terminal region; they have a conserved intramembra-

Table 1. Structural and Functional Properties of CPX-type ATPases

Name	*Function*	*Organism*	*Code*	*Residues*	*M-spans*1	*Cys*	*Cys-Box*	*His-Box*	*CPX*	*EMBL#*	*Reference*
CadA	Cd^{2+}-export	*S. aureus* (pI258)	Sau_Cada	727	6(8)	4	1	0	CPC	J04551	(Nucifora et al., 1989)
CadA	Cd^{2+}-export	*S. aureus* (Tn554)	Tn554_Cada	804	(8)	6	2	0	CPC	L10909	(Dubin et al., 1992)
CadA	Cd^{2+}-export	*B. firmus* (plasmid)	Bfi_Cada	723	(8)	4	1	0	CPC	M90750	(Ivey et al., 1992)
CadA	Cd^{2+}-export	*L. monocytogenes*	Lmo_Cada	711	(8)	4	1	0	CPC	L28104	(Lebrun et al., 1994a)
CopA	Cu^{+}/Ag^{+}-import	*E. hirae*	Ehi_Copa	727	8	4	1	0	CPC	L13292	(Odermatt et al., 1993)
CopB	Cu^{+}/Ag^{+}-export	*E. hirae*	Ehi_Copb	745	8	1	0	2	CPH	L13292	(Odermatt et al., 1993)
CtaA	Cu-transport	*Synechococcus*	Syn_Ctaa	790	(8)	14	1	0	CPC	U04356	(Phung et al., 1994)
PacS	Cu-transport	*Synechococcus*	Syn_Pacs	747	(8)	6	1	0	CPC	D16437	(Kanamaru et al., 1994)
hpCopA	?	*H. pylori*	Hpy_Copa	741	8	9	1	0	CPC	U59625	(Ge et al., 1995)
o732	?	*E. coli*	Eco_O732	732	(8)	9	1	0	CPC	U00039	(Sofia et al., 1994)
HRA-1	?	*E. coli*	Eco_Hra1	731	(8)	2	0	2	CPH	U16658	(Trenor III et al., 1994)
HRA-2	?	*E. coli*	Eco_Hra2	721	(8)	1	0	2	CPH	U16659	(Trenor III et al., 1994)
FixI	?	*R. meliloti*	Rme_Fixi	757	6(8)	9	1	0	CPC	Z21854	(Kahn et al., 1989)
Pca1	Cu-transport	*S. cerevisiae*	Sce_Pca1	1216	6(8)	42	1	0	CPC	Z29332	(Rad et al., 1994)
CCC2	Cu-transport	*S. cerevisiae*	Sce_Ccc2	1004	8	28	2	0	CPC	L36317	(Fu et al., 1995)
Menkes	Cu-transport	Human	Hum_Menkes	1500	8	26	6	0	CPC	L06133	(Vulpe et al., 1993a)
Wilson	Cu-transport	Human	Hum_Wilson	1465	8	33	6	0	CPC	U03464	(Bull et al., 1993)

Note: [1] The digits give the number of membrane spans predicted in the literature while the digits in parenthesis give the number of membrane spans we predict.

nous CPC or CPH motif; they have a conserved HP motif 34 to 43 amino acids C-terminal to the CPC motif; they have two additional predicted transmembranous helices on the N-terminal end of the protein, and (v) they only have two membrane spans at the C-terminal end, thus lacking two to four membrane spans present at this position in other P-type ATPases. Clearly, these ATPases form a subgroup of the P-type ATPases that is quite distinct. Based on the conspicuous feature of the intramembranous CPC or CPH, it has been proposed to call enzymes belonging to this subgroup CPX-type ATPases (Solioz and Vulpe, 1996). The location of the specific features of P-type and CPX-type ATPases in the enzyme molecule are shown in Figure 1. The special features of CPX-type ATPases will be discussed in more detail below.

B. CPX-Type ATPases

A most apparent feature of CPX-type ATPases is the occurrence of one to six copies of a conserved repeat element in the polar N-terminus preceding the first predicted membrane span. The involvement of this sequence in heavy metal binding has first been proposed by Silver et al. (1989) who pointed to the presence of such a conserved sequence in the cadmium ATPase of *S. aureus*, in the periplasmic

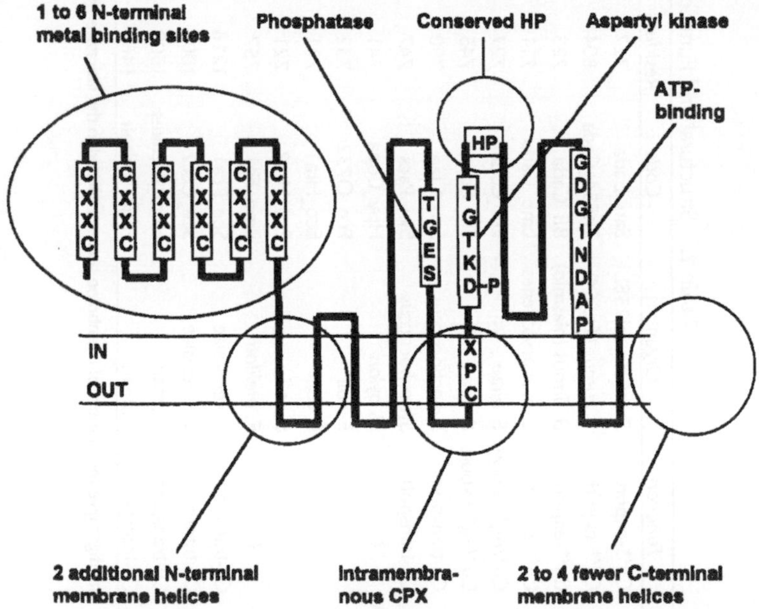

Figure 1. Features of P-type and CPX-type ATPases. The structural elements labeled Phosphatase, Aspartyl Kinase and ATP binding are common to all P-type ATPases; the features that are circled are specific for CPX-type ATPases. See text for further details.

mercury binding protein MerP, and in three different MerA mercuric reductases. This suggested that the sequence is a feature of proteins that interact with heavy metal ions. In the meantime, new proteins containing this sequence element have been found, including many putative copper ATPases and associated proteins. The conserved heavy metal binding motif is about 40 residues long and its most apparent feature are two cysteines, arranged as CXXC. This cysteine-motif is also abundant in metallothioneins and binds copper in the reduced copper(I) state (Karlin, 1993; Presta et al., 1995). The conserved sequence containing the CXXC motif has been called "heavy metal associated" (HMA) sequence (Bull and Cox, 1994), but the term is not specific for this sequence. It will thus be called "cys-box" here.

The Cys-Box

Figure 2 shows an alignment of 37 cys-boxes while Figure 3 provides a plot of the sequence similarity, which is not particularly apparent in Figure 1. It can be seen from this profile that, aside of the CXXC motif, there are conserved amino acids over a stretch of 40 residues. The cys-boxes fall into two categories: those constituting the major element of a small protein (cys-box proteins), and those cys-boxes present in one to six copies in the N-terminal region of a much larger protein. Figure 4 graphically shows the occurrence of cys-boxes in a representative selection of proteins. On the basis of the purported functions of the different proteins, it can be assumed that cys-boxes can interact with a variety of metals, such as copper, mercury or cadmium. However, metal binding has convincingly been shown only for MerP. It is a component of bacterial mercury resistance mechanisms in which Hg^{2+} is taken up by the cells and reduced intracellularly to volatile Hg^0 by the mercuric reductase MerA. The process involves five to six proteins whose genes are arranged in an operon (Brown et al., 1991; Silver and Ji, 1994). Mature MerP proteins are around 70 amino acids in size, not membrane bound, and located in the bacterial periplasm. It has been shown that MerP serves in binding mercuric ions and delivering them to the MerT transport protein. That the conserved cysteines in the CXXC motif are important for metal binding was demonstrated by site-directed mutagenesis (Sahlman and Skarfstad, 1993). A preliminary structure of MerP derived by NMR indicates an "open-faced sandwich" arrangement of the polypeptide, with a β-sheet formed by four antiparallel β-strands, and two α-helices positioned on one side of the β-sheet (Eriksson and Sahlman, 1993).

From the phylogenetic analysis of the cys-boxes (Figure 5), no conclusion can be drawn as to the function of these motifs. Some of the relationships are too distant for a reliable phylogenetic analysis, but the clustering of closely related sequences is significant. The three known MerP sequences are well clustered, but in the same group are the two cys-boxes of the putative yeast copper ATPase CCC2. Also, the cys-box of the copper translocating CopA ATPase of *E. hirae* is most closely related to the corresponding structures in MerA mercuric reductases. Some other relationships apparent in Figure 5 will be alluded to later in this chapter.

```
                           *  *
   Bac_Mera1  KKYRVNVQGM TCSGCEQHVA VALENM.GAK AIEVDFRRGE AVFELPDDVK  50
   Bac_Mera2  KKYRLNVEGM TCTGCEEHIA VALENA.GAK GIEVDFRRGE ALFELPYDVD  129
  Pi258_Mera  NSYKIPIQGM TCTGCEEHVT EALEQA.GAK DVSADFRRGE AIFELSDDQI  52
    Eco_Hra0  ENKSFAIEGM TCAYCAQTVE KAAKKVRGVT QASANLATEK LSIEYNEPTF  55
    Ehi_Copa  KMETFVITGM TCANCSARIE KELNEQPGVM SATVNLATEK ASVKYTDTTT  55
 Tn554_Cada1  DKQVYRVEGF SCANCAGKFE KNVKELSGVH DAKVNFGASK IDVFGSATVE  60
 Tn554_Cada2  DKNVYRVEGF SCANCAGKFE KNVKQLAGVQ DAKVNFGASK IDVYGNASVE  138
    Sau_Cada  EMNVYRVQGF TCANCAGKFE KNVKKIPGVQ DAKVNFGASK IDVYGNASVE  61
    Bfi_Cada  EMKAYRVQGF TCANCAGKFE KNVKQLSGVE DAKVNFGASK IAVYGNATIE  61
    Lmo_Cada  EKTVYRVDGL SCTNCAAKFE RNVKEIEGVT EAIVNFGASK ITVTGEASIQ  52
    Syn_Pacs  NQQTLTLRGM GCAACAGRIE ALIQALPGVQ ECSVNFGAEQ AQVCYDPALT  52
    Eco_O732  TRYSWKVSGM DCAACARKVE NAVRQLAGVN QVQVLFATEK LVVDADNDIR  97
  Tn501_Mera  .MTHLKITGM TCDSCAAHVK EALEKVPGVQ SALVSYPKGT AQLAIVPGTS  49
    Pdu_Mera  .MTHLKITGM TCDSCAAHVK EALEKVPGVQ SAIVSYAKGA AQLALDPGTA  49
   Tn21_Mera  .MSTLKITGM TCDSCAVHVK DALEKVPGVQ SADVSYAKGS AKLAIEVGTS  49
Hum_Menkes3   STATFIIDGM HCKSCVSNIE STLSALQYVS SIVVSLENRS AIVKYNASSV  326
Hum_Wilson3   VTLQLRIDGM HCKSCVLNIE ENIGQLLGVQ SIQVSLENKT AQVKYDPSCI  274
Hum_Menkes2   VVLKMKVEGM TCHSCTSTIE GKIGKLQGVQ RIKVSLDNQE ATIVYQPHLI  220
Hum_Wilson2   AVVKLRVEGM TCQSCVSSIE GKVRKLQGVV RVKVSLSNQE AVITYQPYLI  160
Hum_Menkes1   NSVTISVEGM TCNSCVWTIE QQIGKVNGVH HIKVSLEEKN ATIIYDPKLQ  57
Hum_Menkes4   QETVINIDGM TCNSCVQSIE GVISKKPGVK SIRVSLANSN GTVEYDPLLT  426
Hum_Wilson4   STTLIAIAGM TCASCVHSIE GMISQLEGVQ QISVSLAEGT ATVLYNPAVI  377
Hum_Wilson1   ATSTVRILGM TCQSCVKSIE DRISNLKGII SMKVSLEQGS ATVKYVPSVV  75
Hum_Menkes5   SKCYIQVTGM TCASCVANIE RNLRREEGIY SILVALMAGK AEVRYNPAVI  537
Hum_Wilson5   QKCFLQIKGM TCASCVSNIE RNLQKEAGVL SVLVALMAGK AEIKYDPEVI  506
Hum_Menkes6   GVLELVVRGM TCASCVHKIE SSLTKHRGIL YCSVALATNK AHIKYDPEII  613
Hum_Wilson6   GSIELTITGM TCASCVHNIE SKLTRTNGIT YASVALATSK ALVKFDPEII  582
    Syn_Ctaa  TSILVEVEGM KCAGCVAAVE RRLQQTAGVE AVSVNLITRL AKVDYDAALI  63
  Tn501_Merp  QTVTLSVPGM TCSACPITVK KAISEVEGVS KVDVTFETRQ AVVTFDDAKT  71
    Pdu_Merp  QTVTLSVPGM TCSACPITVK KAISKVEGVS KVNVTFETRE AVVTFDDAKT  71
   R100_Merp  QTVTLAVPGM TCAACPITVK KALSKVEGVS KVDVGFEKRE AVVTFDDTKA  71
   Sce_Ccc21  REVILAVHGM TCSACTNTIN TQLRALKGVT KCDISLVTNE CQVTYDNEVT  51
   Sce_Ccc22  KEGLLSVQGM TCGSCVSTVT KQVEGIEGVE SVVVSLVTEE CHVIYEPSKT  129
    Hpy_Copp  MKVTFQVPSI TCSHCVDKIE KFVGEIEGVS FIDANVEKKS VVVEFDAPAT  50
    Ehi_Copz  MKQEFSVKGM SCNHCVARIE EAVGRISGVK KVKVQLKKEK AVVKFDEANV  50
    Sce_Pca1  EHIVLSVSGM SCTGCESKLK KSFGALKCVH GLKTSLILSQ AEFNLDLAQG  459
    Rme_Fixi  RQTELSVPNA YCGTCIATIE GALRAKPEVE RARVNLSSRR VSIVWKEEVG  86
```

Figure 2. Alignment of the amino acid sequences of cys-boxes. 50 amino acids are represented, although only the central 42 show significant conservation among the sequences. The numbers at the end of each line give the position of the last amino acid of the cys-box in the corresponding protein. The universally conserved cysteines are indicated by an *asterisk*. The EMBL/GenBank accession numbers for the sequences not given in Table 1 are: Bac_Mera (P16171), pl258_Mera (P08663), Eco_Hra-0 (U16658), Tn501_Mera (A00406), pDU_Mera (P08662), Tn21_Mera (P08332), Tn501_Merp (P04131), pDU_Merp (P13113), R100_Merp (P04129), Hpy_Copp (L33259), and Ehi_Copz (Z46807). An extra number at the end of these codes enumerates multiple cys-boxes in the same protein, starting with the most N-terminal one. The alignment was generated with the program Pileup of the GCG software (Devereux et al., 1984).

The three-dimensional structure of MerA mercuric reductase, a protein of 70 kDa, has been solved. It was found that the N-terminal 166 amino acids containing the cys-boxes retained considerable mobility in the crystal and could not be modeled in the X-ray structure (Schiering et al., 1991). Also, this region could be proteolytically cleaved off without loss of enzymatic activity by the mercuric reductase

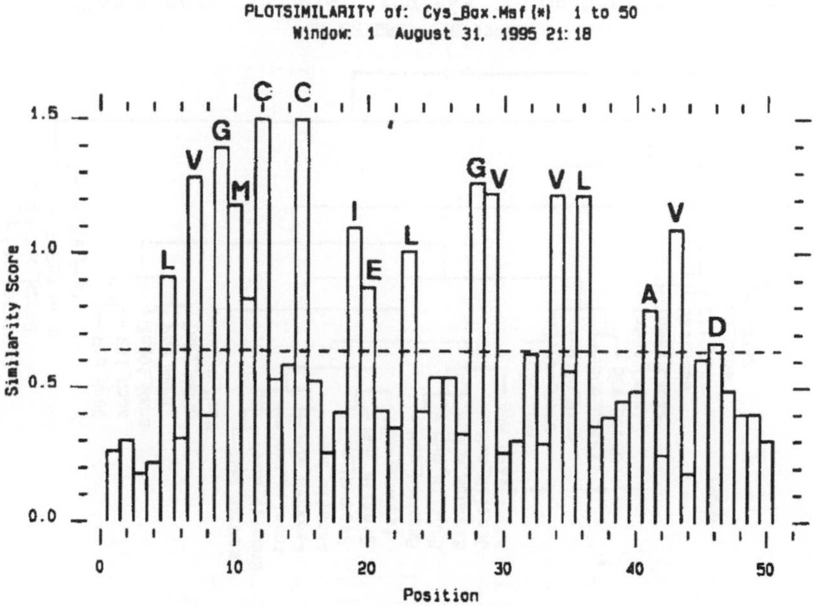

Figure 3. Plot of the sequence similarity profile for the cys-boxes indicated in Figure 2, as calculated by the program Plot Similarity of the GCG software. The peaks corresponding to conserved amino acids are labeled with the respective amino acid. The *dashed line* indicates the average similarity of all the sequences, which is 32%.

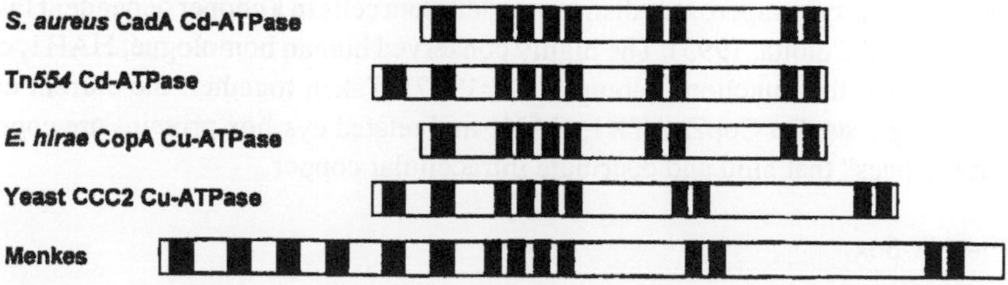

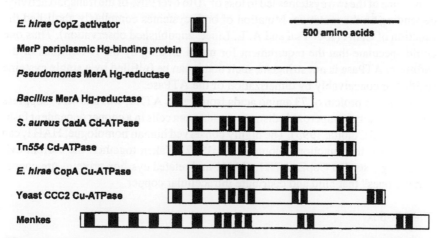

Figure 4. Schematic representation of the occurrence of cys-boxes in various proteins. The polypeptide chains are drawn to scale as boxes. Transmembranous helices are indicated by solid rectangles and cys-boxes by hatched rectangles.

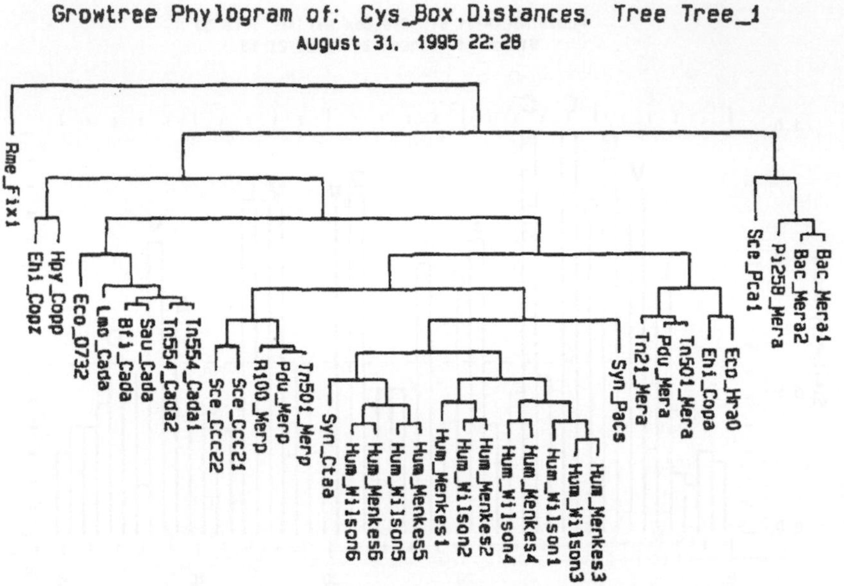

Figure 5. Tree of sequence similarity between cys-boxes of different origin. The alignment in Figure 2 was used to calculate distances by the Jukes-Cantor method (Swofford and Olson, 1990). The sequence designations are as in Figure 2 and Table 1.

(Fox and Walsh, 1982). This indicates that the cys-box has a nonessential, auxiliary function. However, for the cadmium ATPase of *S. aureus*, it was found that mutation of one of the two cysteines led to loss of 70 to over 90% of the transport activity, depending on the mutation. Mutation of both cysteines completely abolished the function of CadA (K-J. Tsai and A. L. Linet, unpublished observation). Thus one could speculate that the requirement for metal binding by the N-terminus of the cadmium ATPase is not stringent, such that it can be fulfilled by a single cysteine residue or conceivably by dimerization of the ATPase.

A cys-box protein of 73 amino acids from yeast, ATX1, was shown to suppress oxygen toxicity in superoxide dismutase-deficient cells in a copper dependent fashion (Lin and Culotta, 1995). The highly conserved human homologue, HAH1, can complement this function (Klomp et al., 1997). Taken together, the current evidence suggests that CopZ, ATX1, HAH1 and related cys-box proteins are copper "chaperones" that bind and distribute intracellular copper.

The His-Box

Three of the CPX-ATPases listed in Table 1 do not possess an N-terminal cys-box. Instead, they have methionine- and histidine-rich motifs that also are metal binding domains. Because of their high histidine content, we call them his-boxes, as listed in Figure 6. Both of the *E. coli* CPX-type ATPases of unknown function, HRA-1 and HRA-2, display a pair of such histidine-rich repeats (Trenor III et al., 1994). In HRA-1, the two his-boxes are nearly perfect repeats, while in HRA-2, they are fairly divergent. The CopB copper ATPase of *E. hirae* also displays two his-boxes (since these repeat elements have overlapping repeats within them, the division into repeat structures is somewhat arbitrary). Similar structures were found in two *Pseudomonas syringae* proteins, CopA and CopB, that are components of a plasmid-encoded copper resistance system (Mellano and Cooksey, 1988; Cooksey, 1994). *P. syringae* CopA was shown to be a periplasmic protein that can bind 11 atoms of copper/monomer and was shown to serve in copper resistance by sequestering large amounts of copper in the periplasm (Fox and Walsh, 1982).

The Intramembranous CPX

The second distinguishing feature of CPX-type ATPases is the intramembranous CPC motif that is present in most CPX-type ATPases. As a variation to the theme, three enzymes have a CPH motif (see Table 1). These motifs are located in the middle of a predicted membrane helix in the most conserved core structure of these ATPases (see Figure 1). The proline residue of the CPX-motif is a conserved feature of all known P-type ATPases. But it is flanked by cysteine(s) only in CPX-type ATPases. The corresponding motifs in the $Na^{+}K^{+}$-ATPase and the Ca^{2+}-ATPase are VPE and IPE, respectively.

For the Ca^{2+}-ATPase of the sarcoplasmic reticulum, it was shown that all mutations of the proline residue (P308) led to ATPases with reduced affinity for Ca^{2+}, suggesting that this proline constitutes a part of the high-affinity calcium binding site (Vilsen et al., 1989). Through its intramembranous location, this region is assumed to form the ion transduction domain or ion channel. These findings indirectly support the idea that the CPX-motif plays a role in heavy metal transduction.

Eco_Hra21	HQHQDHTAMS	GHNMHHEHHE	46
Eco_Hra22	TMRHDHAAMA	HHHMHMSDDP	78
Eco_Hra11	HSAMGHCAMG	GHAHHHHGDM	54
Eco_Hra12	YSEMDHGAMG	GHAHHHHGSF	95
Ehi_Copb	HQQMDHGHMS	GMDHSHMDHE	76
Ehi_Copb2	HSHMDHEDMS	GMNHSHMGHE	89
Psy_Copa	MQGMDHGAMS	GMDHGAMGGM	434
Psy_Copb	MQGMDHSQMQ	GMDHSKMQGM	67

Figure 6. Alignment of the amino acid sequences of his-boxes. The numbers at the end of each line give the position of the last amino acid of the his-box in the corresponding protein. Psy_Copa, CopA of *P. syringae*; Psy_Copb, CopB of *P. syringae*. Other sequence codes are as in Table 1. An extra number at the end of these codes enumerates multiple his-boxes in the same protein, starting with the most N-terminal one. The alignment was generated manually.

The HP Locus

A hitherto unrecognized structural element, a histidine-proline dipeptide, occurs in the CPX-type ATPases and is located 34 to 43 amino acids C-terminal to the phosphorylated aspartic acid residue. Its role in ATPase function is unknown. A frequently occurring mutation in patients with Wilson disease is a change of this histidine to glutamic acid (H714E). Presumably, this renders the Wilson copper ATPase nonfunctional, thus causing the abnormalities in copper metabolism typical of Wilson disease (see also below). No conserved motif is apparent at the equivalent position in other P-type ATPases. Interestingly, an HP locus is also conserved between a group of related copper proteins, including CopA from *P. syringae*, a similar protein from *Xanthomonas campestris*, laccase from four different fungi, and ascorbate oxidase from cucumber (Lee et al., 1994). No mutant studies are as yet available that show whether these histidine and proline residues fulfill an essential role in any of these proteins.

Cysteine Content

When looking at the number of cysteine residues in the different CPX-type ATPases, large differences are apparent. Most bacterial enzymes have very few cysteine residues, with CopB of *E. hirae* at the extreme: It has a single cysteine. At the other end of the scale are the putative copper ATPases of eukaryotic cells that are as rich in cysteines as other P-type ATPases. It has been argued that it might be advantageous for a copper ATPase to have as few cysteines as possible, so as not to be inactivated by nonspecific interaction with copper (Odermatt et al., 1993). If this were true, it would imply that higher cells have more sophisticated systems than bacteria for dealing with reactive heavy metal ions, which is not an unlikely possibility.

Membrane Topology

CPX-type ATPases set themselves apart from the rest of the P-type ATPases not only at the level of the primary structure, but also at the level of the membrane topology. The eight transmembranous helices predicted for CPX-type ATPases by hydropathy analysis (Figure 7) have experimentally been confirmed for the ATPase 439 of *Helicobacter pylori* (Melchers et al., 1996). Following the polar N-terminus, the polypeptide chain of CPX-type ATPases traverses the membrane four times before the first cytoplasmic loop. For other P-type ATPases, there seems to be a consensus that there are only two membrane helices in that region. Furthermore, at the C-terminal end, following the major cytoplasmic loop, two membrane helices are present in CPX-type ATPases. In all other P-type ATPases, there are four to six membrane spans at the C-terminal end.

Without knowing the function of all the CPX-type ATPases under discussion here, it is clear that they have common structural features that differ from those of

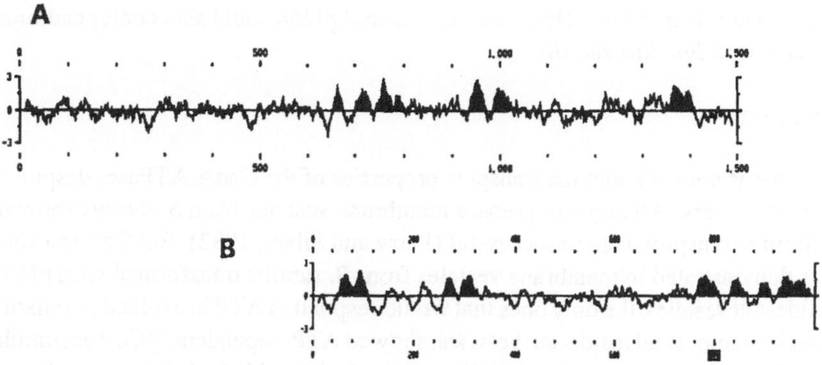

Figure 7. Hydropathy profiles of a P-type and a CPX-type ATPase. Hydropathy profiles of the human Menkes ATPase (A) and the rabbit sarcoplasmatic Ca^{2+}-ATPase (B) were calculated by the method of Kyte and Doolittle, using a window of 21. Peaks of hydrophobicity believed to correspond to membrane helices are filled in black.

other P-type ATPases. It therefore seems reasonable to assign them to a subclass of the P-type ATPases.

IV. PROPERTIES OF KNOWN CPX-TYPE ATPases

A. Bacterial Cd^{2+}-ATPases

The cadmium ATPases will be discussed in some detail since they are well studied examples of the CPX-type ATPases. The first cadmium ATPase was discovered on plasmid pI258 of *Staphylococcus aureus* (Nucifora et al., 1989). It confers cadmium (and possibly zinc) resistance by extruding the cadmium that has fortuitously entered the cells through the Mn^{2+} transport system (Perry and Silver, 1982). Similar cadmium ATPases were later found on transposon Tn*554* of *S. aureus* (Dubin et al., 1992), and on transposon Tn*5422* in the cadmium-resistance plasmid pLm74 of *Listeria monocytogenes* (Lebrun et al., 1994a; Lebrun et al., 1994b). A closely similar Cd^{2+}-ATPase was also found by serendipity on a plasmid of *Bacillus firmus* (Ivey et al., 1992). All these cadmium ATPases form a group of closely related enzymes, as can be seen in Figure 10. The recent finding that at least part of the *S. aureus* plasmid pI258 is also derived from a transposon (Rowland and Dyke, 1990) gives rise to the speculation that all the cadmium ATPases might originate from the same ancestral transposon.

The most extensively studied Cd^{2+}-ATPase, CadA of *S. aureus,* consists of 727 amino acids, has an N-terminal cys-box and an intramembranous CPC motif. For growth of *S. aureus* strains harboring plasmid pI258, the mean inhibitory cadmium concentration was 2.5 mM, or about 10^3 times more than for plasmid free strains

(Yoon and Silver, 1991). The *S. aureus* plasmid pI258 could also confer cadmium resistance in *Bacillus subtilis.*

Transport

Little is known about the transport properties of the CadA ATPase, despite its early discovery. Attempts to prepare membrane vesicles from *S. aureus* showing cadmium transport were unsuccessful (Perry and Silver, 1982). But Cd^{2+} transport was demonstrated in membrane vesicles from *B. subtilis* transformed with pI258. Inside-out vesicles, the only ones that would respond to ATP in a mixed population of right side-out and inside-out vesicles, showed ATP-dependent $^{109}Cd^{2+}$ accumulation. The rate of transport, as estimated from the published data, was about 4 nmol/min/mg at 50 μM Cd^{2+}. No estimate of the K_m was given (Tsai et al., 1992). Transport was not inhibited by vanadate, which is considered a universal inhibitor of P-type ATPases. However, inhibition by VO_4^{3-} is known to be very dependent on the assay conditions and results obtained with crude preparations can give misleading results.

S. aureus (pI258) also exhibits inducible resistance to Zn^{2+} and Zn^{2+} extrusion by whole cells has been observed. However, Zn^{2+} transport by CadA has never been convincingly demonstrated and appears doubtful in the light of two observations: First, no zinc resistance could be observed with the highly similar cadmium ATPases of *L. monocytogenes* and the Tn*554*-encoded *S. aureus* Cd^{2+}-ATPase. And second, CadA of *S. aureus* (pI258) formed an acylphosphate intermediate in the presence of Cd^{2+}, but not in the presence of Zn^{2+} (Tsai and Linet, 1993). This indicates that zinc is not a substrate of this ATPase.

Another notion needs critical evaluation. In some data-base entries and older reports, the cadmium ATPases are referred to as Cd^{2+}/H^+-ATPases, implying that they catalyze cadmium-proton exchange. However, there is no evidence for this in the literature. In early work on cadmium transport, it was suggested that cadmium is extruded electroneutrally in exchange for two protons, but by an antiport mechanism rather than an ATPase (Tynecka et al., 1981). Later work on ATP-driven cadmium transport failed to distinguish between Cd^{2+}/H^+ antiport and Cd^{2+} uniport by CadA. Unambiguous identification of the transport properties of CadA (as well as of any other transport system) will require transport studies with a purified, reconstituted system. However, there can be no doubt that the cadmium ATPases are cadmium translocating pumps. Their similarity to other CPX-type ATPases points to a common evolutionary origin.

Regulation

Cadmium resistance is inducible by cadmium and other heavy metal ions. Maximal induction of CadA by 2 μM Cd^{2+} was demonstrated for *S. aureus* (pI258) and *L. monocytogenes*. In *S. aureus*, induction was shown to also be triggered by 30 μM Pb^{2+} or Bi^{3+}, as measured by the expression of β-lactamase fused to CadA. 100 μM Zn^{2+} or Co^{2+} resulted in low, but significant induction (Yoon et al., 1991).

Upstream of all *cad*A genes cloned thus far is a regulatory gene, *cad*C, that encodes a protein of 119 to 122 amino acids. CadC has a putative cadmium binding motif and exhibits sequence similarity to ArsR, a regulator of bacterial arsenic resistance determinants (Lebrun et al., 1994a). CadC has a predicted helix-turn-helix motif typical of the LysR-type transcriptional activators (Schell, 1993). Overexpressed and partially purified CadC was shown to specifically bind to an inverted repeat in the operator/promotor region of the *cad*-operon. Cd^{2+} and Pb^{2+}, and more efficiently, Bi^{3+} and Hg^{2+}, caused release of the CadC protein from the DNA, in line with the observed *in vivo* induction of the cadmium resistance operon. CadC thus appears to act as a metal responsive repressor of transcription (Endo and Silver, 1995).

B. *Enterococcus hirae* Copper ATPases

The *E. hirae cop*-operon is located on the chromosome and consists of four closely spaced genes in the order: *cop*Y, *copZ*, *cop*A, and *cop*B. *Cop*Y and *copZ* encode regulatory proteins, whose function is described below, while *cop*A and *cop*B encode CPX-type ATPases of 727 and 745 amino acids, respectively (Odermatt et al., 1993). They were the first genes encoding copper ATPases to be described (Odermatt et al., 1992). Their cloning was fortuitous when trying to clone a potassium ATPase using an antibody of low specificity. The sequence similarity of the his-box in the N-terminus of CopB to a 120 amino acid periplasmic copper binding protein of *P. syringae*, CopP, initially gave the clue to involvement of CopB in copper homeostasis.

The evidence for the function of CopA in copper and silver transport is indirect. Cells disrupted in *cop*A cease to grow in media in which the copper has been complexed with 8-hydroxyquinoline or o-phenanthroline. This growth inhibition could be overcome by adding copper to the growth media. Interestingly, null-mutants in *cop*A could grow in the presence of 5 μM $AgNO_3$, conditions that fully inhibit the growth of wild-type cells. Thus, the CopA ATPase appears to be a route for the entry of copper as well as silver into the cell (Odermatt et al., 1993). The transport of Ag(I) by CopA is an indication that Cu(I) rather than Cu(II) is transported by CopA of *E. hirae*.

The CopB ATPase of *E. hirae* was found to be required for copper resistant growth of the cells. Wild-type cells can grow in the presence of up to 6 mM $CuSO_4$.. Null mutants in *cop*B became sensitive to copper, while null-mutation of *cop*A had no significant effect on copper tolerance. This suggested that the CopB ATPase is a copper export ATPase, extruding excess copper from the cytoplasm of *E. hirae* and thus conferring copper resistance.

Transport

CopB is the only copper ATPase, for which copper transport has been demonstrated directly. CopB was shown to catalyze ATP-driven accumulation of copper(I) and silver(I) in native membrane vesicles of *E. hirae*. These vesicles showed ATP-dependent accumulation of radiolabeled copper, but only under strongly reducing conditions. ^{64}Cu(I) was thus the transported species. Uptake of copper by these vesicles would correspond to copper extrusion in whole cells. Use of null-mutants in either *copA*, *copB*, or *copA* and *copB* made it possible to attribute the observed transport to the activity of the CopB ATPase. Copper transport exhibited an apparent K_m for Cu^+ of 1 μM and a V_{max}, of 0.07 nmol/min/mg of membrane protein. $^{110m}Ag^+$ was transported with a similar affinity and rate (Solioz and Odermatt, 1995). However, since Cu^+ and Ag^+ were not free in solution but complexed to Tris-buffer and dithiothreitol under the experimental conditions, the K_m values must be considered as relative only. The results obtained with membrane vesicles were further supported by evidence of $^{110m}Ag^+$ extrusion from whole cells pre-loaded with this isotope. Again, transport depended on the presence of functional CopB (Odermatt et al., 1994).

Vanadate showed an interesting biphasic pattern of inhibition of ATP-driven copper and silver transport: maximal inhibition of Cu^+ transport was observed at 40 μM VO_4^{-3} and of Ag^+ transport at 60 μM VO_4^{3-}. Higher concentrations relieved the inhibition of transport. This behavior is unexplained at present, but may relate to the complex chemistry of vanadate involving many oxidation states (Pope and Dale, 1968).

These results show that CopA functions as a Cu^+/Ag^+-ATPase for the import of Cu^+ and Ag^+ and CopB as a Cu^+/Ag^+-ATPase for the export of Cu^+ and Ag^+. Although CopA and CopB are capable of transporting silver, this is probably fortuitous and has no physiological significance. Ag^+ is very toxic and has no known biological role in living cells. Also, in wild-type cells, Ag^+ imported by CopA cannot be extruded efficiently enough by CopB to prevent cell death and therefore constitutes a suicide mechanism.

Regulation

The two copper ATPases of *E. hirae* exhibit biphasic regulation: induction of the genes is lowest in standard growth media (average copper content = 10 μM). If media copper is increased, an up to 50-fold induction is observed at 2 mM extracellular copper. Full induction is also obtained by 5 μM Ag^+ or 5 μM Cd^{2+}. The induction by silver and cadmium is in all likelihood fortuitous, since it does not confer resistance to these highly toxic metal ions. Interestingly, high induction is also observed when copper is depleted from the media. Since CopA serves in copper uptake and CopB in its extrusion, this co-induction of CopA and CopB by high and low copper seems rather puzzling. However, it could be a safety mechanism: if cells express, under copper-limiting conditions, only the import ATPase, they would become highly vulnerable to copper poisoning in the event of a sudden increase in ambient copper.

The two regulatory genes, *cop*Y and *cop*Z upstream of the genes encoding the CopA and CopB ATPases were cloned by chromosome walking (Ochman et al., 1988). *Cop*Y encodes a repressor protein of 145 amino acids (Odermatt and Solioz, 1995). The N-terminal half of CopY exhibits around 30% sequence identity to the bacterial repressors of β-lactamases MecI, PenI and BlaI (Himeno et al., 1986; Suzuki et al., 1993; Hackbarth and Chambers, 1993). In the best studied of these, PenI, this N-terminal portion appears to be the domain that recognizes the operator (Wittman and Wong, 1988). In the C-terminal half of CopY, there are multiple cysteine residues, arranged as $CXCX_4CXC$. The consensus motif $CXCX_{4,5}CXC$ is also found in the three yeast copper responsive transcriptional activators, ACE1, AMT1 (Zhou and Thiele, 1991; Dobi et al., 1995), and MAC1 (Jungmann et al., 1993). Disruption of the *E. hirae cop*Y gene results in constitutive overexpression of the *cop*-operon (Odermatt and Solioz, 1995). Binding of CopY to an inverted repeat sequence upstream of the *cop*Y gene has been demonstrated *in vitro* (D. Strausak, unpublished). Thus, CopY appears to be a copper responsive repressor protein with an N-terminal DNA binding domain and a C-terminal copper binding domain (Figure 8).

CopZ, the second regulatory gene of the *E. hirae cop*-operon, displays the cys-box motif and probably functions as a copper carrier. Null-mutants in CopZ show greatly reduced expression of the *cop*-operon. The interplay of CopY and other component of the *cop*-operon is not entirely clear. Figure 9 provides a possible scheme of the interaction of CopY, CopZ and copper in the biphasic regulation of the *cop*-operon by copper. Although final proof of this concept requires further

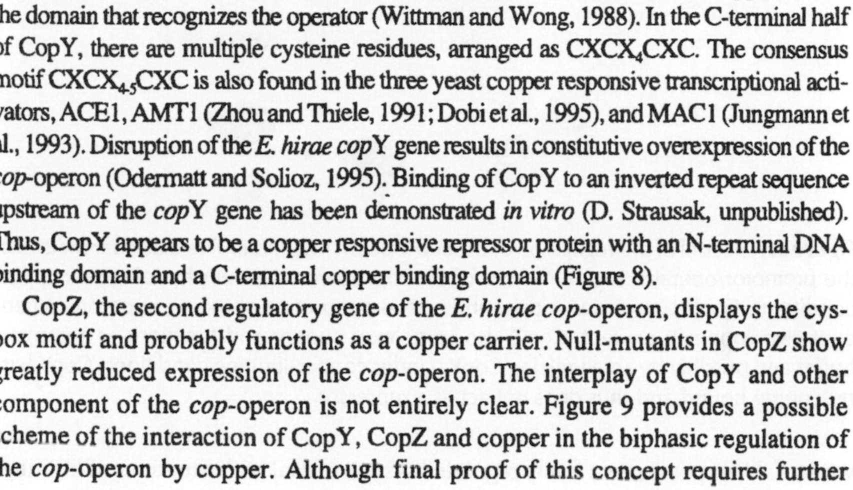

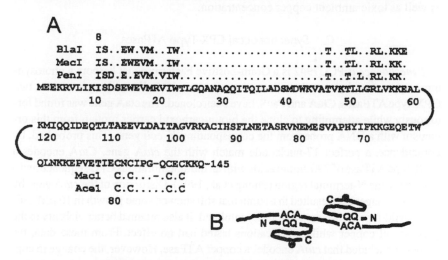

Figure 8. A. schematic representation of the CopY repressor. The N-terminal sequence similarity with other repressor proteins is shown, as are the cysteine patterns in the C-terminus. B. Model of the interaction of two CopY monomers with the inverted repeat sequence of the *cop*-operon.. From the X-ray structure of the lambda 434 suppresser, it is known that a pair of glutamine residues forms a tight interaction with an ACA base triplet (Anderson et al., 1987).

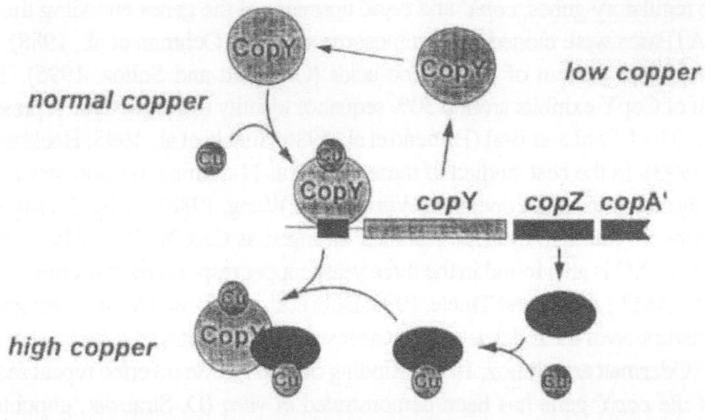

Figure 9. Model of the regulation of the *cop*-operon in *E. hirae*. The black bar indicates the promotor/operator region in front of the *cop*-operon. Under physiological copper conditions, CopY complexes copper and acts as a repressor of the *cop*-operon. When excessive copper is present, CopZ also complexes copper and functions as a copper buffer and possibly by antagonizing CopY. Under copper-limiting conditions, CopY has no copper bound and thus does not act as a repressor.

work, *E. hirae* possesses a finely regulated system to deal with situations of limiting as well as toxic ambient copper concentration.

C. *Synechococcal* CPX-Type ATPases

Synechococcus PCC7942 is a Gram-positive bacterium that harbors a photosynthetic apparatus (thylakoid) similar to that of chloroplasts. From *Synechococcus*, two CPX-type ATPases, CtaA and PacS, have been cloned. The ctaA gene was found fortuitously while attempting to clone the biotin-carboxyl carrier protein from this organism with a DNA probe from the corresponding *E. coli* gene. This probe had, by coincidence, a perfect 17-nucleotide match with the *cta*A gene. *Cta*A encodes a CPX-type ATPase of 790 amino acids with an intramembranous CPC sequence and a cys-box in the N-terminal region (Phung et al., 1994). Disruption of the *cta*A gene by cassette mutagenesis resulted in a strain that still showed some growth in 10 μM Cu^{2+} while wild-type cells were completely inhibited. It also retained better viability in the presence of copper while other cations tested had no effect. From these data, the authors concluded that *cta*A encodes a copper ATPase. However, the change in copper resistance is marginal, with wild-type and mutant showing nearly the same growth behavior in 3 μM Cu^{2+}. The sequence similarity of CtaA, which is greatest to the CCC2 copper ATPase of yeast, also suggests that CtaA is a copper pump. However, additional evidence for its function as a copper pump is required.

The *pac*S gene was found in a systematic search for P-type ATPases in *Synechococcus*. With degenerate primers corresponding to the phosphorylation domain and the ATP binding region, respectively, of P-type ATPases, PCR fragments were generated from total *Synechococcus* DNA and these fragments then used to screen a library. One of the genes so cloned is the *pac*S, the second known CPX-type ATPase gene of *Synechococcus*. It encodes a protein of 747 amino acids that also has intramembranous CPC and an N-terminal cys-box. A membrane topology with eight membrane spans has been predicted for this ATPase (Kanamaru et al., 1994). Several lines of evidence suggest that PacS is a copper translocating ATPase located in the thylakoid membrane. PacS mRNA was induced 20- to 30-fold by 5 μM Cu^{2+} or 40 μM Ag^{+} but not by metal-depleted growth media. Deletion of the *pac*S gene resulted in hypersensitivity to growth in copper. Growth of these cells was inhibited by 5 μM Cu^{2+} or 25 μM Ag^{+}, conditions that did not significantly affect the wild-type. With an antibody against PacS, the protein was localized in the thylakoid membranes. Taken together, these results point to a role of PacS in intracellular copper distribution in *Synechococcus*.

D. *Helicobacter pylori* CPX-Type ATPases

Helicobacter pylori is a curved, microaerophilic Gram-negative bacterium. It is currently attracting enormous attention since it is now recognized as a causative agent of chronic active type B gastritis in humans (Dick, 1990). Efficient treatment of *H. pylori* infections can be accomplished with a combination of the antibiotic roxithromycin and omeprazole (Cellini et al., 1991). Omeprazole is a pro-drug that is converted to the active form in the acid environment of the stomach. When activated, it reacts with sulfhydryls and strongly inhibits the gastric P-type $K^{+}H^{+}$-ATPase, but also the growth of *H. pylori*. Because of this dual effect, it seemed reasonable to search for an essential, omeprazole-sensitive ATPase in *H. pylori* and many laboratories set out to clone P-type ATPases from this organism.

The *H. pylori* CPX-type ATPase designated hpCopA was cloned by isolating a PCR product derived from genomic *H. pylori* DNA with two primers to regions conserved in the *E. coli* KdpB K^{+}-ATPase subunit and the *E. hirae* CopB ATPase. This PCR fragment was then used to screen a library. The resultant gene, *hpcop*A, was initially reported incomplete (Ge et al., 1995). The complete hpCopA protein has a cys-box in the N-terminus and an intramembranous CPC motif (cf. Table 1).

Knock-out mutants in the *hpcop*A gene were constructed in two unrelated isolates of *H. pylori*. Two of these constructs exhibited a mean inhibitory concentration for growth in Cu^{2+} that was raised from 1.1 and 0.1 mM in the wild-type to 2 mM and 0.5 mM in the mutants, respectively. Two other mutants did not show a change in sensitivity to copper. All strains showed the same inhibition by Cd^{2+}, Hg^{2+} and Ni^{2+}. Based on these data and the sequence similarity of hpCopA to the Menkes ATPase and other putative copper ATPases, Ge et al. (1995) proposed that hpCopA is a copper exporting ATPase. However, the data are not entirely consistent and further experimentation will be required for a definitive assignment of a function to the hpCopA ATPase. Also, this ATPase seemed not to be the target for inhibition of *H. pylori* by omeprazole.

A second gene downstream of *hpcop*A, termed *hpcop*P, encodes a cys-box protein of 66 amino acids. hpCopP has the most extensive amino acid similarity to *E. hirae* CopZ protein. We like to hypothesize that hpCopP and related proteins act as a type of bacterial metallothioneins and are required for intracellular complexation and/or distribution of copper.

More recently, another CPX-type ATPase gene from *H. pylori* was reported (Melchers et al., 1996). The gene was cloned by screening an *H. pylori* library with pools of degenerate 20mer oligonucleotides corresponding to the phosphorylation motif in P-type ATPases. The derived protein sequence of the gene cloned in this way is 686 amino acids long and has 50% sequence similarity to CadA of *S. aureus*, the copper ATPases of *E. hirae*, and the Menkes ATPase. It exhibits an N-terminal cys-box and an intramembranous CPC motif. Using an *in vitro* transcription/translation/glycosylation system to identify transmembranous helices (Bamberg and Sachs, 1994) predict a membrane topology with eight membrane spans, similar to the ones proposed for CopA and CopB of *E. hirae* and the Menkes ATPase (Vulpe et al., 1993a; Vulpe et al., 1993b; Solioz et al., 1994). Preliminary data indicate that the enzyme could be involved in the loading of urease with nickel.

E. *Escherichia coli* CPX-Type ATPases

Three sequences for CPX-type ATPases from *E. coli* are known. An open reading frame, o732, that is predicted to encode an ATPase of this type was obtained in an *E. coli* genome sequencing project (Sofia et al., 1994). The predicted protein has 732 amino acids, an intramembranous CPC and an N-terminal cys-box. However, phylogenetic analysis places this enzyme as a whole (see below), as well as the cys-box alone, in a branch with the cadmium ATPases. Since there is no biochemical information available, the function of this ATPase remains obscure.

Two more *E. coli* CPX-type ATPase genes, *hra*-1 and *hra*-2 (histidine rich ATPase), were cloned accidentally when screening a human small intestine cDNA library from Clonetech. Using oligonucleotides homologous to conserved sequences of the *E. hirae* and human CPX-type ATPases, multiple isolates of the two new ATPase genes were obtained. Both genes were represented as human expressed sequence tags in the database. It later turned out that these genes were not of human origin but from *E. coli* (Trenor III et al., 1994). *E. coli* DNA obviously contaminated the human cDNA library and *E. coli* DNA sequences had found their way into the human expressed sequence tag database. The problem of library and database contamination has been noted earlier (Anderson, 1993).

The two new predicted proteins, HRA-1 and HRA-2, are interesting in two respects: First, they contain an intramembranous CPH rather than CPC and thus resemble the *E. hirae* CopB ATPase. And second, they do not have cys-boxes, but his-boxes in the polar N-terminal region. HRA-1 and HRA-2 constitute a separate branch on the phylogenetic tree, together with the *E. hirae* CopB ATPase (Figure 10). Since CopB has been shown to be a copper exporting ATPase, this would suggest a similar function for the two *E. coli* CPX-type ATPases.

Upstream of the *hra*-1 gene is a small open reading frame that encodes a cys-box protein. This open reading frame was considered to be a remnant of an ancestral gene for an ATPase with a cys-box, probably because there is no ATG initiator methionine codon (Trenor III et al., 1994). However, *E. coli* has the unique ability to also initiate protein synthesis at TTG, coding for leucine, or GTG, coding for valine (Mangroo et al., 1995). There is a consensus GGAGA ribosome binding site 16 bases upstream of a TTG codon and thus this open reading frame has all the features of a gene, which we call *hra*-0 in this discussion. *Hra*-0 would encode a cys-box protein that could, like other such proteins, act as a metal chelating protein, fulfilling a regulatory role in *E. coli*. Phylogenetically, HRA-0 clusters with the cys-box of *E. hirae* CopA ATPase and the cys-boxes of a group of mercuric reductases.

F. *Rhizobium meliloti* CPX-Type ATPase

Rhizobia fix nitrogen when living in symbiotic association with a host plant. This process takes place in specialized cells, the bacteroids, inside the root nodules. The FixI ATPase of *R. meliloti*, the alfalfa symbiont, was discovered as part of the gene cluster

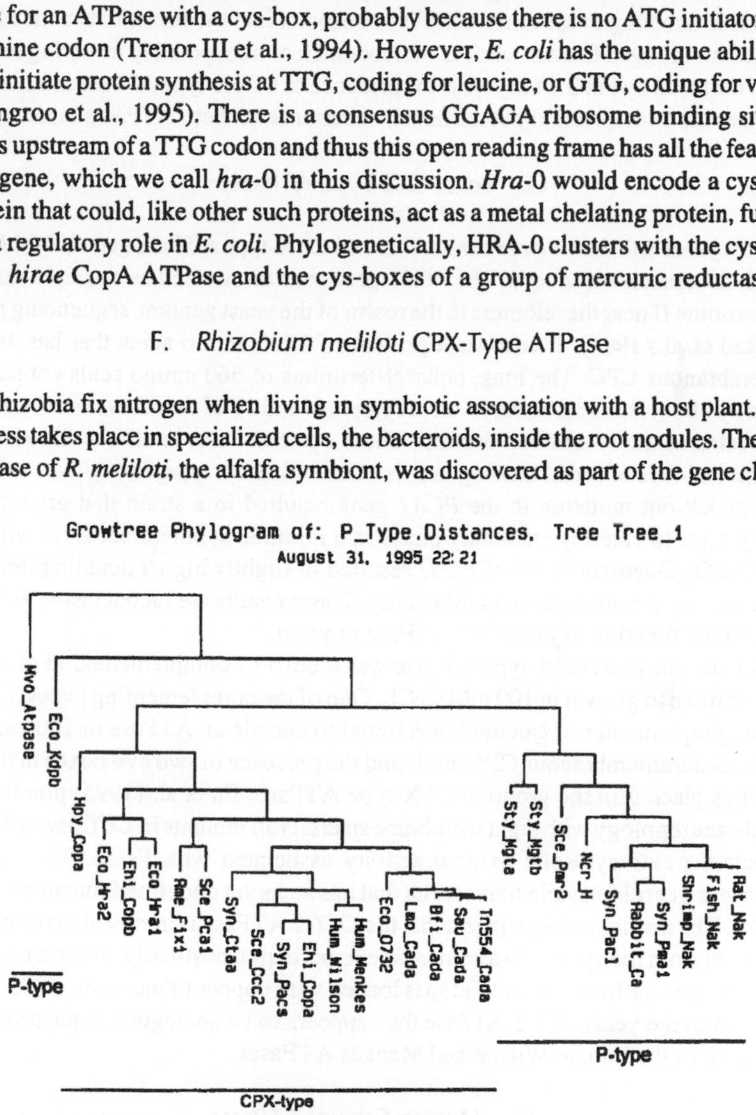

Figure 10. Phylogram of the CPX-type ATPases with a selected sample of P-type ATPases. Divergence was scored by the Jukes-Cantor method (Swofford and Olson, 1990). Relationships between distant branches are not reliable.

*fix*GHI involved in nitrogen fixation (Kahn et al., 1989). *FixI* encodes a protein of 757 amino acids that has a clear cys-box in the N-terminus and an intramembranous CPC. Based on the structural features, FixI is clearly a CPX-type ATPase. It may have a role in symbiotic nitrogen fixation, but evidence of its transport specificity is not yet available. *Fix*G, a gene upstream of *fix*I, probably encodes a membrane bound redox enzyme. It is tempting to speculate that FixI transports into the bacterioids a metal ion required for nitrogen fixation, such as molybdenum or iron, possibly involving reduction of the metal by FixG. In any case, the presence of a CPX-type ATPase in Rhizobia illustrates the widespread, if not universal, occurrence of these ATPases in nature.

G. Yeast CPX-Type ATPases

In the yeast *Saccharomyces cerevisiae*, two CPX-type ATPase genes, *PCA1* and *CCC2*, have been sequenced. The *PCA1* gene was discovered on the right arm of chromosome II near the telomere in the realm of the yeast genome sequencing project (Rad et al., 1994). It encodes a protein of 1216 amino acids that has an intramembranous CPC. The long, polar N-terminus of 560 amino acids contains a single cys-box. But Pca1 is very rich in cysteines and 33 of the 42 cysteine residues are located in the N-terminal domain. These cysteines are, in all likelihood, constituent residues of metal binding motifs of an unknown type.

A knock-out mutation in the *PCA1* gene resulted in a strain that grew more slowly prior to entering stationary phase in a minimal synthetic medium with >1 mM $CuSO_4$. Overexpression of *PCA1* resulted in slightly higher doubling times in minimal media containing 0.3 mM $CuSO_4$. These results are inconclusive with respect to the function of the *PCA1* ATPase in yeast.

The second yeast CPX-type ATPase was cloned by complementation of a mutant sensitive to growth in 100 mM $CaCl_2$. One of the complementing genes, *CCC2* (cross complementor of calcium) was found to encode an ATPase of 1004 amino acids. Its intramembranous CPC motif and the presence of two cys-boxes in the N-terminus place it in the group of CPX-type ATPase. Fu et al. (1995) predicted a membrane topology with eight membrane spans. Null mutants in *CCC2* were found to lack the extracytosolic oxidase activity associated with Fet3p. This cuproenzyme is a ceruloplasmin homologue that has an as yet undefined function in iron uptake. The available data suggest that the *CCC2* ATPase is involved in the export of copper from the cytosol to a compartment, perhaps a post-Golgi compartment of the endocytic pathway, where Fet3p is loaded with copper (Yuan et al., 1995). The function of the yeast *CCC2* ATPase thus appears to be analogous to the proposed functions of the human Wilson and Menkes ATPases.

H. Human Copper ATPases

The cloning of two putative human copper ATPases, one which is defective in Menkes disease and the other in Wilson disease, is a milestone in the study of heavy metal homeostasis in higher cells. It substantiates the concept that the regulation of heavy metal ions involves active transport systems, as had been suggested earlier for bacteria.

Menkes disease as described in 1962 by John Menkes is a familial condition resulting in brain deterioration, white, twisted, and brittle hair and death within one to two years of birth (Menkes et al., 1962). Ten years later, David Danks noted that the unusual hair in Menkes disease (also called kinky hair disease) resembled the wool of sheep in Australia that suffered from copper deficiency (the crosslinking of amino acids in keratins and collagens requires the copper enzyme lysyl oxidase). Copper deficiency in Menkes disease results from the trapping of copper in intestinal mucosa, kidney and connective tissue, and failure of its distribution to other tissues. At the cellular level, most types of cultured cells from Menkes patients exhibit excessive copper accumulation, with the exception of hepatocytes and chorionic villus cells (Danks, 1989).

Menkes disease is caused by mutations that have been mapped to the human X chromosome (Xq13.3) by linkage analysis and deletion mapping. As expected with an X-linked recessive disease, patients are normally male. However, a female patient with a balanced X:2 chromosome translocation had symptoms of Menkes disease and it was speculated that the translocation had disrupted the Menkes gene (Kapur et al., 1987). By cloning the DNA encompassing the site of the chromosome translocation, the Menkes gene could eventually be isolated. Its sequence was reported independently by three groups (Chelly et al., 1993; Vulpe et al., 1993a; Mercer et al., 1993). The Menkes gene is officially designated as *ATP7A,* but is also referred to as *MNK* or *Mc1*. It consists of 23 exons spanning a 150 kb genomic region (Dierick et al., 1995; Tumer et al., 1995). The predicted gene product is a CPX-type ATPase of 1500 amino acids with an intramembranous CPC and containing six cys-boxes in the very long polar N-terminus that precedes the first predicted membrane span. A membrane topology involving eight transmembranous helices was proposed (Vulpe et al., 1993b). Expression of an 8.5 kb transcript from the Menkes gene was detected in heart, brain, lung, muscle, kidney, pancreas, fibroblasts, lymphoblasts and, to a lesser extent, in placenta. No Menkes mRNA was detected in liver.

Copper resistant variants of Chinese hamster ovary cells were found to have undergone genomic amplification of the Menkes gene, accompanied by increased levels of Menkes mRNA and protein. Also, the copper resistant cells exhibited less copper accumulation due to enhanced copper efflux (Camakaris et al., 1995). These findings, combined with the clinical and cellular phenotype in Menkes disease, favor a role of the Menkes ATPase in transmembrane copper efflux.

A number of mutations have been characterized in Menkes patients. mRNA isolated from cultured fibroblasts was reverse transcribed, followed by PCR amplification of selected regions of the Menkes gene. Splice site mutations and deletions could be detected by the resultant PCR products of different sizes, while point mutations were characterized by chemical cleavage analysis and sequencing of the amplified DNA (Das et al., 1994). Interestingly, a V767L polymorphism in the third membrane helix was detected. Ten of 23 patients analyzed had leucine at this position. This polymorphism was recognized because it seemed very unlikely that a valine-to-leucine mutation in a membrane helix would affect the function of the ATPase and additional mutations were sought for in these cases. But other polymorphisms may be mistaken for the disease-causing mutations. Polymorphisms are of course useful markers in genetic analysis.

Some of the mutations thought to cause Menkes disease are listed in Figure 11. The R645S mutation in patient 1286 is a point mutation that still yields a full size Menkes gene product. But at the same time, a smaller transcript resulting from exon skipping can also be detected. R645 is the first arginine in the sequence REIRQWRR that precedes the first proposed transmembrane helix. Positively charged amino acids in front of a helix pair that inserts into the membrane act as stop-transfer signals (Sipos and Von Heijne, 1993; Gafvelin and Von Heijne, 1994). However, a methionine replaces R645 in the Wilson ATPase. It is thus not obvious how an R645S mutation could render the Menkes ATPase inactive.

G1302 in the ATP binding domain is conserved in all P-type ATPases and must have an essential role. In patient 1645, it is mutated to arginine and in all likelihood leads to a non-functional enzyme. But the mutation also induces the generation of a second mRNA species in which the mutated exon is skipped. A point mutation in patient 1488 in the intron just preceding this exon also leads to two aberrant splice products, both resulting in premature chain termination. A point mutation in an intron in patient 1190 leads to exon skipping, resulting in a frameshift and premature chain termination. Finally, patient 1095 had suffered a two basepair deletion in an

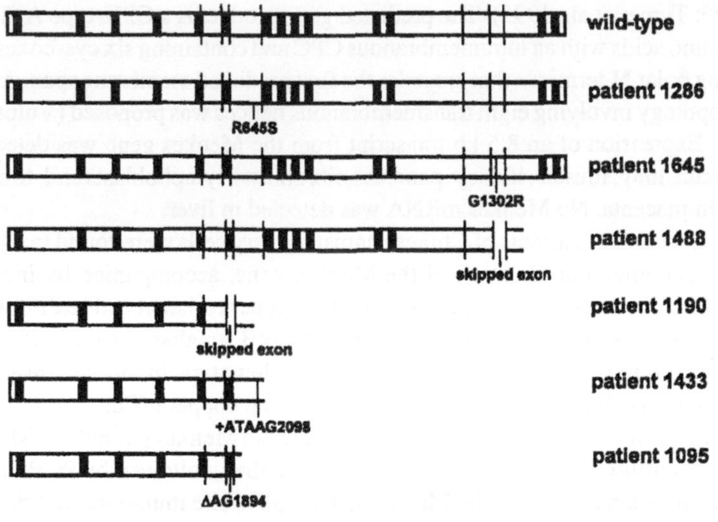

Figure 11. Schematic representation of the Menkes ATPase products expected to be generated in various Menkes patients. See text for details. The Figure is drawn to scale. Cys-boxes are delineated by shaded bars and membrane spans by black bars.

exon encoding the last cys-box. This resulted in normally spliced mRNA and frameshift/chain termination, but also two alternatively spliced mRNA species with one and three skipped exons, respectively. Exon-skipping thus appears to be a frequent consequence of mutations, even when they are located in exons. But exon-skipping and aberrant splicing has also been observed to occur with wild-type transcripts and may have a regulatory role. Examples include the plasma membrane Ca^{2+}-ATPase and the $Na^{+}H^{+}$-antiporter (Collins et al., 1993; Enyedi et al., 1994).

What can we learn from mutation analysis? Obviously the identification of mutations and their association with clinical observations is of great diagnostic value. On a molecular level, point mutations will help to identify functionally important amino acid residues for future studies on structure-function relationships in the Menkes ATPase.

The mottled mice strains are regarded as murine models of Menkes disease. Using a cDNA probe of the human gene, the mouse homologue of the Menkes gene was isolated (Levinson et al., 1994; Mercer et al., 1994). The derived protein is a CPX-type ATPase with 90% sequence identity to the Menkes enzyme. It could be shown that mottled mice have mutations in the Menkes gene homologue (Cecchi et al., 1997; Reed and Boyd, 1997). These results prove that the mottled mouse is the murine model for Menkes disease, providing the basis for future biochemical and therapeutic studies.

The occipital horn syndrome in humans appears to be a milder variant of the Menkes disease. In this disorder, connective tissue defects similar to those in the blotchy strain of mottled mice are observed, such as hyperelastic skin, bladder diverticulae and skeletal abnormalities, including bony exostoses of the occiput (Lazoff et al., 1975). It was shown that in these two potentially homologous disorders, similar splicing mutations of the Menkes mRNA occurred. Some expression of normal mRNA was also detectable in the mutant tissues, suggesting that the synthesis of low amounts of normal Menkes ATPase could result in the relatively mild phenotype of occipital horn syndrome and the blotchy mouse mutation (Das et al., 1995).

The primary connective tissue defects observed in these diseases can be ascribed to a deficiency in the copper enzyme lysyl oxidase that cross-links keratins and collagens. This secreted cuproenzyme appears to be much more sensitive to defects in the Menkes ATPase than intracellular copper proteins, such as cytochrome *c* oxidase or tyrosinase. This suggests that the Menkes copper ATPase is responsible for transporting copper into a compartment where copper is delivered to apolysyl oxidase, and probably other cuproenzymes. Clearly, further studies are required to determine the intracellular location and the function of the Menkes copper ATPase.

Wilson's disease was discovered in 1912 by Kinnear-Wilson and in the 1940s, the disease was found to be connected to the toxic effects of copper accumulation. Copper accumulates primarily in the liver, but also in the brain, kidney, cornea, and other tissues. This is accompanied by a failure of incorporation of copper into ceruloplasmin in the liver and the excretion of copper into the bile. The accumulated copper induces liver cirrhosis and/or progressive neurological damage and has an age of 20-40 years (Danks, 1989). While an effective treatment with the chelating agent penicillamine has been available for many years, there has been little progress in understanding the defect in copper transport and metabolism. The Wilson disease gene has been mapped to a single locus on chromosome 13 (13q14.3) by genetic analysis, and more recently, cloned by two groups using similar approaches. Based on the hypothesis that the Wilson gene encodes a copper ATPase similar to the Menkes protein, but active in the liver, Bull et al. (1993) used a 1 kb DNA probe from the cys-box region of the Menkes ATPase as a hybridization probe to clone the Wilson ATPase gene *Wc1* (official designation: *ATP7B*). At the same time, Tanzi et al. used a degenerate oligonucleotide corresponding to a heavy metal binding site of amyloid β-protein precursor. This motif is not present in the Wilson ATPase, but their DNA probe fortuitously hybridized (out of frame) to the Wilson gene.

The cloned Wilson gene encodes a predicted protein of 1465 amino acids with 57% sequence identity to the Menkes ATPase. Its overall primary structure is very similar to that of the Menkes ATPase: It is a CPX-type ATPase with an intramembranous CPC sequence and 6 cys-boxes in the polar N-terminus. There are eight putative transmembranous helices and it has all the conserved sites common to P-type ATPases. All 21 intron-exon boundaries (=22 exons) have been characterized (Petrukhin et al., 1994).

Yeast mutants deficient in *CCC2* ATPase cannot deliver copper to Fet3p, a multicopper oxidase related to ceruloplasmin (Yuan et al., 1995). This defect can be functionally complemented by human Wilson ATPase (Hung et al., 1997). The Wilson protein would thus deliver copper to a subcellular compartment where copper gets incorporated into ceruloplasmin and possibly other cupro-proteins. Failure to do so would then result in the observed deficiency in ceruloplasmin, a lack of copper secretion and the concomitant accumulation of copper in tissue.

Depending on the tissue, alternative splicing may lead to variants of the Wilson gene product. Two shorter transcripts were described in brain and one in the liver (Tanzi et al., 1993; Petrukhin et al., 1994). They are generated by exon-skipping and are predicted to lead to gross alterations of the gene product or a truncated protein. Exon skipping was also observed in kidney and placenta. These variants are unlikely to be active in transport. Whether the splicing aberrations are a regulatory mechanism *per se* or whether the products of alternative splicing play a role in the cell remains to be established.

A number of mutations causing Wilson's disease have been characterized using methods similar to those applied to Menkes patients (Cellini et al., 1991; Petrukhin et al., 1994). In contrast to the genetic lesions in Menkes patients, which are often gross deletions in the chromosome, Wilson patients predominantly exhibit point mutations, small deletions, or insertions. Over 16 mutations have already been characterized that, either directly or through altered splicing, lead to frameshifts and truncated gene products, predicted not to be active. A number of mutations caused amino acid substitutions that apparently impaired the function of the Wilson ATPase. A mutation detected in samples from American and Russian patients lead to an H714E mutation in the conserved HP-locus typical of CPX-type ATPases, thus supporting the idea that this residue is functionally important (Tanzi et al., 1993). A mutation found in a single patient converts asparagine-915 to serine (N915S). N915 is conserved in all P-type ATPases and is part of the proposed ATP-binding domain.

Many of the clinical and biochemical features of the Long-Evans Cinnamon (LEC) rat are similar to those seen in Wilson disease. Cloning of the rat homologue of the Wilson gene has identified the nature of the genetic defect in the LEC rat. A deletion that removes at least 900 bp of the coding region at the 3' end of the gene removes the ATP binding domain from the encoded copper ATPase and extends downstream of the gene (Wu et al., 1994). This supports the long-held view that the LEC rat is an animal model for Wilson disease. An animal model is important for studying liver pathophysiology, developing therapy for Wilson disease and studying the pathway of copper transport and its possible interaction with other heavy metals.

The cloning of the human Menkes and Wilson CPX-type ATPases is unlikely to lead in the near future to a cure of these copper metabolic diseases. However, it has greatly stimulate the investigation of copper homeostasis and paved the way to efficient prenatal diagnosis and genetic counseling. Many of the characterized mutations will help to identify functionally important residues in these ATPases and thus will contribute to the study of the mechanism of copper transport by CPX-type ATPases.

V. EVOLUTION OF P-TYPE ATPases

The chief features that distinguish CPX-type ATPases from the rest of the P-type ATPases are: heavy metal binding motifs in the N-terminus, two additional membrane spans preceding the first cytoplasmic loop, an intramembranous CPX motif, and only two predicted membrane helices at the C-terminus, compared to four or six in most P-type ATPases (see Figure 1). What is the evolutionary relationship of the CPX-type ATPases to the rest of the P-type ATPases (reviewed by Fagan and Saier Jr. 1994)?

For parsimony analysis across this broad spectrum of enzymes, only regions most conserved among all the P-type ATPases are useful. The most conserved domain is a region of 70 amino acids in the aspartyl kinase domain, including the membrane helix with the ion transduction domain (amino acids 369 to 438 in CopA of *E. hirae*). Using the algorithm of Jukes and Cantor, the phylogram shown in Figure 10 was obtained. Some of the ATPases were too unrelated (comparison score >100) to yield unambiguous assignments of the branching points. However, the clustering is in all cases significant and it becomes apparent that the CPX-type ATPases are on a separate evolutionary branch. Also on distinct branches are the KdpB subunit of the *E. coli* K^+-ATPase and a *Methanococcus voltae* ATPase that is only marginally related to the family of P-type ATPases (Dharmavaram et al., 1991). It even lacks the universally conserved DKTGT phosphorylation site, possibly represented by EKDTT. In fact, the *M. voltae* ATPase has been claimed to be unrelated to other ATPases, yet analysis of only the most conserved domains reveals significant sequence similarity to P-type ATPases.

The results of the analysis in Figure 10 are represented as an unrooted evolutionary tree in Figure 12. Again the distances of the branches are only approximate, as some of the sequences are too distant to allow a reliable phylogenetic analysis. Several points of interest deserve discussion. First, all the present day P-type ATPases appear to be descendants of the same primordial enzyme. The precursor enzyme could have resembled the *M. voltae* ATPase [the latter enzyme has been suggested to only be membrane associated, without any transmembranous helices (Dharmavaram et al., 1991)]. From such a primordial ATPase, a core structure with six membrane helices could have evolved. Known representatives of this branch are the KdpB subunits of bacterial K^+-ATPase. The Kdp-ATPases do, however, require two additional subunits for function. Other prokaryotic and all the eukaryotic ATPases are well grouped in another evolutionary lineage. These ATPases have acquired two to four additional membrane helices at the C-terminus (the number of helices at the C-terminus is still an issue of debate, but six seem to be favored by most workers in the field). This lineage is also marked by the presence of an invariant cysteine preceding the phosphorylated aspartic acid, resulting in the consensus sequence CXDKTGT. A separate branch of the evolutionary tree is occupied by the CPX-type ATPases. They have, as the result of evolutionary change, acquired two additional membrane helices at the N-terminus. The bacterial CPX-type ATPases

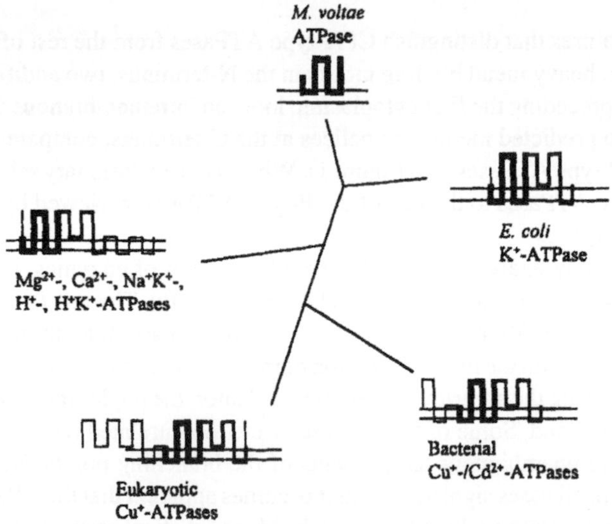

Figure 12. Phylogenetic, unrooted tree of P-type ATPases. The part of the ATPases thought to be the oldest is drawn in bold lines. Newer additions to the ATPases are drawn in thinner lines.

have diverged significantly from eukaryotic-type ATPases in that the former only developed a small extension of the N-terminus, in contrast to the very long N-termini that have been added to some of these eukayotic enzymes. Thus, starting from an archetype ATPase, a class of modern P-type ATPases for various transport functions have evolved. This evolution took place around a fairly invariant core structure, chiefly involving sites interacting with ATP.

VI. SUMMARY

Copper in biological systems presents a formidable problem due to the high reactivity of this transition metal. The recent discovery of ATP-driven copper pumps in bacteria and the identification of the human genes associated with the two copper metabolic disorders, Menkes disease and Wilson disease, have created new interest in this field. Over two dozen putative copper ATPases have already been described. While the function of most of these ATPases has not yet been elucidated, they clearly form a subclass of the P-type ATPases. Members of this subclass have a number of distinctive features, the most important of which are: a membrane topology deviating from that of other P-type ATPases, cysteine- or histidine-rich putative metal binding domains in the N-terminus, and a conserved intramembranous cysteine-proline-cysteine or cysteine-proline-histidine motif. Based on this latter signature, we propose to call these enzymes CPX-type ATPases. Bacterial cadmium ATPases are also members of this subclass. The structural and functional properties of known and putative copper ATPases and the cadmium ATPases are described in this chapter. The special role of copper in biological systems is also touched upon.

ACKNOWLEDGMENT

I am indebted to P. Wylev-Duda, D. Strausak, C. Vulpe, S. Silver, J. Gitschier, S. Packman, K. Melchers and K. Schäfer for valuable discussions and for providing various data prior to publication. Some of the work described here was supported by grant 3200-046804 from the Swiss National Foundation.

REFERENCES

Anderson, C. (1993). Genome databases worry about yeast (and other) infections. Science 259, 1685.

Anderson, J.E., Ptashne, M., & Harrison, S.C. (1987). Structure of the repressor-operator complex of bacteriophage 434. Nature 326, 846-852.

Ascone, I., Longo, A., Dexpert, H., Ciriolo, M.R., Rotilio, G., & Desideri, A. (1993). An X-ray absorption study of the reconstitution process of bovine Cu, Zn superoxide dismutase by Cu(I)-glutathione complex. FEBS Lett. 322, 165-167.

Askwith, C., Eide, D., Vanho, A., Bernard, P.S., Li, L.T., Daviskaplan, S., Sipe, D.M., & Kaplan, J. (1994). The FET3 Gene of S. cerevisiae encodes a multicopper oxidase required for ferrous iron uptake. Cell 76, 403-410.

Bamberg, K., & Sachs, G. (1994). Topological analysis of H^+,K^+-ATPase using in vitro translation. J. Biol. Chem. 269, 16909-16919.

Bell, P.F., Chen, Y., Potts, W.E., Chaney, R.L., & Angle, J.S. (1991). A reevaluation of the Fe(III), Ca(II), Zn(II), and proton formation constants of 4,7-diphenyl-1,10-phenanthrolinedisulfonate. Biol. Trace Elem. Res. 30, 125-144.

Brown, N.L. et al. (1991). Bacterial resistances to mercury and copper. Cell. Biochem. 46, 106-114.

Brown, N.L., Camakaris, J., Lee, B.T., Williams, T., Morby, A.P., Parkhill, J., & Rouch, D.A. (1991). Bacterial resistances to mercury and copper. J. Cell Biochem. 46, 106-114.

Bull, P.C., Thomas, G.R., Rommens, J.M., Forbes, J.R., & Cox, D.W. (1993). The Wilson disease gene is a putative copper transporting P-type ATPase similar to the Menkes gene. Nature Genet. 5, 327-337.

Bull, P.C., & Cox, D.W. (1994). Wilson disease and Menkes disease: new handles on heavy-metal transport. Trends Genet. 10, 246-252.

Camakaris, J., Petris, M.J., Bailey, L., Shen, P., Lockhart, P., Glover, T.W., Barcroft, C., Patton, J., & Mercer, J.F.B. (1995). Hum. Molec. Gen. 4, 2117-2123.

Cecchi, C., Biasotto, M., Tosi, M., & Avner, P. (1997). The *mottled* mouse as a model for human Menkes disease: identification of mutations in the *Atp7a* gene. Hum. Mol. Genet. 6, 425-433.

Cellini, L., Marzio, L., Di Girolamo, A., Allocati, N., Grossi, L., & Dainelli, B. (1991). Enhanced clearing of Helicobacter pylori after omeprazole plus roxithromycin treatment. FEMS Microbiol. Lett. 68, 255-257.

Cervantes, C., & Gutierrez-Corona, F. (1994). Copper resistance mechanisms in bacteria and fungi. FEMS Microbiol. Rev. 14, 121-138.

Cha, J.S., & Cooksey, D.A. (1991). Copper resistance in Pseudomonas syringae mediated by periplasmic and outer membrane proteins. Proc. Natl. Acad. Sci. USA 88, 8915-8919.

Chang, A., & Fink, G.R. (1994). Metal ion metabolism: The copper-iron connection. Curr. Biol. 4, 532-533.

Chelly, J., Tumer, Z., Tonnesen, T., Petterson, A., Ishikawa Brush, Y., Tommerup, N., Horn, N., & Monaco, A.P. (1993). Isolation of a candidate gene for Menkes disease that encodes a potential heavy metal binding protein (see comments). Nature Genet. 3, 14-19.

Ciriolo, M.R., Desideri, A., Paci, M., & Rotilio, G. (1990). Reconstitution of Cu, Zn-superoxide dismutase by the Cu(I) .glutathione complex. J. Biol. Chem. 265, 11030-11034.

Collins, J.F., Honda, T., Knobel, S., Bulus, N.M., Conary, J., Dubois, R., & Ghishan, F.K. (1994). Molecular cloning, sequencing, tissue distribution, and functional expression of a Na^+/H^+ Exchanger (NHE-2). Proc. Natl. Acad. Sci. USA 90, 3938-3942.

Cooksey, D.A. (1994). Molecular mechanisms of copper resistance and accumulation in bacteria. FEMS Microbiol. Rev. 14, 381-386.

Culotta, V.C., Hsu, T., Hu, S., Furst, P., & Hamer, D. (1989). Copper and the ACE1 regulatory protein reversibly induce yeast metallothionein gene transcription in a mouse extract. Proc. Natl. Acad. Sci. USA 86, 8377-8381.

Dancis, A., Yuan, D.S., Haile, D., Askwith, C., Eide, D., Moehle, C., Kaplan, J., & Klausner, R.D. (1994). Molecular characterization of a copper transport protein in s. cerevisiae - an unexpected role for copper in iron transport. Cell 76, 393-402.

Danks, D.M. (1989). Basis of Inherited Disease (Scirves, C., Beaudet, A., Sly, W. and Valle, D., Eds.), pp. 1411-1432. McGraw-Hill, New York.

Das, S., Levinson, B., Whitney, S., Vulpe, C., Packman, S., & Gitschier, J. (1994). Diverse mutations in patients with Menkes disease often lead to exon skipping. Am. J. Hum. Genet. 55, 883-889.

Das, S., Levinson, B., Vulpe, C., Whitney, S., Gitschier, J., & Packman, S. (1995). Similar splicing mutations of the Menkes/mottled copper-transporting ATPase gene in occipital horn syndrome and the blotchy mouse. Am. J. Hum. Genet. 56, 570-576.

Devereux, J., Haeberli, P., & Smithies, O. (1984). A comprehensive set of sequence analysis programs for the VAX. Nucleic. Acids. Res. 12, 387-395.

Dharmavaram, R., Gillevet, P., & Konisky, J. (1991). Nucleotide sequence of the gene encoding the vanadate-sensitive membrane-associated ATPase of methanococcus-voltae. J. Bacteriol. 173, 2131-2133.

Dick, J.D. (1990). Helicobacter (Campylobacter) pylori: a new twist to an old disease Enhanced clearing of Helicobacter pylori after omeprazole plus roxithromycin treatment. Annu. Rev. Microbiol. 44, 249-269.

Dierick, H. A., Ambrosini, L., Spencer, J., Glover, T. W., & Mercer, J. F. B. (1995). Genomics, in press.

Dierick, H.A., Ambrosini, L., Spencer, J., Glover, T.W., & Mercer, J.F. (1995). Molecular structure of the Menkes disease gene (ATP7A). Genomics 289, 462-469.

Dobi, A., Dameron, C.T., Hu, S., Hamer, D., & Winge, D.R. (1995). Distinct regions of Cu(I)×ACE1 contact two spatially resolved DNA major groove sites. J. Biol. Chem. 270, 10171-10178.

Dubin, D.T., Chikramane, S.G., Inglis, B., Matthews, P.R., & Stewart, P.R. (1992). Physical mapping of the mec region of an Australian methicillin-resistant Staphylococcus aureus lineage and a closely related American strain (published erratum appears in J Gen Microbiol 1992 Mar; 138(Pt 3):657). J. Gen. Microbiol. 138, 169-180.

Endo, G., & Silver, S. (1995). CadC, the transcriptional regulatory protein of the cadmium resistance system of Staphylococcus aureus plasmid pI258. J. Bacteriol. 177, 4437-4441.

Enyedi, A., Verma, A.K., Heim, R., Adamo, H.P., Filoteo, A.G., Strehler, E.E., & Penniston, J.T. (1994). The Ca^{2+} affinity of the plasma membrane Ca^{2+} pump is controlled by alternative splicing. J. Biol. Chem. 269, 41-43.

Eriksson, P.O., & Sahlman, L. (1993). 1H NMR studies of the mercuric ion binding protein MerP: sequential assignment, secondary structure and global fold of oxidized MerP. J. Biomol. NMR. 3, 613-626.

Fagan, M.J., & Saier, M.H., Jr. (1994). P-type ATPases of eukaryotes and bacteria: sequence analyses and construction of phylogenetic trees. J. Mol. Evol. 38, 57-99.

Ferreira, A.M., Ciriolo, M.R., Marcocci, L., & Rotilio, G. (1993). Copper(I) transfer into metallothionein mediated by glutathione. Biochem. J. 292, 673-676.

Fox, B., & Walsh, C.T. (1982). Mercuric reductase. Purification and characterization of a transposon-encoded flavoprotein containing an oxidation-reduction-active disulfide. J. Biol. Chem. 257, 2498-2503.

Freedman, J.H., Ciriolo, M.R., & Peisach, J. (1989). The role of glutathione in copper metabolism and toxicity. J. Biol. Chem. 264, 5598-5605.

Freedman, J.H., & Peisach, J. (1989). Intracellular copper transport in cultured hepatoma cells. Biochem. Biophys. Res. Commun. 164, 134-140.

Fu, D., Beeler, T.J., & Dunn, T.M. (1995). Sequence, mapping and disruption of CCC2, a gene that cross-complements the Ca^{2+}-sensitive phenotype of csg1 mutants and encodes a P-type ATPase belonging to the Cu^{2+}-ATPase subfamily. Yeast 11, 283-292.

Gafvelin, G., & Von Heijne, G. (1994). Topological "frustration" in multispanning *E. coli* inner membrane proteins. Cell 77, 401-412.

Ge, Z., Hiratsuka, K., & Taylor, D.E. (1995). Nucleotide sequence and mutational analysis indicate that two Helicobacter pylori genes encode a P-type ATPase and a cation-binding protein associated with copper transport. Molec. Microbiol. 15, 97-106.

Hackbarth, C.J., & Chambers, H.F. (1993). blaI and blaR1 regulate beta-lactamase and PBP 2a production in methicillin-resistant Staphylococcus aureus. Antimicrob. Agents Chemother. 37, 1144-1149.

Harris, E.D., & Percival, S.S. (1989). Copper transport: insights into a ceruloplasmin-based delivery system. Adv. Exp. Med. Biol. 258, 95-102.

Hassett, R., & Kosman, D.J. (1995). Evidence for Cu(II) reduction as a component of copper uptake by Saccharomyces cerevisiae. J. Biol. Chem. 270, 128-134.

Himeno, T., Imanaka, T., & Aiba, S. (1986). Nucleotide sequence of the penicillinase repressor gene penI of Bacillus licheniformis and regulation of penP and penI by the repressor (published erratum appears in J Bacteriol 1987 Jul; 169(7) :3392). J. Bacteriol. 168, 1128-1132.

Huckle, J.W., Morby, A.P., Turner, J.S., & Robinson, N.J. (1993). Isolation of a prokaryotic metallothionein locus and analysis of transcriptional control by trace metal ions. Mol. Microbiol. 7, 177-187.

Hung, I.H., Suzuki, M., Yamaguchi, Y., Yuan, D.S., Klausner, R.D., & Gitlin, J.D. (1997). Biochemical characterization of the Wilson disease protein and functional expression in the yeast *Saccharomyces cerevisiae*. J. Biol. Chem. 272, 21461-21466.

Ivey, D.M., Guffanti, A.A., Shen, Z., Kudyan, N., & Krulwich, T.A. (1992). The cadC gene product of alkaliphilic Bacillus firmus OF4 partially restores Na^+ resistance to an *Escherichia coli* strain lacking an Na^+/H^+ antiporter (NhaA). J. Bacteriol. 174, 4878-4884.

Jungmann, J., Reins, H.A., Lee, J.W., Romeo, A., Hassett, R., Kosman, D., & Jentsch, S. (1993). MAC1, a nuclear regulatory protein related to cu-dependent transcription factors is involved in Cu/Fe utilization and stress resistance in yeast. EMBO J. 12, 5051-5056.

Kahn, D., David, M., Domergue, O., Daveran, M.L., Ghai, J., Hirsch, P.R., & Batut, J. (1989). Rhizobium meliloti fixGHI sequence predicts involvement of a specific cation pump in symbiotic nitrogen fixation. J. Bacteriol. 171, 929-939.

Kanamaru, K., Kashiwagi, S., & Mizuno, T. (1994). A copper-transporting P-type ATPase found in the thylakoid membrane of the cyanobacterium *Synechococcus* species PCC7942. Mol. Microbiol. 13, 369-377.

Kapur, S., Higgins, J.V., Delp, K., & Rogers, B. (1987). Menkes syndrome in a girl with X-autosome translocation. Am. J. Med. Genet. 26, 503-510.

Karlin, K.D. (1993). Metalloenzymes, structural motifs, and inorganic models. Science 261, 701-708.

Kägi, J.H., & Schaffer, A. (1988). Biochemistry of metallothionein. Biochemistry 27, 8509-8515.

Klomp, L.W., Lin, S.J., Yuan, D.S., Klausner, R.D., Culotta, V.C., & Gitlin, J.D. (1997). Identification and functional expression of HAH1, a novel human gene involved in copper homeostasis. J. Biol. Chem. 272, 9221-9226.

Kyte, J., & Doolittle, R.F. (1982). A simple method for displaying the hydropathic character of a protein. J. Mol. Biol. 157, 105-132.

Lazoff, S.G., Rybak, J.J., Parker, B.R., & Luzzatti, L. (1975). Skeletal dysplasia, occipital horns, diarrhea and obstructive uropathy- a new hereditary syndrome. Birth Defects 11, 71-74.

Lebrun, M., Audurier, A., & Cossart, P. (1994a). Plasmid-borne cadmium resistance genes in Listeria monocytogenes are present on Tn5422, a novel transposon closely related to Tn917. J. Bacteriol. 176, 3049-3061.

Lebrun, M., Audurier, A., & Cossart, P. (1994b). Plasmid-borne cadmium resistance genes in Listeria monocytogenes are similar to cadA and cadC of Staphylococcus aureus and are induced by cadmium. J. Bacteriol. 176, 3040-3048.

Lee, Y.A., Hendson, M., Panopoulos, N.J., & Schroth, M.N. (1994). Molecular cloning, chromosomal mapping, and sequence analysis of copper resistance genes from xanthomonas campestris PV Juglandis - homology with small blue copper proteins and multicopper oxidase. J. Bacteriol. 176, 173-188.

Levinson, B., Vulpe, C., Elder, B., Martin, C., Verley, F., Packman, S., & Gitschier, J. (1994). The mottled gene is the mouse homologue of the Menkes disease gene. Nature Genet. 6, 369-373.

Lin, S.J., & Culotta, V.C. (1995). The ATX1 gene of *Saccharomyces cerevissiae* encodes a small metal homeostasis factor that protects cells against reactive oxygen toxicity. Proc. Natl. Acad. Sci. USA 92, 3784-3788.

Linder, M.C., & Hazegh Azam, M. (1996). Copper biochemistry and molecular biology. Am. J. Clin. Nutr. 63, 797S-811S.

Mangroo, D., Limbach, P.A., McCloskey, J.A., & RajBhandary, U.L. (1995). An anticodon sequence mutant of *Escherichia coli* initiator tRNA: possible importance of a newly acquired base modification next to the anticodon on its activity in initiation. J. Bacteriol. 177, 2858-2862.

McLaggan, D., Logan, T.M., Lynn, D.G., & Epstein, W. (1990). Involvement of gamma-glutamyl peptides in osmoadaptation of *Escherichia coli*. J. Bacteriol. 172, 3631-3636.

McPhail, D.B., & Goodman, B.A. (1984). Tris buffer—a case for caution in its use in copper-containing systems (letter). Biochem. J. 221, 559-560.

Melchers, K., Weitzenegger, T., Buhmann, A., Steinhilber, W., Sachs, G., & Schäfer, K.P. (1996). Cloning and membrane topology of a P type ATPase from *Helicobacter pylori*. J. Biol. Chem. 271, 446-457.

Mellano, M.A., & Cooksey, D.A. (1988). Nucleotide sequence and organization of copper resistance genes from Pseudomonas syringae pv. tomato. J. Bacteriol. 170, 2879-2883.

Menkes, J.H. et al. (1962). Pediatrics 29, 764-779.

Menkes, J.H., Milton, A., Steigleder, G.K., Weakley, D.R., & Sung, J.H. (1962). A sex-linked recessive disorder with retardation of growth, peculiar hair, and focal cerebral and cerebellar degeneration. Pediatrics 29, 764-779.

Mercer, J.F.B., Livingston, J., Hall, B., Paynter, J.A., Begy, C., Chandrasekharappa, S., Lockhart, P., Grimes, A., Bhave, M., Siemieniak, D., & Glover, T.W. (1993). Isolation of a partial candidate gene for Menkes disease by positional cloning (see comments). Nature Genet. 3, 20-25.

Mercer, J.F.B., Grimes, A., Ambrosini, L., Lockhart, P., Paynter, J.A., Dierick, H., & Glover, T.W. (1994). Mutations in the murine homologue of the Menkes gene in dappled and blotchy mice. Nature Genet. 6, 374-378.

Morby, A.P., Turner, J.S., Huckle, J.W., & Robinson, N.J. (1993). SmtB is a metal-dependent repressor of the cyanobacterial metallothionein gene smtA: identification of a Zn inhibited DNA-protein complex. Nucleic. Acids. Res. 21, 921-925.

Nielson, K.B., & Winge, D.R. (1985). Independence of the domains of metallothionein in metal binding. J. Biol. Chem. 260, 8698-8701.

Nucifora, G., Chu, L., Misra, T.K., & Silver, S. (1989). Cadmium resistance from Staphylococcus aureus plasmid pI258 cadA gene results from a cadmium-efflux ATPase. Proc. Natl. Acad. Sci. USA 86, 3544-3548.

Ochman, H., Gerber, A.S., & Hartl, D.L. (1988). Genetic applications of an inverse polymerase chain reaction. Genetics 120, 621-623.

Odermatt, A., Suter, H., Krapf, R., & Solioz, M. (1992). An ATPase operon involved in copper resistance by Enterococcus hirae. Ann. N. Y. Acad. Sci. 671, 484-486.

Odermatt, A., Suter, H., Krapf, R., & Solioz, M. (1993). Primary structure of 2 P-Type ATPases involved in copper Homeostasis in Enterococcus hirae. J. Biol. Chem. 268, 12775-12779.

Odermatt, A., Krapf, R., & Solioz, M. (1994). Induction of the putative copper ATPases, CopA and CopB, of Enterococcus hirae by Ag^{+} and Cu^{2+}, and Ag^{+} extrusion by CopB. Biochem. Biophys. Res. Commun. 202, 44-48.

Odermatt, A., & Solioz, M. (1995). Two trans-acting metalloregulatory proteins controlling expression of the copper-ATPases of Enterococcus hirae. J. Biol. Chem. 270, 4349-4354.

Pedersen, P.L., & Carafoli, E. (1987). Ion motive ATPases. Trends Biochem. Sci. 12, 186-189.

Percival, S.S., & Harris, E.D. (1989). Ascorbate enhances copper transport from ceruloplasmin into human K562 cells. J. Nutr. 119, 779-784.

Perry, R.D., & Silver, S. (1982). Cadmium and manganese transport in Staphylococcus aureus membrane vesicles. J. Bacteriol. 150, 973-976.

Petrukhin, K., Lutsenko, S., Chernov, I., Ross, B.M., Kaplan, J.H., & Gilliam, T.C. (1994). Characterization of the Wilson disease gene encoding a P-type copper transporting ATPase: genomic organization, alternative splicing, and structure/function predictions. Hum. Molec. Gen. 3, 1647-1656.

Phung, L.T., Ajlani, G., & Haselkorn, R. (1994). P-type ATPase from the cyanobacterium *Synechococcus* 7942 related to the human Menkes and Wilson disease gene products. Proc. Natl. Acad. Sci. USA 91, 9651-9654.

Pope, M.T., & Dale, B.W. (1968). Isopoly-vanadates, -niobates, and tantalates. Rev. Chem. Soc. 22, 527-545.

Presta, A., Green, A.R., Zelazowski, A., & Stillman, M.J. (1995). Copper binding to rabbit liver metallothionein—Formation of a continuum of copper(I)-thiolate stoichiometric species. Eur. J. Biochem. 227, 226-240.

Rad, M.R., Kirchrath, L., & Hollenberg, C.P. (1994). A putative P-type Cu^{2+}-transporting ATPase gene on chromosome II of Saccharomyces cerevisiae. Yeast 10, 1217-1225.

Reed, V., & Boyd, Y. (1997). Mutation analysis provides additional proof that mottled is the mouse homologue of Menkes' disease. Hum. Mol. Genet. 6, 417-423.

Rowland, S.J., & Dyke, K.G. (1990). Tn552, a novel transposable element from Staphylococcus aureus. Mol. Microbiol. 4, 961-975.

Sahlman, L., & Skarfstad, E.G. (1993). Mercuric ion binding abilities of MerP variants containing only one cysteine. Biochem. Biophys. Res. Commun. 196, 583-588.

Samson, S.L.-A., & Gedamu, L. (1995). Metal-responsive elements of the rainbow trout metallothionein-B gene function for basal and metal-induced activity. J. Biol. Chem. 270, 6864-6871.

Samuni, A., Chevion, M., & Czapski, G. (1981). Unusual copper-induced sensitization of the biological damage due to superoxide radicals. J. Biol. Chem. 256, 12632-12635.

Schell, M.A. (1993). Molecular biology of the LysR family of transcriptional regulators. Annu. Rev. Microbiol. 47, 597-626.

Schiering, N., Kabsch, W., Moore, M.J., Distefano, M.D., Walsh, C.T., & Pai, E.F. (1991). Structure of the detoxification catalyst mercuric ion reductase from Bacillus sp. strain RC607. Nature 352, 168-172.

Silver, S., Nucifora, G., Chu, L., & Misra, T.K. (1989). Bacterial resistance ATPases: primary pumps for exporting toxic cations and anions. Trends Biochem. Sci. 14, 76-80.

Silver, S., & Ji, G. (1994). Newer systems for bacterial resistances to toxic heavy metals. Environ. Health Perspect. 102 Suppl. 3, 107-113.

Silver, S., & Phung, L.T. (1996). Bacterial heavy metal resistance: New surprises. Annu. Rev. Microbiol. 50, 753-789.

Sipos, L., & Von Heijne, G. (1993). Predicting the Topology of Eukaryotic Membrane Proteins. Eur. J. Biochem. 213, 1333-1340.

Sofia, H.J., Burland, V., Daniels, D.L., Plunkett, G., & Blattner, F.R. (1994). Analysis of the *Escherichia coli* genome. V. DNA sequence of the region from 76.0 to 81.5 minutes. Nucleic. Acids. Res. 22, 2576-2586.

Solioz, M., Odermatt, A., & Krapf, R. (1994). Copper pumping ATPases: Common concepts in bacteria and man. FEBS Lett. 346, 44-47.

Solioz, M., & Odermatt, A. (1995). Copper and silver transport by CopB-ATPase in membrane vesicles of Enterococcus hirae. J. Biol. Chem. 270, 9217-9221.

Solioz, M., & Vulpe, C. (1996). CPX-type ATPases: a class of P-type ATPases that pump heavy metal ions. Trends Biochem. Sci., in press.

Suzuki, E., Kuwahara Arai, K., Richardson, J.F., & Hiramatsu, K. (1993). Distribution of mec regulator genes in methicillin-resistant Staphylococcus clinical strains. Antimicrob. Agents Chemother. 37, 1219-1226.

Swofford, D.L., & Olson, R.B. (1990). Phylogeny reconstruction. In Molecular Systematics (Hillis, D.M., and Moritz, C. Eds.), pp. 21-132. Sinauer, Sunderland, MA.

Tanzi, R.E., Petrukhin, K., Chernov, I., Pellequer, J.L., Wasco, W., Ross, B., Romano, D.M., Parano, E., Pavone, L., & Brzustowicz, L.M. (1993). The Wilson disease gene is a copper transporting ATPase with homology to the Menkes disease gene. Nature Genet. 5, 344-350.

Trenor, C., III, Lin, W., & Andrews, N.C. (1994). Novel bacterial P-type ATPases with histidine-rich heavy-metal-associated sequences. Biochem. Biophys. Res. Commun. 205, 1644-1650.

Tsai, K.J., Yoon, K.P., & Lynn, A.R. (1992). ATP-dependent cadmium transport by the cadA cadmium resistance determinant in everted membrane vesicles of Bacillus subtilis. J. Bacteriol. 174, 116-121.

Tsai, K.J., & Linet, A.L. (1993). Formation of a phosphorylated enzyme intermediate by the cadA Cd^{2+}-ATPase. Arch. Biochem. Biophys. 305, 267-270.

Tumer, Z., Vural, B., Tonnesen, T., Chelly, J., Monaco, A. P., & Horn, N. (1995) Characterization of the exon structure of the Menkes disease gene using vectorette PCR. Genomics 26, 437-442.

Tylecote, R.F. (1992). A History of Metallurgy. Institute of Materials, London.

Tynecka, Z., Gos, Z., & Zajac, J. (1981). Energy-dependent efflux of cadmium coded by a plasmid resistance determinant in Staphylococcus aureus. J. Bacteriol. 147, 313-319.

Vilsen, B., Andersen, J.P., Clarke, D.M., & MacLennan, D.H. (1989). Functional consequences of proline mutations in the cytoplasmic and transmembrane sectors of the Ca^{2+}-ATPase of sarcoplasmic reticulum. J. Biol. Chem. 264, 21024-21030.

Vulpe, C., Levinson, B., Whitney, S., Packman, S., & Gitschier, J. (1993a). Isolation of a candidate gene for Menkes disease and evidence that it encodes a copper-transporting ATPase. Nature Genet. 3, 7-13.

Vulpe, C., Levinson, B., Whitney, S., Packman, S., & Gitschier, J. (1993b). Isolation of a candidate gene for Menkes disease and evidence that it encodes a copper-transporting ATPase (Erratum). Nature Genet. 3, 273.

Vulpe, C.D., & Packman, S. (1995). Cellular copper transport. Annu. Rev. Nutr. 15, 293-322.

Wittman, V., & Wong, H.C. (1988). Regulation of the penicillinase genes of Bacillus licheniformis: interaction of the pen repressor with its operators. J. Bacteriol. 170, 3206-3212.

Wu, J., Forbes, J. R., Chen, H. S., & Cox, D. W. (1994). The LEC rat has a deletion in the copper transporting ATPase gene homologous to the Wilson disease géne. Nature Genet. 7, 541-545.

Yoon, K.P., Misra, T.K., & Silver, S. (1991). Regulation of the cada cadmium resistance determinant of staphylococcus aureus plasmid-pI258. J. Bacteriol. 173, 7643-7649.

Yoon, K.P., & Silver, S. (1991). A second gene in the Staphylococcus aureus-cadA cadmium resistance determinant of plasmid-pI258. J. Bacteriol. 173, 7636-7642.

Yoshida, Y., Furuta, S., & Niki, E. (1993). Effects of metal chelating agents on the oxidation of lipids induced by copper and iron. Biochim. Biophys. Acta 1210, 81-88.

Yuan, D.S., Stearman, R., Dancins, A., Dunn, T., Beeler, T., & Klausner, R.D. (1995). The Menkes/Wilson disease gene homologue in yeast provides copper to a ceruloplasmin-like oxidase required for iron uptake. Proc. Natl. Acad. Sci. U. S. A. 92, 2632-2636.

Zhou, J., & Goldsbrough, P.B. (1994). Functional homologs of fungal metallothionein genes from Arabidopsis. Plant Cell 6, 875-884.

Zhou, P.B., & Thiele, D.J. (1991). Isolation of a metal-activated transcription factor gene from Candida glabrata by complementation in Saccharomyces cerevisiae. Proc. Natl. Acad. Sci. USA 88, 6112-6116.

ISOFORM DIVERSITY AND REGULATION OF ORGANELLAR-TYPE Ca^{2+}-TRANSPORT ATPases

Frank Wuytack, Luc Raeymaekers, Jan Eggermont,

Ludo Van Den Bosch, Hilde Verboomen, and

Luc Mertens

Advances in Molecular and Cell Biology
Volume 23A, pages 205-248.

ISBN: 0-7623-0287-9

I. INTRODUCTION

One of the themes emerging in the field of cell biology, mainly as a result of the application of powerful molecular biology techniques, concerns the existence of an often bewildering protein isoform diversity. In this respect, the Ca^{2+}-transport ATPases are no exception. This chapter covers the study of the properties, the differential expression and the physiological meaning of the Ca^{2+}-transport ATPase isoforms involved in intracellular Ca^{2+} accumulation.

In the first part an update will be presented on what is currently known about the diversity of the eukaryotic Ca^{2+}-transport ATPases in intracellular Ca^{2+}-storage sites. Data on vertebrates, as well as on invertebrates and non-animal species will be included. Since alternative processing of primary gene transcripts contributes to the isoform diversity, special attention will be devoted to alternative transcript processing in Ca^{2+}-transport ATPases and to the mechanisms controlling them. The second part deals with the regulatory proteins of the Ca^{2+}-transport ATPases, through which the different signal-transduction pathways control the ATPase activity. And in the third part, factors controlling tissue-dependent expression of Ca^{2+}-transport ATPases both in health and disease will be addressed.

Further background information can be found in recent reviews on Ca^{2+}-transport ATPase isoform diversity and gene expression (Grover and Khan, 1992; Wuytack et al., 1992; Dux, 1993; Lompré et al., 1994).

II. THE FUNCTIONAL ROLE OF INTRACELLULAR Ca^{2+} STORAGE

A rise in cytosolic free Ca^{2+} above the resting level of 50-100 nM serves in most animal (Dawson, 1990) and plant cells (Evans et al., 1991) as a major intracellular messenger controlling a wide range of cellular processes including contraction, secretion and cell cycling. Ca^{2+} release from intracellular storage sites holds a key position in the intracellular signaling system even in small cells (reviewed by Pozzan et al., 1994). The presence of intracellular Ca^{2+} stores deeper in the larger cell-types allows the rapid establishment of Ca^{2+} gradients originating from regions far removed from the original site of cell stimulation at the plasma membrane. But even superficially located Ca^{2+} stores may exert in several ways an important influence over Ca^{2+} entering the cell. They might establish a Ca^{2+} compartment separate from the bulk cytosol which functions as a superficial buffer barrier (van Breemen and Saida, 1989; Chen and van Breemen, 1992). Although the loading state of the store is signaled to the plasma membrane (Berridge and Irvine, 1989; Putney, 1993), the nature of the retrograde messenger(s) and the channels in the plasma membrane mediating the ensuing increase in Ca^{2+} influx remains largely unknown (reviewed by Berridge, 1995).

Intracellular Ca^{2+} stores do not only accumulate and release Ca^{2+} in order to control cytosolic Ca^{2+} levels. The importance of luminal Ca^{2+} in many other cellular functions is only now becoming increasingly evident. Luminal Ca^{2+}, for example, intervenes in the regulation of the synthesis, folding and sorting of proteins in the endoplasmic reticulum (Brostrom and Brostrom, 1990; Sambrook, 1990; Suzuki et al., 1991; Wileman et al., 1991); it also controls signal-mediated transport and passive diffusion through the nuclear pore complex (Greber and Gerace, 1995). In view of the large diversity of intracellular Ca^{2+}-storage organelles (Krause, 1991; Sitia and Meldolesi, 1992) and the multiple functions of these stores in cellular physiology, the large isoform diversity of the Ca^{2+}-accumulation pumps may not be surprising.

III. THE SERCA GENE FAMILY

In vertebrates, three separate genes (SERCA1-3)[1], all members of a multigene family, are known to encode at least seven different Ca^{2+}-transport ATPase isoforms (SERCAla/b, SERCA2a/b and SERCA3a/b/c). The SERCA proteins are targeted to the endoplasmic reticulum (ER) or to specialized ER subdomains, like the sarcoplasmic reticulum (SR). Although they are termed SERCA (from Sarco/Endoplasmic Reticulum Ca^{2+}-ATPases), a more defined subcellular localization is often not known with certainty (see Table 1).

The SERCA Ca^{2+}-transport ATPases belong to the group of ATP-driven ion-motive ATPases or P-type ion-motive ATPases, which are characterized by the phosphorylation of an aspartic acid residue at the active site (that is, formation of a

Table 1. Overview of the Established Organellar-Type Ca^{2+}-Transport ATPases and of Sequences Denoted as Ca^{2+}-Transport ATPases Based on Similarity

Gene	*Organismal species*	*Acc. No.*	*Reference*
SERCA1	*Oryctolagus cuniculus* (rabbit)	M12898	Brandl et al., 1986
SERCA1	*Rattus norvegicus* (rat)	M99223	Wu and Lytton, 1993
SERCA1	*Homo sapiens* (human)	U96773-80	Zhang et al., 1995
SERCA1	*Gallus gallus* (chicken)	M26064	Karin et al., 1989
SERCA1	*Rana esculenta* (frog)	S18884	Vilsen and Andersen, 1992
SERCA1	*Makaira nigricans* (fish)	U65228	Block et al., unpublished
SERCA2	*Homo sapiens* (human)	M23114	Lytton and MacLennan, 1988
SERCA2	*Rattus norvegicus* (rat)	J04022	Gunteski-Hamblin et al., 1988
SERCA2	*Oryctolagus cuniculus* (rabbit)	J04703	Lytton et al., 1989
SERCA2	*Sus scrofa* (pig)	X15073	Eggermont et al., 1989
SERCA2	*Gallus gallus* (chicken)	M66385	Campbell et al., 1991
SERCA2	*Felis catus* (cat)	Z11500	Gambel et al., 1992
SERCA3	*Rattus norvegicus* (rat)	M30581	Burk et al., 1989
SERCA3	*Mus musculus* (mouse)	U49393-4	Tokuyama et al., unpublished
SERCA3	*Homo sapiens* (human)	Z69881	Dode et al., 1996
SERCA	*Artemia franciscana* (brine shrimp)	X51674	Palmero and Sastre, 1989
SERCA	*Procamburus clarkii* (crab)	AF025848-9	Chen et al., unpublished
SERCA	*Drosophila melanogaster* (fruit fly)	M62892	Magyar and Váradi, 1990
SPCA1	*Rattus norvegicus* (rat)	M93017	Gunteski-Hamblin et al., 1992
SMA1	*Schistosoma mansoni* (flatworm)	L40328	de Mendonça et al., 1995
PEA1	*Arabidopsis thaliana* (tale cress)	L08469	Huang et al., 1993
ECA1	*Arabidopsis thaliana* (tale cress)	U96455	Liang and Sze, unpublished
LCA1	*Lycopersicon esculentum* (tomato)	M96324	Wimmers et al., 1992
BCA1	*Brassica oleracea* (cabbage)	X99972	Malmstrom et al., 1997
CA1	*Dunaliella bioculata* (algae)	X93592	Raschke and Wolf, 1996
PMR1	*Saccharomyces cerevisiae* (yeast)	M25488	Rudolph et al., 1989
PMC1	*Saccharomyces cerevisiae* (yeast)	U03060	Cunningham and Fink, 1994
YER166w	*Saccharomyces cerevisiae* (yeast)	P32660*	Dietrich et al., unpublished
YEL031w	*Saccharomyces cerevisiae* (yeast)	P39986*	Dietrich et al., unpublished
YIL048w	*Saccharomyces cerevisiae* (yeast)	P40527*	Barrell et al., unpublished
YM8520.11c	*Saccharomyces cerevisiae* (yeast)	Z49705	Hunt and Bowman, unpublished
CTA3	*Schizosaccharomyces pombe* (yeast)	J05634	Ghislain et al., 1990
SPAC29A4.19c	*Schizosaccharomyces pombe* (yeast)	Z97210	Brown and Churcher, unpublished

continued

Table 1. (Continued)

Gene	*Organismal species*	*Acc. No.*	*Reference*
patA	*Dictyostelium discoideum* (slime mold)	X89369	Moniakis et al., 1995
TVCA1	*Trichomonas vaginalis*	U65066	Meade et al., unpublished
-	*Plasmodium falciparum* (protozoa)	X71765	Kimura et al., 1993
PfATPase4	*Plasmodium falciparum* (protozoa)	U16995	Dyer et al., 1995
YEL6	*Plasmodium yoelii* (protozoa)	X55197	Murakami et al., 1990
TBA1	*Trypanosoma brucei* (protozoa)	M73769	Revelard and Pays, 1991
PMA1	*Synechocystis* sp. (cyanobacteria)	X71022	Geisler et al., 1993
cda	*Flavobacterium odoratum* (bacteria)	L42816	Peiffer et al., 1996

Note: The accession number refers to the GenBank-EMBL accession number of the nucleotide sequence, except for the sequence indicated by * which refers to the corresponding protein in the SWISS-PROT database.

phosphoprotein intermediate) as an obligatory part of their catalytic cycle (Pedersen and Carafoli, 1987). A comprehensive review on the structural organization, ion transport and energy transduction of the superfamily of P-type ATPases appeared recently (Møller et al., 1996). Other eukaryotic members of the P-type ion-motive ATPases include the plasma membrane Ca^{2+}-transport ATPases (PMCA), the Na^+-K^+-ATPases, the gastric H^+-K^+-ATPases, the plant and fungal plasma membrane H^+-ATPases and even some putative copper-ion transporters like the Menkes and Wilson ATPases (reviewed by Bull and Cox, 1994). More distantly related members of the P-type ion motive ATPases which include putative Ca^{2+} pumps (Geisler et al., 1993; Peiffer et al., 1996), have also been described in prokaryotes (see Green, 1992). Song and Fambrough (1994) recently reviewed the evolutionary relationship among the different Ca^{2+}-transport ATPase isoforms.

The SERCA Ca^{2+}-transport ATPases can be distinghuised from their plasma-membrane counterparts (PMCA) and from the other ion-motive ATPases by the SERCA-selective inhibitors: thapsigargin (Thastrup et al., 1990), cyclopiazonic acid (CPA) (Seidler et al., 1989), and 2,5-di(*tert*-butyl)-1,4-benzohydroquinone (tBu-BHQ) (Oldershaw and Taylor, 1990).

A. SERCA1: The Fast-Twitch Skeletal Muscle Isoform

SERCA1 cDNA was originally cloned from rabbit skeletal muscle. The SERCA1 gene is only expressed in significant amounts in fast-twitch skeletal muscle fibers. It encodes two different protein isoforms: SERCAlb or the neonatal isoform (1001 amino-acids long, Brandl et al., 1986) and the adult isoform SERCAla (994 amino-acids long, Brandl et al., 1987). The neonatal polypeptide chain contains a highly charged carboxyl-terminal extension -Asp-Pro-Glu-Asp-Glu-Arg-Arg--Lys. In contrast, SERCAla encodes a protein isoform in which a single Gly residue substitutes for the neonatal octapeptide terminus, but which is otherwise identical. The difference between both isoforms results from a developmentally-regulated alternative splicing in which a 42-bp optional exon (exon 22 in rat and human) is retained in the adult mRNA but is excised in the neonatal one (Korczak et al., 1988).

Early in development, the optional adult exon remains largely unrecognized by the nuclear splicing machinery. The reason for this is at present unknown. Splicing could be inhibited through binding of inhibitory factors. Alternatively, specific stimulatory splice factors might be lacking in fetal and neonatal conditions. Gradually during development the exon is retained in the mature transcript. Brandl et al. (1987) reported that in the rabbit the retention of the exon increased from 28% just prior to birth to 83% by the age of two weeks, but never becomes 100 %. Even in muscle from one-year-old rabbits, some 4% of the mature SERCA1 messengers still lack the exon. Muscle regeneration following necrosis is also accompanied by a transient expression of the neonatal SERCA1b (Zador et al., 1996).

Remarkably, alternative splicing of exons encoding C-terminal sequences appears to be common among Ca^{2+}-transport ATPase genes. It is also observed in the case of SERCA2 and SERCA3 (*cfr infra*), the plasma membrane Ca^{2+}-transport ATPases (PMCA1-4, Strehler, 1991; Carafoli, 1994) and the mammalian homologue of the yeast secretory-pathway Ca^{2+}-transport ATPase (Gunteski-Hamblin et al., 1992). Alignment of SERCA1 , SERCA2 and SERCA3 cDNAs indicates that for SERCA2 the splice donor site (5'D1; *cfr infra*) involved in this alternative splice process is positioned 3 nucleotides more downstream from the corresponding splice donor site in SERCA1 and SERCA3. The splicing modes differ. Whereas for SERCA1 and SERCA3 splicing occurs via an exon skipping-type of mechanism, processing of SERCA2 transcripts is much more complicated and involves an internal splice donor site (see Figure 1).

SERCAla cDNAs have also been cloned from rat diaphragm (Wu and Lytton, 1993), from human fetal *psoas* muscle (Zhang et al., 1995) and from chicken embryo thigh muscle (Karin et al., 1989). In the chicken, the C-terminal amino acid is Ala instead of Gly as in the other species. Analysis of the 3'-untranslated end of rat, human and chicken cDNAs suggests the possibility of there being similar alternative processing patterns to the one documented in the rabbit to give rise to the SERCAlb isoform. In the rat and human, this would also require the removal of a 42-basepair (bp) exon and the new extended tail would then become exactly the same as in the rabbit. In the chicken, however, the optional exon would count only 37 bp resulting in a decapeptide tail (-Asp-Ala-Glu-Asp-Leu-Arg-Lys-Lys-Arg--Lys) highly homologous to the mammalian octapeptide.

The functional significance of the highly charged polar carboxy-terminal extension found in the neonatal forms of SERCA1 is not yet clear. When expressed in COS cells, adult and neonatal forms exhibit a nearly identical maximal Ca^{2+}-turnover rate, Ca^{2+}-affinity and ATP-dependency of Ca^{2+} transport (Maruyama and MacLennan, 1988). One should keep in mind however that the alternative carboxyl termini could act as a binding site for regulatory factors which might be absent from the heterologous COS cell system, and hence, a potential functional difference in muscle might thereby be overlooked. Another functional meaning of the different tails might be sought in the targeting of the individual isoforms to distinct cellular organelles or organellar subdomains. The tail might also be an adaptation to changes in the characteristics of the sorting machinery during development. Finally it may merely represent an epiphenomenon resulting from a developmentally dependent general change in splicing activity.

An older form of the SERCA1 gene may be found in the amphibia. A SERCA cDNA was cloned from frog *Rana esculenta* thigh muscle (Vilsen and Andersen, 1992). This type of pump is at the DNA and protein level more similar to SERCA1a than to the SERCA2 isoforms of rabbit or chicken. In addition, at low stringency, no cross hybridization to the RNA from other frog tissues, including cardiac muscle, could be observed. This suggests that it represents the frog equivalent of SERCA1a. However, in contrast to mammals and birds, the frog SERCA1 transcript does not allow the formation of a SERCA1b variant.

Frog SERCA1a enzyme, when expressed in COS cells, displayed an approximately 2.5-fold lower apparent affinity for Ca^{2+} and an increased apparent affinity for the inhibitors vanadate and thapsigargin, relative to the mammalian enzyme (Vilsen and Andersen, 1992). Since the proposed ligands for Ca^{2+} residing in M4-M8 are conserved in frog and rabbit, the structural basis for this functional difference must be sought elsewhere. The observations can be explained by assuming that minor structural differences in the frog enzyme compared to its rabbit counterpart allow the frog enzyme to reside for a relatively longer duration of its catalytic cycle in the E_2 conformation (where the enzyme shows low-affinity for Ca^{2+}).

The human SERCA1 gene (ATP2A1) is mapped to chromosome 16p12.1 (MacLennan et al., 1987; Callen et al., 1991), and its exon/intron organization is the same as that of the homologous rabbit gene, with the exception of the boundary between exons 12 and 13, where splicing occurs in the human transcript one nucleotide downstream of the splicing site in the rabbit transcript (Zhang et al., 1995). Both genes comprise 23 exons of which exon 22 represents the 42-bp optional exon. The sizes of the corresponding introns vary slightly between the human and the rabbit. The rabbit SERCA1 gene is about 23 kb long (Korczak et al., 1988), its human counterpart 26 kb. In general no close correlation could be found between the exon/intron boundaries and the boundaries of predicted functional domains in the pump protein. Thus, for example, the parts of the gene encoding three of the ten membrane-spanning segments (M1, M5, and M7) are each interrupted by an intron/exon boundary. Since it can be supposed that each of the multigene families of P-type ATPases (that is, the SERCA family, the PMCA family, and the Na^+-K^+-ATPase family) originates from a common ancestral gene, this could be reflected in a common intron/exon layout. For the majority of the introns however such a conservation of the positions is not observed among these three families of P-type ATPases. This might indicate that the evolutionary event of separation of the gene families is very old. Within a given family, however, the intron positions are generally very well conserved. For the SERCA family this is clearly the case between SERCA1, SERCA2 and SERCA3 (Dode et al., 1997b). The only exception is the boundary between exon 8 and 9 in SERCA1, which is absent from SERCA2 and SERCA3 (Dode et al., 1997b).Even in the crustacean SERCA gene (*Artemia franciscana*), 12 out of 17 introns are inserted in the same position as in rabbit SERCA1 (Escalante and Sastre, 1994a). The *Drosophila melanogaster* SERCA gene has only 9 exons and 8 introns (Magyar et al., 1995). Five of the 8 introns *in Drosophila* are found in identical positions as in *Artemia* and in the vertebrate SERCA1, but intron 2 of the *Drosophila* gene, which is preserved in the rabbit and human is absent from the *Artemia* gene. The rat PMCA3 (Burk and Shull, 1992) and the human PMCA1 gene (Hilfiker et al., 1993) belong together to a separate family. Both genes present an almost identical intron/exon layout which differs from that in SERCA. Finally the human α2 and α3 Na^+-K^+ATPase (Shull et al., 1989) and the human gastric H^+-K^+-ATPase (Maeda et al., 1990) form together a third independent family, again characterized by a specific intron layout.

B. SERCA2 is Expressed in Muscle and Non-Muscle Cells

The human SERCA2 gene maps to chromosome 12q23-q24.1 (Otsu et al., 1993), *i.e.*, on the same chromosome as the plasma membrane Ca^{2+}-transport ATPase gene PMCA1 (12q21-q23) (Olson et al., 1991). Up till now no complete structure of a vertebrate SERCA2 gene has been published. However, partial sequences comprising the 5'end including the promotor and the 3'end suggest that the exon/intron layout is conserved between SERCA1 and SERCA2 (Zarain-Herzberg et al., 1990; Eggermont et al., 1991).

Characterization of the 3' end of the SERCA2 gene revealed the presence of four optional exons, which can be alternatively processed, resulting in four distinct classes of mRNAs (*cfr. infra*). Class 1 encodes the SERCA2a pump, and classes 2, 3 & 4 encode SERCA2b. Both Ca^{2+}-transport ATPases are completely identical up to amino acid 993. Four more amino acids are present in SERCA2a, whereas an extension of 49 amino acids is found in SERCA2b. This prolonged SERCA2b tail contains a highly hydrophobic stretch which has been suggested to represent an eleventh membrane-spanning segment (Lytton & MacLennan, 1988; Gunteski-Hamblin et al., 1988; Eggermont et al., 1989; Campbell et al. 1991). Immunocytochemical epitope mapping (Campbell et al., 1992) localized the extreme C-termini of SERCA2a and SERCA2b on opposite sides of the ER membrane (with the SERCA2b tail protruding in the lumen). Further support for the existence of an eleventh membrane-spanning segment in SERCA2b is given by *in vitro* translation scanning experiments where the insertion into the membrane of *in vitro* expressed SERCA2 peptide fragments is assessed (Bayle et al., 1995). Despite the topological difference between SERCA2a and SERCA2b, Campbell et al. (1991) working with the COS-1 cell expression system, have not been able to measure any functional difference between SERCA2a and SERCA2b. However, other groups (Lytton et al.,

1992; Verboomen et al., 1992) have observed a difference in Ca^{2+} affinity (SERCA2a < SERCA2b) and in turnover rate of the phosphoprotein intermediate (SERCA2a > SERCA2b), but not in phospholamban (PLB) and thapsigargin sensitivity. Moreover, work with truncated SERCA2b mutants shows that the last 12 amino acids of SERCA2b are of critical importance in this respect (Verboomen et al., 1994). By means of site-directed mutagenesis, three SERCA2b deletion mutants are constructed, lacking progressively larger stretches of 12, 31, and 49 amino acids at their C-terminus. These truncated variants are expressed in COS cells and analyzed for functional characteristics relative to the parental ATPases. SERCA2a and the truncated mutants show a 2-fold higher turnover rate, a 2-fold lower Ca^{2+} affinity and a 10-fold higher vanadate sensitivity compared to SERCA2b (Verboomen et al., 1994).

The characteristic structural SERCA2 duality and the underlying RNA-processing patterns exist in mammals (Lytton & MacLennan, 1988; Gunteski-Hamblin et al. 1988; Eggermont et al., 1989), birds (Campbell et al., 1991; Campbell et al. 1992) and even in invertebrates (*Artemia franciscana*) (Escalante & Sastre, 1993). Analysis of the structure of the SERCA gene in *Drosophila* also indicates that as the result of alternative processing of the 3' end of the primary transcript, two different protein isoforms could be expressed, but the occurrence of this alternative splicing remains to be established (Magyar et al., 1995). The full physiological significance of these highly conserved isoforms in the different species is not yet understood. Both vertebrate SERCA2 isoforms are expressed in a tissue-dependent way and their expression pattern is not only qualitatively but also quantitatively different (Eggermont et al., 1990a). Cardiac muscle expresses 5- (Lompré et al., 1994) to 20-fold (Eggermont et al., 1990a) higher levels of SERCA2 than smooth muscle. Whereas slow-twitch skeletal and cardiac muscle only express SERCA2a, SERCA2b represents 70 to 80% of the Ca^{2+}-transporting ATPases in smooth-muscle tissues. In addition, SERCA2b which is also referred to as the 'housekeeping' isoform, is expressed in most—if not all—non-muscle tissues. In view of this expression pattern, it is tempting to speculate that those tissues which have to pump relatively larger amounts of Ca^{2+}, for instance, cardiac muscle, slow-twitch skeletal muscle, and, to a lesser extent, smooth muscle, can economize on the number of Ca^{2+}-pump molecules by expressing a SERCA2 variant with a roughly 2-fold higher turnover rate, despite a lower Ca^{2+} affinity. For nonmuscle tissues, on the other hand, a slower pump variant with a twofold higher Ca^{2+}-transport rate at submicromolar Ca^{2+} concentrations will do.

Early claims that the Ca^{2+}-transport ATPase in non-muscle (SERCA2b) is more sensitive to thapsigargin than the cardiac-muscle isoform (SERCA2a), and that the fast-twitch skeletal-muscle isoform (SERCA1) is almost insensitive (Thastrup et al., 1990) are based on an artefact. Indeed Lytton et al. (1991) showed that when expressed in COS cells, all SERCA isozymes are inhibited by thapsigargin with the same potency. Thapsigargin binds with such a high affinity in a stoichiometric complex to the ATPases, presumably in the membrane domain in a crevice between segments M3 and M4 (Nørregaard, et al., 1994; Andersen, 1995), that it effectively titrates the number of pump molecules. Therefore, those membranes showing the lowest density of pump molecules show the highest apparent affinity for the inhibitor.

The level of Ca^{2+}-pump protein in a cell depends on several factors including efficiency of gene transcription, message stability, efficiency of translation and protein stability. This might result in different protein-to-message ratios for SERCA2a and SERCA2b. According to Khan et al. (1990), cardiac muscle which expresses SERCA2a, contains 70 times more protein, but only 7 times more mRNA compared to stomach smooth muscle which expresses predominantly SERCA2b. Wu and Lytton (1993), on the other hand, report for cardiac muscle and stomach or aorta smooth muscle ratios of 2×10^5 protein molecules / mRNA molecules. Evidence for additional control of SERCA2 expression at the translational level in canine *latissimus dorsi* muscle is presented by Hu et al. (1995).

A major unresolved question concerns the subcellular localization of the SERCA2a and SERCA2b isozymes. In some cell types like smooth-muscle cells and neuronal cells, Ca^{2+} release can be mediated by the inositol trisphosphate receptor (IP_3R) Ca^{2+}-release channel, and by the ryanodine receptor (RYR) Ca^{2+}-release channel. These cells also co-express nonmuscle type and muscle type Ca^{2+}-binding proteins in the lumen of their endomembranes (Wuytack et al., 1987; Michelangeli et al., 1991; Takei et al., 1992; Johnson et al., 1993; Raeymaekers et al., 1993). The possible association of the distinct isoforms of Ca^{2+} pumps, luminal Ca^{2+}-binding proteins, and Ca^{2+}-release channels in morphologically and functionally distinct stores with a specific intracellular sublocalization remains largely unexplored. *In situ* hybridization and immunocytochemical studies show that some types of neuronal cells in the brain, in particular those with high Ca^{2+} metabolism, for example, Purkinje neurons, express high levels of SERCA2 pumps (Plessers et al., 1991; Miller et al., 1991; Takei et al., 1992). SERCA2b is in this case the major isoform. Immunocytochemical analysis shows that low levels of SERCA2a might be present in avian brain (Campbell et al., 1993). In this respect, the alternative termini of SERCA2a and 2b have been suggested to contain targeting information for a different subcellular localization (Campbell et al., 1993). Reports of the occurrence of SERCA2a based on *in situ* hybridization must be interpreted with caution because the major neuronal SERCA2 transcript is class 4. This mRNA is translated in SERCA2b, but contains the SERCA2a-specific exon 25 in its untranslated 3'end (Plessers et al., 1991; *cfr. infra* and Figure 1).

Immunocytochemical staining shows that, at least when overexpressed in COS-1 cells, both Ca^{2+}-pump isoforms as well as PLB are targeted to the same store (Campbell et al., 1991; Verboomen et al., 1992).

In *Artemia,* SERCA expression is developmentally controlled. Palmero and Sastre (1989) describe SERCA messengers of 5.2 and 4.5 kb in the cryptobiotic embryos. In the nauplii, the 5.2 kb mRNA is undetectable, but the level of the 4.5 kb mRNA increases during early development. Later Escalante and Sastre (1993) reported that transcript processing would allow the formation of two different protein isoforms. Since the last six amino acids of the isoform translated from the 4.5-kb mRNA would be substituted by a variant tail of 30 amino acids in the isoform corresponding to the 5.2-kb mRNA, one form would be short with 1003 amino acids and the other longer with 1027 amino acids. The extended tail contains a hydrophobic segment with the propensity of forming an additional transmembrane segment comparable to the one observed in vertebrate SERCA2b. It should be added that besides their hydrophobic character, the vertebrate and crustacean stretches show no further significant homology. Furthermore, whereas in vertebrate SERCA2b, an extension of about 10 hydrophilic amino acids follows the hydrophobic stretch, no such extension is seen in *Artemia.* It is a matter of special interest that alternative transcript processing responsible for this isoform diversity is the same in vertebrates and *Artemia* and that the processing sites occur in an homologous position. The two isoforms of *Artemia* also present tissue-specific expression. The 4.5 kb mRNA, and the corresponding 1003 amino-acids long protein is expressed in muscle, whereas the 5.2-kb mRNA and the 1027 amino-acid protein is constitutively expressed in all tissues (Escalante and Sastre, 1995). The tissue-specific expression results from the presence of two alternative first exons, each with its own promotor. The existence of alternative promotors in *Artemia* is made possible because of the presence of an intron in the 5'-untranslated region that is not present in the vertebrate genes. A muscle-specific exon 1 and promotor are found 11 kb upstream from exon 2 which contains the ATG start codon. When this promotor is used, this leads to the transcription of the 4.5-kb mRNA. Within the intron separating exon 1 from exon 2, an alternative exon (exon 1') is found in combination with its own promotor from which the 5.2-mRNA is transcribed. Crustaceans are therefore as yet the only taxon for which isoform diversity is controlled both by the use of tissue-specific alternative SERCA transcript processing and tissue-specific use of alternative SERCA promotors. Although the former process is retained in vertebrates and possibly also in insects, there is no indication for the existence of more than one promotor in vertebrates or in *Drosophila* (Magyar et al., 1995).

Recent reports on the possible expression of two sub-isoforms of SERCA2b in pancreatic ER (Dormer et al., 1993; Webb and Dormer, 1995) are interesting but need confirmation.

C. SERCA3 is a Non-Muscle Isoform

The human SERCA3 gene maps to chromosome 17p13.3 (Dode et al., 1996). The SERCA3 cDNA, which was first cloned as a rat kidney cDNA by Burk et al. (1989), encodes a 999 amino acid-long protein which shows 75-77% amino acid identity with the SERCA1 and SERCA2 proteins. Expression in the COS-cell system indicates that SERCA3 acts as a Ca^{2+} pump but, at least in this heterologous system, it shows an approximately 5-fold lower affinity for Ca^{2+} and a slightly higher pH optimum compared to the SERCA 1 and SERCA2 pumps. Work with SERCA2/SERCA3 chimeras shows that the specific SERCA3 nucleotide-binding/hinge domain through its interaction with the COOH-terminal transmembrane domain is responsible for these deviating properties (Toyofuku et al., 1992). SERCA3 also lacks the putative cytosolic interaction domain for phospholamban, and does not respond to this modulator (Toyofuku et al., 1993). Remarkably, a truncated form of phospholamban lacking partially or entirely its cytosolic domain, inhibits Ca^{2+} transport by lowering the apparent Ca^{2+} affinity. These results suggest an interaction between SERCA3 and phospholamban via intramembrane interactions (Kimura et al., 1996). In platelets, mast cells and lymphoid cells, SERCA3 is co-expressed with the housekeeping SERCA2b (Wuytack et al., 1994; Bobe et al., 1994; Wuytack et al., 1995). This can be shown at the protein level by means of its phosphoprotein intermediate and by the use of SERCA3 and SERCA2b-specific antibodies (Wuytack et al., 1989; 1994). Recently sequences for two mouse SERCA3 splice variants (SERCA3a and SERCA3b) were entered in EMBL/Genbank (see Table 1). Moreover unpublished observations indicate that mouse and human might even express a SERCA3c variant (Dode et al., 1997a). A mouse monoclonal antibody, PL/IM430, raised against human platelet intracellular membranes and known to inhibit Ca^{2+} uptake in platelet organelles (Hack et al., 1988), was also reported to recognize SERCA3 (Wuytack et al., 1994; Bokkala et al., 1995). However, subsequent observations by Kovács et al. (1994) cast some doubt on this matter; these authors conclude that PL/IM430 reacts with another SERCA pump co-migrating with SERCA3 (*cfr. Infra*). The SERCA3 phosphoprotein intermediate migrates with a slightly lower apparent M_r than SERCA2b (that is, 97 kDa instead of 100 kDa). The phosphoprotein intermediates of both pumps can be easily distinguished by their different tryptic-fragmentation pattern, resulting from the absence of the T_1-cleavage site in SERCA3. Ratios of SERCA3/SERCA2 mRNA which are measured using a quantitative ratio reverse transcriptase PCR confirm the protein data (Wuytack et al., 1994; 1995). *In situ* hybridization experiments indicate the expression of SERCA3 in some arterial endothelial cells and in early (10 days *post coitum)* developing rat heart (Anger et al., 1993). SERCA3 is also expressed in some secretory epithelial cells of endodermal origin like those in the crypts of the colon (Wu et al., 1995) and in the cerebellar Purkinje neurons (Wu et al. 1995; Baba-Aissa et al., 1996). SERCA3 is upregulated in platelets of spontaneously hypertensive rats (Papp et al., 1993) and downregulated in a non-insulin-dependent diabetes mellitus rat model (Varadi et al., 1996). T cell activation is accompanied by a very rapid downregulation of SERCA3 and a concomitant upregulation of SERCA2b (Launay et al., 1997).

D. Evidence for a Novel Non-Muscle SERCA Isoform?

Studies based on controlled proteolysis of SERCA phosphoprotein intermediates in human platelets, megakaryoblastoid and lymphoblastoid cell lines led Kovács et al. (1994) to conclude that these cells express another SERCA isoform besides SERCA2b and SERCA3. In the unproteolyzed condition both SERCA3 and the novel SERCA isoform are found to co-migrate at 97 kDa. But whereas trypsin slowly degrades SERCA3 into an 80 kDa C-terminal fragment containing the autophosphorylation site, and an 25 kDa amino-terminal fragment, the other form is more trypsin sensitive and is quickly split into fragments of 73, 68, and 40 kDa, all of which react, like the parent protein, with monoclonal antibody PL/IM430. The 73 and 68 kDa fragments are precursors of the 40 kDa one while the 73 kDa fragment shows Ca^{2+}-dependent phospholabeling in the presence of La^{3+}. The relationship between both 97 kDa isoforms is not yet clear. That the PL/IM430-recognized 97 kDa pump might belong to the SERCA family is deduced from its reaction with an anti-SERCA1 antibody and from the tBu-BHQ-sensitivity of the autophosphorylated 73 kDa fragment. The expression of the PL/IM430-recognized 97 kDa pump in platelets might explain the observed increase in the steady-state level of the 97 kDa Ca^{2+}-dependent phosphoprotein intermediate in the presence of La^{3+} (Papp et al., 1991). Such an increase is not true of the other SERCA pumps, but is characteristic of the PMCA ATPases (Wuytack et al., 1984). Furthermore, the autophosphorylation of the 97 kDa protein is relatively more sensitive to tBu-BHQ (Papp et al., 1992) whereas that of SERCA2b is more sensitive to thapsigargin (Papp et al., 1991).

A correlated expression is observed in platelets and various cell lines between the PL/IM430-immunostained 97 kDa protein and RaplB, a Ras-related protein which is a target for cAMP-dependent kinase (Magnier et al., 1994). A controversy over the functional relationship between phosphorylation of RaplB and alleged stimulation of Ca^{2+} transport in platelet organelles still exists (see Magnier et al., 1994).

E. Targeting of the SERCA Isoforms to their Subcellular Compartments

The SERCA ATPases are synthesized on membrane-bound polysomes (Greenway and MacLennan, 1978; Reithmeier et al., 1980) and co-translationally inserted into the ER membrane without cleavage of signal sequences (Chyn et al., 1979; Mostov et al., 1981; Anderson et al., 1983). However the mechanisms for membrane insertion are not yet fully understood. *In vitro* translation experiments done in the presence of microsomes show that a truncated SERCAla variant lacking the first two membrane-spanning segments (M1-M2) is not inserted into the membrane. Putative membrane segments M8, M9 or M10, however, are not required for stable association of the ATPase with the membrane (Skerjanc et al., 1993). In striking contrast, truncated variants of the PMCA ATPase, missing the two first membrane-spanning segments, although functionally inactive, are still inserted into the membrane and delivered to the plasma membrane (Helm et al., 1992).

Another major unresolved issue concerns the mechanism underlying retention of the SERCA pumps in intracellular membranes and their targeting to the right subcellular compartment (ER in general, SR, or one of the other subcompartments of the ER). Among the P-type ion-motive ATPases, only the SERCA pumps and the yeast-type organellar Ca^{2+} pump are retained in intracellular organelles. All the others are targeted to the plasma membrane. It is thus clear that the SERCA pumps must contain sorting signals for intracellular membranes but the nature of these signals remains obscure. Structurally simple retention signals like the short linear C-terminal sequence *viz.*, -Lys-Asp-Glu-Leu (Munro and Pelham, 1987) or a related sequence (Andres et al., 1991) found in many luminal ER residents are not present in SERCA. For SERCA3, the sequence -Lys-Lys-Asp--Leu-Lys at its C terminus, has been viewed as retaining the pump in the ER (Burk et al., 1989). This is so, because it resembles the ER-targeting signal found at the C-terminus of some ER membrane proteins (Jackson et al., 1990). The exact subcellular localization of SERCA3 is however not known. In expression studies with COS cells it follows an ER distribution pattern similar to that of SERCA1 and SERCA2 which both lack this signal for protein retention (Lytton, et al., 1992). Moreover, the signal for retention of membrane proteins in the ER might be more complex than originally suggested (Gabathuler and Kvist, 1990). A pentapeptide motif Arg/Lys-Ile-Leu-Leu-Leu, found in the first transmembrane segment of the SERCAs and in the α chain of the T cell antigen receptor may as suggested act as another putative ER-retention signal (Magyar and Váradi, 1990). Zones in the SERCA proteins carrying ER-targeting sequences could in principle be traced by constructing chimeras between SERCA and one of the ATPases normally targeted to the PM (for example, the Na^+-K^+-ATPase, the gastric H^+-K^+ ATPase or PMCA), but an overview of the many chimeras which have been made so far has failed to shed some light on the problem of ER retention of SERCA. One complicating factor in studies with recombinant proteins is that the ER harbors a chaperone system checking the structural integrity of newly made membrane proteins and retaining or even degrading those that have not yet reached their correct mature tertiary or quaternary structure (see Klausner and Sitia, 1990; Bonifacino and Lippincott-Schwartz, 1991). In the case of Na^+-K^+-ATPase and gastric H^+-K^+-ATPase (but not PMCA), the complexation of the main α subunit with an independent heterologous β subunit soon after their synthesis is required for exit of the holoenzyme from the ER and its correct insertion into the plasma membrane. A 26 aminoacyl long stretch in the extracytosolic loop connecting putative membrane segments 7 and 8 in the α subunit of the Na^+-K^+-ATPase mediates the interaction with the β subunit (see Fambrough et al., 1994).

SERCAla is a typical major component of the SR of fast-twitch skeletal muscle which accounts for more than 70% of the membrane protein. It is synthesized in the ER and then moves to the SR possibly through membrane continuities between both organelles. Immunostaining in chick skeletal-muscle myotubes shows its presence in the SR and the nuclear envelope, which belongs to the ER (Kaprielian and Fambrough, 1987; Karin and Settle, 1992). The fact that ER staining persists after cycloheximide treatment indicates that the SERCA1 found in the ER is a permanent resident of this compartment and not merely in transit to the SR.

Finally it is interesting to note that the Ca^{2+} stores may undergo spatial reorganization during cell activation. For example, immunostaining of SERCA2 pumps in neutrophils shows that during phagocytosis the stores congregate in a ring-like structure around the phagosomes. Since actin also shows a similar redistribution, the actin cytoskeleton might participate in this process (Stendahl et al., 1994).

IV. OTHER ORGANELLAR Ca^{2+}-TRANSPORT ATPASES

A. A New Class of Ca^{2+} Pumps in Mammals

Besides the SERCA and PMCA classes of mammalian P-type Ca^{2+} pumps, an enzyme of what could be a third class of Ca^{2+}-transport ATPases has been described by Gunteski-Hamblin et al. (1992). These workers cloned a cDNA from rat stomach encoding a 919-amino acid P-type ion-motive ATPase (M_r 100500) exhibiting 23% amino-acid identity to PMCAla, 33% identity to SERCA2a and, remarkably, a 50% identity to PMR1, the enzyme in yeast which is considered to function as a secretory pathway Ca^{2+}-transport ATPase (Rudolph et al., 1989; Halachmi and Eilam, 1996; *cfr. infra*). The PMR1 pump is most likely a Golgi associated Ca^{2+} pump (Antebi and Fink, 1992, Sorin et al., 1997). The identity between this novel ATPase and the yeast enzyme exceeds 80% within the predicted transmembrane domains comprising the six Ca^{2+}-binding residues in the SR Ca^{2+}-transport ATPase (Clarke et al., 1989). All six of these residues are identical in both ATPases, but only 3 out of the 6 residues have been conserved between these pumps and the SERCAs, a property which sets both classes of pumps apart. The residues which are not conserved are most likely forming part of the Ca^{2+}-binding site 1 which is also absent from the PMCA family (MacLennan et al., 1997). Therefore these pumps might transport only a single Ca^{2+} per catalytic cycle. These pumps are phylogenetically most closely related to the products of the PEA1 gene from *Arabidopsis* (Huang et al., 1993), the BCA1 from *Brassica* (Malmstrom et al., 1997) and the patA from *Dictyostelium* (Moniakis et al., 1995). Northern-blot analysis shows an almost uniform expression of two mRNAs (3.9 and 5 kb) of the PMR1 homologue in all rat tissues, suggesting that the encoded enzyme performs a housekeeping function. Rat testis expresses a splice variant of this pump, in which a short peptide Phe-Tyr-Pro-Lys-Ile replaces the last residue (Val 919) of the putative housekeeping isoform.

B. Expression of SERCA-Related Genes in Non-Vertebrate Taxa

SERCA-type ATPases presenting about 70% amino-acid identity with the mammalian SERCA enzymes have been cloned and sequenced in invertebrates, including *Drosophila melanogaster* (Magyar and Váradi, 1990), the crustacean *Artemia franciscana* (Palmero and Sastre, 1989) and the parasitic flatworm *Schistosoma mansoni* (de Mendonça et al., 1995). The six residues in the transmembrane segments M4-M8 forming the Ca^{2+}-binding sites in mammalian SERCAs are conserved in these invertebrate enzymes. The *Drosophila* gene encoding a 1002 amino-acid peptide maps at the end of the right arm of chromosome 2, in band 60A. Developmental control of expression of the gene is observed (Váradi et al., 1989). In contrast to vertebrates, there is no evidence for a multi SERCA gene family in *Drosophila* or in *Artemia*, but for *Artemia*, a retropseudogene is known (Escalante and Sastre, 1994b). In *Schistosoma*, fragments have been sequenced of what could be another SERCA-type of ATPase and of a putative homologue of the yeast secretory pathway Ca^{2+}-transporting ATPase (de Mendonça et al., 1995). SERCA homologues have been cloned from higher plants. Partial cDNA sequences are obtained from tobacco (*Nicotiana tabacum* - clone pH27) (Perez-Pet et al., 1992) and tomato (*Lycopersicon* - clone LCA1). In the latter case, a complete genomic clone (gLCA13) is described (Wimmers et al., 1992). The ECA1 gene product from *Arabidopsis* (Liang et al., 1997; see also Table 1) complements yeast mutants defective in Ca^{2+} pumps and appears mainly associated with the endoplasmic reticulum. These plant homologues of SERCA share about 50% amino-acid identity with the mammalian SERCAs, and they too conserve each of the six amino-acid residues constituting the Ca^{2+}-binding sites. Interestingly, the mRNA levels vary in the different plant tissues and the presence of multiple transcripts suggests the possibility of differential mRNA splicing. Expression of both the tobacco and the tomato genes is induced in NaCl-adapted cells. Remarkably specific polyclonal antibodies raised against a fusion protein encoding a portion of the LCA polypeptide reacted with two polypeptides of 116 and 120 kDa, which in tomato root cells, are localized in the vacuolar (tonoplast) and plasma membranes, respectively (Ferrol and Bennett, 1996). The LCA pump therefore seems not to be targeted to the endoplasmic reticulum. The antibodies partially inhibited the corresponding ATPase activity. Genomic blot analysis failed to demonstrate another related Ca^{2+}-transport ATPase gene in the tomato. The tomato gene stretches over a length of some 6 kilo basepairs and shows 7 exons and 6 short introns. The gene structure in plants is therefore much simpler than in animals where rabbit SERCA1 counts 22 exons (Korczak et al., 1988) and *Artemia* SERCA counts 17 (Escalante and Sastre, 1994a), but more complicated than in the protists where the Ca^{2+} pump of *Plasmodium yoelii* shows only 3 exons (Murakami et al., 1990) and where that of *Trypanosoma brucei* has no introns (Revelard and Pays, 1991).

SERCA-like ATPases have been found in protista. In general they show close to 50% amino-acid identity with the mammalian SERCAs. Analysis of a Ca^{2+}-transport ATPase *(tba1)* genomic DNA from *Trypanosoma brucei* predicts a 110 kDa protein showing 49% sequence conservation compared with the rabbit SERCA2a (Revelard and Pays, 1991). Overexpression of this gene in transfected procyclic forms of *T. brucei* indicated that it encodes a Ca^{2+}-transport ATPase which is recovered in the microsomes and inhibited by thapsigargin (Nolan et al., 1994).

Several presumably organellar-type Ca^{2+}-transport ATPase genes have been cloned from malaria parasites: both from *Plasmodium yoelii,* a rodent malaria species (Murakami et al., 1990) and from *Plasmodium falciparum* (Kimura et al., 1993; Trottein et al., 1985; Dyer et al., 1996). The *Plasmodiium* proteins show 46-63% amino-acid identity with the vertebrate SERCA ATPases in the conserved domains, but the most remarkable difference between vertebrate and some of the malarial ATPases is the presence in the latter of several long peptide inserts which render these ATPases 100 to 200 amino acids longer (Kimura et al., 1993). The longest of these inserts is found immediately downstream of the autophosphorylation site and replaces the domains which in SERCA1 and SERCA2 function as interaction sites for phospholamban. Pasmodium ATPases might have their own regulatory system. PfATPase4 was first described by Trottein et al. (1995) but later Dyer et al. (1996) described a sequence which diverged from the former in that the last 5 amino acids of the translated sequence were replaced by 130 amino acids. They suggested that the Trottein sequence was a cloning artefact. The PfATPase 4 gives rise to an 8kb mRNA shortly after erythrocyte invasion. In the later stages this mRNA disappears, but the protein of 190 kDa persists.

C. Non-SERCA Organellar Ca^{2+} Pumps in Fungi and Protista

Putative organellar Ca^{2+}-transport ATPases are also cloned from yeast (see also Catty and Goffeeau, 1996 for a recent review). All of them seem to belong to classes separate from SERCA. *PMR1* in *Saccharomyces cerevisiae* has already been mentioned. It encodes a P-type ion-motive ATPase which is thought to act as a Ca^{2+}-transport ATPase controlling the secretory pathway (Rudolph et al., 1989). It is targeted to a Golgi-like compartment (Antebi and Fink, 1992). A null mutation of the gene displays Ca^{2+}-related growth defects, incomplete outer-chain glycosylation of invertase and mislocalization of proteins in the secretory pathway. The PMR1 pump, however, seems to differ from another recently described Ca^{2+}-transport ATPase in the yeast Golgi membranes (Okorokov et al., 1993). A related gene *PMR2, is* most likely not involved in Ca^{2+} transport and might encode an Na^{+} pump. In *Schizosaccharomyces pombe* the *cta3* gene (Ghislain et al., 1990) appears to encode an intracellular Ca^{2+} pump and shows 45% identity with the *S. cerevisiae* PMR2 (Halachmi et al., 1992).

In yeast large amounts of Ca^{2+} are accumulated in the vacuole. This Ca^{2+} accumulation is partially catalyzed by a P-type ion transport ATPase encoded by the PMC 1 gene, an homologue of the mammalian PMCA rather than SERCA (Cunningham and Fink, 1994) and partially by a low affinity H^{+}/Ca^{2+} exchanger which depends for its function on a vacuolar P-type H^{+}-ATPase. A similar secondary Ca^{2+}-accumulation system is found in the slime mold *Dictyostelium* (Rooney et al., 1994).

The suggestion that both the malarial and the yeast ATPase function as Ca^{2+}-transporters is attractive but needs to be confirmed. Neither ATPases contain all six amino acid residues involved in high-affinity Ca^{2+} binding.

V. ALTERNATIVE TRANSCRIPT PROCESSING IN SERCA2

Northern blots and S1 nuclease assays have revealed four different SERCA2 mRNAs (de la Bastie et al., 1988; Eggermont et al., 1990b; Plessers et al., 1991) which differ in their 3'end and are tissue specific (Figure 1). The class 1 mRNA is generated in cardiac muscle, slow-twitch skeletal muscle and smooth muscle by a splicing process, removing an optional intron at the 3'end of the SERCA2 messenger. Polyadenylation at a site (Pa_u) that lies in between the muscle-specific 5' donor (5'D1) and 3' acceptor (3'A) site generates the non-muscle class 2 mRNA. The class 4 mRNA is brain specific and formed after splicing of a 5' donor splice site (5'D2) downstream of the muscle-specific one (5'D1) to the same acceptor (3'A) as that used in muscle cells. Class 1 encodes the muscle-specific SERCA2a isoform, while classes 2, 3, and 4 all encode the nonmuscle SERCA2b isoform (Figure 1).

In cultured pig aorta smooth-muscle cells, platelet derived growth factor (PDGF) is able to induce the formation of class 1 messengers at the cost of class 2/3. The na-

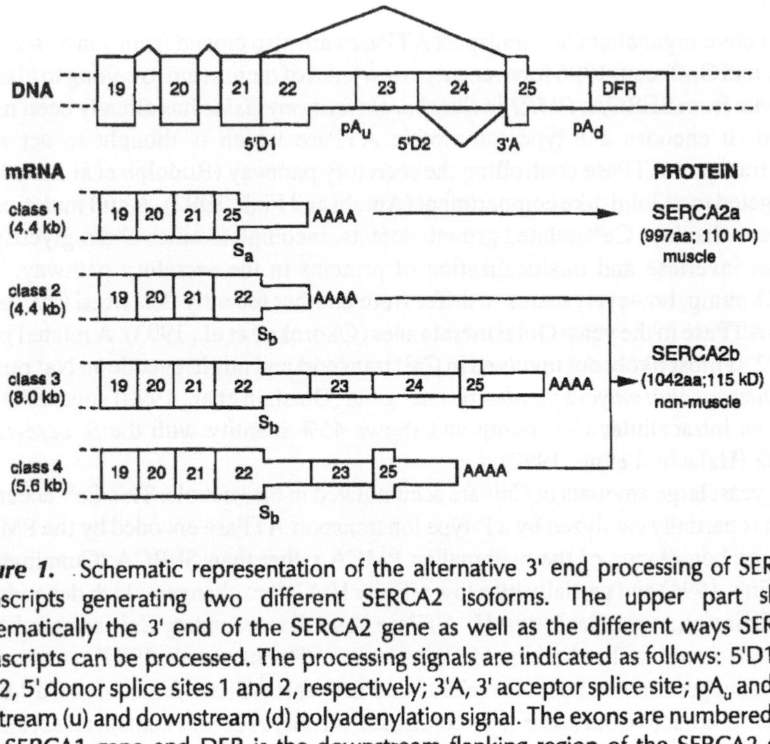

Figure 1. Schematic representation of the alternative 3' end processing of SERCA2 transcripts generating two different SERCA2 isoforms. The upper part shows schematically the 3' end of the SERCA2 gene as well as the different ways SERCA2 transcripts can be processed. The processing signals are indicated as follows: 5'D1 and 5'D2, 5' donor splice sites 1 and 2, respectively; 3'A, 3' acceptor splice site; pA_u and pA_d, upstream (u) and downstream (d) polyadenylation signal. The exons are numbered as in the SERCA1 gene and DFR is the downstream flanking region of the SERCA2 gene. Coding and noncoding sequences are represented by wide and narrow boxes respectively. Beneath the genomic structure, the 3' ends as well as the length of the four different SERCA2 messengers are shown. Exons 19-21 are constitutive exons present in the four classes of mRNA, while exons 22-25 are optionally included or excluded. The class 1 messenger is generated by splicing the 5'D1 to the 3'A. Class 2 is formed by polyadenylation at the pA_u site, while class 3 is polyadenylated at the pA_d site without any further modification. In the class 4 messenger the 5'D2 is spliced to the 3'A. The position of the stop codon used is indicated: S_a in the class 1 messenger and S_b in the three other messengers. The class 1 messenger encodes the muscle-specific SERCA2a isoform, while the three other messengers encode the nonmuscle SERCA2b isoform.

ture of this mechanism remains unknown (Magnier et al., 1992). Other studies use the mouse tumor cell line BC_3H1 as a model system in order to investigate the regulation of SERCA2 RNA processing. The BC_3H1 cell line is widely used as a model for studying muscle-cell differentiation. When these cells are exposed to a mitogen-deficient medium, they develop a large variety of muscle-specific proteins but do not fuse (Taubmann et al., 1989). De Smedt et al. (1991) report that during myogenic differentiation of BC_3H1 cells the muscle-specific isoform is expressed. Northern blots (De Smedt et al., 1991) as well as PCR analysis (De Jaegere et al., 1993) confirm that class 1 mRNA is formed during myogenic differentiation. Furthermore, the muscle-specific processing responsible for class 1 mRNA formation in BC_3H1 cells is dependent on ongoing protein synthesis and is tightly coupled to the myogenic differentiation of the BC_3H1 cells (De Jaegere et al., 1993).

In order to define the *cis*-active elements involved in differential splicing, a mini-gene was constructed (pCMβSERCA2) containing the 3' end of the SERCA2 gene (Van Den Bosch et al., 1994). When stably transfected into BC_3H1 cells, the mini-gene transcripts are differentially processed depending on the differentiation state of the cells. Myogenin, a transcription factor of the helix-loop-helix family, can induce muscle-specific splicing of pCMβSERCA2 transcripts in a fibroblast cell line (10T1/2) after exposing these cells to a mitogen-deficient medium. These results are taken to mean that the *trans*-acting factor(s) responsible for muscle-specific SERCA2 processing can be induced by one of the master regulatory genes of muscle differentiation and that growth factors antagonize this induction (Van Den Bosch et al., 1994).

Activation of an otherwise inefficient splicing process seems to lead to the formation of the class 1 messenger in muscle cells. This is concluded from the fact that the muscle-type processing of pCMβSERCA2 transcripts can be induced in non-muscle cells (10T1/2) by optimizing the muscle-specific donor splice site. A splice induction in 10T1/2 cells can also be obtained by shortening the very long muscle-specific intron. This effect is a result of altered spacing between the 5'D1 and 3'A splice sites, as replacement of the deleted intron fragment by an unrelated DNA sequence of the same length prevents splicing in 10T1/2 cells (Van Den Bosch et al., 1994). Moreover, in this fibroblast cell line an inverse relation is observed between the length of the intron and the efficiency of splicing (Mertens et al., 1995).

The presence of weak processing sites is a prerequisite for regulated splicing of SERCA2 mRNA in BC_3H1 cells (Mertens et al., 1995). The 5' donor splice site (5'D1) is weak as it deviates from the consensus sequence in three positions. Optimizing the 5'D1 splice site can induce muscle-specific splicing in undifferentiated myogenic cells. However, there is still a significant upregulation during differentiation. The neuronal 5' donor splice (5'D2) only deviates from the consensus at one position (-1). Nevertheless, it too seems to be a weak site as neuronal splicing is induced in BC_3H1 cells by converting the neuronal 5' donor into the consensus sequence. Also the polyadenylation site present in the muscle-specific intron (pA_u)

has to be weak. Converting this site into a strong polyadenylation site almost completely prevents muscle-specific splicing.

Decreasing the spacing of the splice donor and acceptor sites and increasing the strength of the 5' donor splice site leads to constitutive splicing and hence prevents further stimulation of splicing during myogenic differentiation. This observation indicates that once a certain efficiency of splicing is reached, the *trans*-acting factors responsible for the stimulation of splicing are no longer effective (Mertens et al., 1995). These *trans*-acting factors are most likely responsible for potentiating the use of the non-consensus 5' donor splice site and/or for overcoming the inhibitory effect imposed by the long terminal intron.

VI. REGULATION OF THE ACTIVITY OF SERCA

A. Regulation by Phosphorylation of the Regulatory Protein Phospholamban

In slow-twitch skeletal muscle, cardiac muscle and smooth-muscle tissues (Fuji et al., 1991), SERCA2 activity is modulated by phosphorylation of the regulatory protein phospholamban (PLB). Primary sequence analysis of PLB reveals the presence of an N-terminal hydrophilic domain (domain I, residues 1-30), protruding in the cytosol, and a C-terminal hydrophobic domain (domain II, residues 31-52), which would be mainly α-helical (Simmerman et al., 1986; Simmerman et al., 1989) and which anchors the protein in the SR membrane. The suggestion has been made that the first 20 residues of domain I form an amphipathic α-helix (Simmerman et al., 1986; Fujii et al., 1987; Simmerman et al., 1989) with a net charge of +2 (2 Glu^-, 3 Arg^+, 1 Lys^+), containing the two phosphorylation sites Ser 16 and Thr 17. In cardiac muscle, *in vivo* phosphorylation of PLB by cAMP- or Ca^{2+}/Calmodulin(CaM)-dependent protein kinase exerts a stimulatory effect on the Ca^{2+}-transport rates at Ca^{2+} concentrations below saturation (Le Peuch et al., 1979; Tada et al., 1979; Davis et al., 1983; Wegener et al., 1989). This stimulatory effect is the result of an enhanced turnover rate. Pre-steady state measurements had shown that phosphorylation of PLB accelerates the rate limiting steps of the catalytic cycle (E2 $\rightarrow$ E1 and ElP $\rightarrow$ E2P, corresponding to the major conformational transition steps), resulting in an increase of both Ca^{2+} affinity and turnover rate of EP (Tada and Katz, 1982; Tada et al., 1988). However, more recent experiments only confirm the effect of PLB phosphorylation on Ca^{2+} affinity. Indeed, the effect of PLB phosphorylation on Ca^{2+}-ATPase and uptake activity can be mimicked by the addition of antibodies against the cytoplasmic domain of PLB (Morris et al., 1991; Cantilina et al., 1993). Antibody binding to the cytoplasmic PLB domain can specifically reduce the activation energy for the slow transition, activated by Ca^{2+}, whereas the turnover of the phosphoenzyme intermediate is not affected. In order to determine the exact role of PLB *in vivo*, Luo et al. (1994) have generated PLB-deficient mice. Ca^{2+}-uptake rates by homogenates of the PLB-deficient hearts reveal a marked effect on Ca^{2+} affinity, as compared with the wild-type hearts, but no effect on V_{max}. The ablation of the PLB gene is associated with increased myocardial contractility, and a loss of the positive inotropic response to β-adrenergic stimulation.

Thus far, the precise molecular mechanism underlying the pump's modulation by PLB is not apparent. In addition to antibody binding, PLB phosphorylation can also be mimicked by tryptic proteolysis of PLB (Kirchberger et al., 1986), suggesting that unphosphorylated PLB inhibits the pump via close interaction between its cytoplasmic domain and the Ca^{2+} pump. In addition, an hydrophobic interaction between the membrane-spanning domains of both proteins occurs. It is actually this interaction that appears to exert the inhibitory effect on the pump (Kimura et al., 1996). Phosphorylation of PLB alters its electrostatic and hydrophobic properties. This affects the interaction of PLB with the pump and in turn the pump's activity. Indeed, cross-linking experiments have shown direct interaction between unphosphorylated PLB and SERCA2a in the absence of Ca^{2+} (James et al., 1989). Cross-linking was inhibited by phosphorylation of PLB. That is to say, the mere addition of a negative phosphoryl group to the cytoplasmic domain of PLB induces a conformation change, thereby relieving its interaction with the pump. This inhibitory effect of PLB has been confirmed in COS-1 cells by its co-expression with SERCA2a (Fuji et al., 1990).

In SDS-polyacrylamide gel electrophoresis (SDS-PAGE), PLB migrates as a pentamer ($M_r \approx 25$ kDa), suggesting a strong tendency of the protein to form homopentameric complexes (Kirchberger and Antonetz, 1982). However, the pentamer is easily decomposed; when boiled prior to gel electrophoresis, PLB migrates as a monomer ($M_r \approx 5$ kDa). The sites on PLB that are important for functional interaction with the pump have been identified by co-expression of the pump with a number of PLB mutants (Toyofuku et al., 1994a; Kimura et al., 1996). The most essential residues appear to be localized in the cytoplasmic domain. Functional association seems to occur through a balance of both electrostatic and hydrophobic interactions. In other words, each of the 5 charged residues (Lys3, Arg9, 13 & 14 and Glu2) in the N-terminal domain, as well as the hydrophobic residues are critical for interaction. In addition, the residues that accept phosphoryl groups play a functional role, even when unphosphorylated. Surprisingly, modification of Pro21, which may be involved in the conformational change of the cytoplasmic domain in response to phosphorylation (Simmerman et al., 1986), fails to diminish the inhibitory effect of PLB on Ca^{2+} transport (Toyofuku et al., 1994a). However, Simmerman et al. have not mentioned whether the observed inhibition is also relieved by phosphorylation. Pentamer formation during SDS-PAGE is abolished by site-directed mutagenesis of the Cys residues (Cys36, Cys41 and Cys46) in the transmembrane segment (Fuji et al., 1989). More recently Simmerman et al. (1996) have proposed that closely packed leucines and isoleucines in the transmembrane domain form a leucine zipper motif that is essential for oligomeric assembly (Simmerman et al., 1996).Upon co-expression of monomeric mutants with SERCA2a in COS-1 cells, inhibition of the Ca^{2+} pump is retained as well (Toyofuku et al., 1994a; Kimura et al., 1997), supporting the idea that the functional unit consists of a monomer, and that pentameric phospholamban represents a less active or inactive reservoir that dissociates to provide inhibitory monomers.

The domain of the SERCA2 ATPase that interacts with PLB is localized by means of cross linking to a stretch immediately C-terminal of the phosphorylation site neighboring Lys397 and Lys400 (James et al., 1989). This putative PLB-binding domain is also found in SERCA1 but not in SERCA3, nor in any other P-type ion pump. Indeed, when co-expressed in COS-1 cells, SERCA1, but not SERCA3, is also inhibited by PLB (Toyofuku et al., 1993). Remarkably this indicates that SERCA1 has the potential to be regulated by PLB in spite of the fact that both proteins are never co-expressed *in vivo*. Co-expression of a number of chimerical SERCA2/SERCA3 constructs with PLB has identified the nucleotide-binding/hinge domain as a second region required for functional interaction with PLB. Additionally, co-expression of a number of SERCA2a mutants with PLB demonstrated that the peptide Lys-Asp-Asp-Lys-Pro-Val402 comprises one of the sites in SERCA2 which associate with PLB, and that both hydrophobicity and charge are important for functional interaction (Toyofuku et al., 1994b).

B. Regulation by Direct Phosphorylation of the Ca^{2+} Pump

Endogenous membrane-bound Ca^{2+}/CaM kinase II can also directly phosphorylate the SR Ca^{2+}-transport ATPase in the heart and in slow-twitch skeletal muscle. Xu et al. (1993) reported that phosphorylation resulted in 2-fold higher maximal velocity without a change in the $K_{0.5}$ value for Ca^{2+} (Xu et al., 1993). However more recent experiments failed to find any evidence for a functional effect of Ca^{2+}/calmodulin-dependent phosphorylation on the activity of the Ca^{2+} pump (Odermatt et al., 1996; Reddy et al., 1996). Site-directed mutagenesis of all three serine residues located within a minimal consensus site for CaM-kinase II phosphorylation (Ser38, Ser167, Ser531), has identified Ser38 as the actual phosphorylation site for CaM kinase II (Toyofuku et al., 1994c). In addition, CaM kinase phosphorylation appears to be specific for the SERCA2 Ca^{2+}-transport ATPases (Hawkins et al., 1994).

VII. CONTROL OF SERCA GENE EXPRESSION

A. Influence of Innervation and Muscle Activity on SERCA Expression in Skeletal Muscle

Expression of SERCA isoforms in skeletal muscle is a tightly regulated process, both during and after muscle development. The predominant isoform in fetal and early neonatal skeletal muscle is the SERCA2 pump. After birth, SERCA1 expression in fast-twitch muscle is upregulated while SERCA2 expression is downregulated to a point where the SERCA2 isoform is nearly completely replaced by SERCA1 in adult fast-twitch muscle. In contrast, adult slow-twitch muscle almost exclusively expresses SERCA2 (Brandl et al., 1987; Arai et al., 1992). Importantly, the expression pattern of SERCA pumps in skeletal muscle is modulated quantitatively (up or downregulation) and qualitatively (isoform switches) by the innervation state and muscle activity state (Leberer et al., 1986; Salvatori et al., 1988; Leberer et al., 1989; Briggs et al., 1990; Kaprielian et al., 1991; Schulte et al., 1993; Schulte et al., 1994).

Intact muscle innervation is required for normal SERCA levels to occur in skeletal muscle. For example, denervation of fast-twitch muscles of rats or rabbits leads to a decrease in SERCA1 mRNA and protein, but is without effect on SERCA2 expression. Similarly, SERCA2 mRNA and protein levels are reduced in denervated slow-twitch muscle of rat and rabbits but there are no changes in SERCA1 expression (Salvatori et al., 1988; Schulte et al., 1994). In contrast, a qualitative change in SERCA-expression pattern is found in chicken where both SERCA1 and SERCA2 Ca^{2+} pumps are present in denervated fast-twitch muscle (Kaprielian et al., 1991). In chicken and rabbit, the denervation-induced changes in SERCA expression are fully reversible. The isoform pattern is also normalized after nerve regeneration and muscle re-innervation (Salvatori et al., 1988; Kaprielian et al., 1991). Muscle innervation is also important for postnatal upregulation of SERCA expression in both slow- and fast-twitch muscles. Thus, for example, SERCA1 and SERCA2 levels in newborn-rabbit muscle gradually increase in the two weeks after birth, whereas in denervated muscles this increase does not occur (Leberer et al., 1986).

Chronic nerve stimulation of fast-twitch muscle results in an isoform switch with a gradual decrease in SERCA1 and a progressive increase in SERCA2 such that after prolonged stimulation the SERCA2 isoform becomes the only detectable Ca^{2+} pump in the originally fast-twitch muscle (Leberer et al., 1989; Briggs et al., 1990). Simultaneously with the SERCA isoform switch, phospholamban mRNAs is induced and reaches levels typical of slow-twitch muscle (Leberer et al., 1989). Similarly, chronic mechanical overload of the rat fast-twitch plantaris muscle by surgical ablation of the soleus and gastrocnemius muscles leads to a 2-fold rise in SERCA2 expression (mRNA and protein) and a small decline in SERCA1 level (Kandarian et al., 1994). Therefore, chronic muscle load either due to rhythmic nerve stimulation or increased mechanical load induces a partial to a complete SERCA1 to SERCA2 switch in fast-twitch muscles. Opposite changes in SERCA expression are observed when the muscle load is artificially reduced. When rats are prevented from using their hindlimbs (hindlimb upweighing with intact innervation), SERCA1 levels (mRNA and protein) in the slow-twitch soleus muscle gradually rise with time whereas SERCA2 protein levels decline (Schulte et al., 1993). At the functional level the isoform switch is accompanied by a greater Ca^{2+}-ATPase activity of the sarcoplasmic reticulum and a shorter relaxation time (Schulte et al., 1993).

The mechanisms by which innervation and muscle activity modulate SERCA expression in skeletal muscle are still unknown, but their effect is largely exerted at the transcriptional level since mRNA and protein levels somewhat vary in parallel, but translational regulation also exists (Hu et al., 1995). What is not clear is how changes in neural stimulation influence SERCA expression. One possibility is that the observed changes in SERCA expression after denervation or chronic nerve stimulation are the mere consequence of changed muscle activity *per se.* Alternatively, the increased or decreased release of neurotransmitters and/or trophic factors at the neuromuscular synapse may directly trigger changes in SERCA expression independently of changes in muscle activity.

B. Influence of Thyroid Hormone on SERCA Expression in Skeletal and Cardiac Muscle

The effect of thyroid hormone on SERCA expression in skeletal muscle is now well established (Simonides et al., 1990; Sayen et al., 1992; van der Linden et al., 1992; Muller et al., 1994). In fact, administration of thyroid hormone (T3 or T4) to euthyroid or hypothyroid animals strongly augments SERCA levels in skeletal muscle. In fast-twitch muscle the SERCA1 levels highly depend on the thyroid state of the animal. For example, SERCA1 mRNAs in gastrocnemius muscle of euthyroid rats rise 5-fold between day 5 and 20 after birth. This increase in SERCA1 is not observed in newborn hypothyroid animals, unless T3 hormone is given to the hypothyroid newborn rats (van der Linden et al., 1992). In older rats, SERCA1 expression in fast-twitch muscle remains sensitive to thyroid hormone as SERCA1 is further upregulated in hyperthyroid animals, but downregulated in hypothyroid animals (Simonides et al., 1990; Sayen et al., 1992). A striking feature is that thyroid hormone exerts the opposite effect on SERCA2 expression in fast-twitch muscle; that is, the low degree of SERCA2 expression in fast-twitch muscle is further decreased in hyperthyroid animals, but upregulated in hypothyroid animals (Sayen et al., 1992; van der Linden et al., 1992).

In slow-twitch skeletal muscle the situation is more complex. Biochemical assessment of SERCA expression (that is, determination of mRNA or protein levels in muscle homogenates) in slow-twitch muscle reveals a prominent induction of SERCA1 and either a moderate increase or a slight decrease in SERCA2 upon thyroid hormone administration to hypothyroid or euthyroid animals (Simonides et al., 1990; Sayen et al., 1992; Muller et al., 1994). Immunocytochemical analysis with SERCA1- or SERCA2-specific antibodies shows that every individual muscle fiber in soleus muscle becomes SERCA1 positive after thyroid hormone treatment, whereas a significant fraction (30%) of the slow fibers becomes SERCA2 negative (Muller et al., 1994). Thus, thyroid hormone seems to have a dual effect on SERCA2 expression in slow-twitch muscle: one is a conversion of SERCA2 positive fibers into SERCA1 positive fibers leading to a reduction of the number of SERCA2 positive fibers; the other is an increase in SERCA2 expression in the remaining slow SERCA2 positive fibers.

Collectively, these data indicate that thyroid hormone exerts a strong upregulating effect on SERCA1 expression in both fast- and slow-twitch muscle. Its effect on

SERCA2 expression depends on the muscle type since in fast-twitch muscle there is downregulation of SERCA2 levels and in slow-twitch muscle thyroid hormone reduces the number of SERCA2-positive fibers. However, the SERCA2 content in the remaining individual slow-twitch fibers may increase.

In contrast to skeletal muscle, thyroid hormone has an unequivocal positive effect on SERCA2 expression (mRNA and protein) in cardiac muscle (Rohrer and Dillmann, 1988; Nagai et al., 1989; Arai et al., 1991; Balkman et al., 1992; Ojamaa et al., 1992; Sayen et al., 1992; Kiss et al., 1994). In hearts of hypothyroid animals SERCA2 levels are 2- to 3-fold lower than in control animals, whereas in hyperthyroid animals SERCA2 levels rise approximately 1.5-fold *versus* the euthyroid control (Rohrer and Dillmann, 1988; Nagai et al., 1989; Kiss et al., 1994). It is noteworthy that phospholamban mRNA levels are decreased in hyperthyroid conditions whereas in hypothyroidism there is a decrease, no change or an increase (Nagai et al., 1989; Arai et al., 1991; Kimura et al., 1994; Kiss et al., 1994). These changes have profound effects on the kinetics of Ca^{2+} uptake: there is in cardiac SR vesicles from hyperthyroid animals an increase in both the maximal Ca^{2+}-uptake rate and the affinity for Ca^{2+} (Kimura et al., 1994; Kiss et al., 1994). The simultaneous upregulation of SERCA2 (that is, higher number of Ca^{2+}-pump units) and downregulation of phospholamban (that is, smaller number of inhibitor molecules) may also explain the two phenomena observed in intact hearts of hyperthyroid animals, namely, an increased speed of diastolic relaxation due to the higher Ca^{2+}-uptake capacity of the SR and a diminished effect of β-adrenergic agonists on the diastolic relaxation rate due to the reduced phospholamban content (Beekman et al., 1989; Mintz et al., 1991; Kiss et al., 1994).

C. Promoter Analysis of SERCA Genes

Positive effects of thyroid hormone on SERCA1 and SERCA2 expression are also observed in cultured cells from skeletal or cardiac muscle (Muller et al., 1991; Rohrer et al., 1991; Zarain-Herzberg et al., 1994). These cell systems have been exploited to delineate the *cis*-acting elements in the SERCA2 promoter that confer thyroid responsiveness to the gene (Rohrer et al., 1991; Hartong et al., 1994; Zarain-Herzberg et al., 1994). A detailed analysis of the rat SERCA2 promoter has led to the identification of 3 thyroid-responsive elements (TRE) between nucleotides -485 and 190 (relative to the start of transcription). These 3 TREs differ in sequence, orientation and spacing of the half sites, that is, the binding sites for the thyroid receptor (TR) complexes. In *in vitro* binding studies the 3 TREs display different binding affinities for TR complexes depending on the subunit composition of the receptor complex (monomeric, homodimeric, or heterodimeric). However, each TRE is equally capable of conferring thyroid hormone responsiveness on a CAT reporter gene when tested in a transient transfection assay (Hartong et al., 1994). At present, the physiological significance of TRE diversity and the *in vivo* contribution of each TRE to thyroid hormone modulation are not known. What is generally accepted is that thyroid-receptor complexes influence gene expression by activating transcriptional initiation. Consistent with this mode of action, nuclear run on experiments on the SERCA1 gene have revealed an increased rate of transcription of the SERCA1 gene after thyroid hormone stimulation (Thelen et al., 1994).

Besides thyroid hormone, two other substances are known to modulate SERCA expression in cultured cells. First, the positive effect of thyroid hormone on SERCA1 expression in the L6 muscle cell line is potentiated by Insulin Growth Factor I (IGFI) suggesting crosstalk between these two factors (Muller et al., 1991; Thelen et al., 1994). Second, SERCA2 expression in primary neonatal cardiac cells is upregulated by retinoic acid, but the retinoic acid response element in the SERCA2 promoter has not yet been identified (Rohrer et al., 1991).

Finally, the SERCA2 promoter has been analyzed in an effort to identify *cis*-acting elements that mediate the upregulation of SERCA2 expression during myogenic differentiation. Indeed, endogenous SERCA2 levels rise when C2C12 myoblasts or myogenin-transfected C3H/10T1/2 fibroblasts are allowed to differentiate into "myocytes" (Zarain-Herzberg et al., 1990; Arai et al., 1992). This search has led to the identification in the rabbit SERCA2 promoter of a 17-bp element (UPE = Upstream Promoter Element) between nucleotides -267 and -284 (relative to the transcriptional start site) that confers high level expression on a CAT reporter gene in C2C12 myoblasts with a further increase after myocyte differentiation. This element however fails to activate reporter gene expression in wild-type 3T3 fibroblasts and myogenin-transfected C3H/10T1/2 fibroblasts (Sukovich et al., 1993). Furthermore, a protein complex that specifically binds to this UPE has been identified in nuclear extracts of C2C12 cells, but it is not yet clear how this factor mediates muscle-specific upregulation, as it is also detected in nuclear extracts of 3T3 fibroblasts (Sukovich et al., 1993). In addition to the positive UPE, the rabbit SERCA2 promoter contains a negative element upstream of nucleotide -658 that represses gene transcription in C2C12 muscle cells (Sukovich et al., 1993).

Cis-acting elements in the rabbit SERCA2 promoter that are responsible for high-level expression in myocardial cells have been localized. In this work gene expression was monitored not only after transient transfection in fetal cardiocytes of a luciferase reporter gene containing SERCA2-promoter fragments, but also after direct injection of the same reporter gene into adult myocardium *in vivo*. Both types of experiments led to evidence that the first 284 nucleotides upstream of the SERCA2 cap site are required for efficient expression in myocardial cells (Fisher et al., 1993).

VIII. SERCA EXPRESSION IN PATHOLOGY

Abnormal cytosolic free Ca^{2+} levels have been shown to be involved in different muscle pathologies (Morgan, 1991; Perreault et al., 1993). Since the SERCA Ca^{2+}-transport ATPases play a crucial role in regulating cytosolic Ca^{2+} levels, changes in their expression levels could result in abnormal Ca^{2+} handling, and hence, in abnormal cellular function. In this context, three different pathological models have been extensively studied: cardiac hypertrophy, heart failure and arterial hypertension (Arai et al., 1994; Lompré et al., 1994).

A. SERCA and Cardiac Hypertrophy

Cardiac hypertrophy is an adaptive response of the cardiac muscle to a hemodynamic overload which precedes most forms of heart failure. Diastolic dysfunction is one of the earliest signs of a pathological hypertrophic response. Relaxation abnormalities may also be implicated in this. Most studies on alterations in Ca^{2+} transport in cardiac hypertrophy have been performed on animal models. In these models a highly significant positive correlation is obtained between end-diastolic cytosolic free Ca^{2+} levels and diastolic relaxation abnormalities. Prolongation of relaxation correlates well with prolongation of the cytosolic free Ca^{2+} transient as measured with fluorescent dyes (Perreault et al., 1993). Changes in SERCA2 gene expression is investigated in different experimental models of the hemodynamically overloaded heart. One frequently used technique is to produce left ventricular pressure overload by aortic banding. However, conflicting results have been obtained using this technique. After one month of aortic banding (Komuro et al., 1989; de la Bastie et al., 1990), a 32% decrease in SERCA2 mRNA levels is measured, which is associated with a decrease in SR Ca^{2+} uptake, and a decrease in SERCA2-protein levels. These changes in gene expression are only observed in severe hypertrophy. Moderate hypertrophy is not associated with significant changes in SERCA2 mRNA expression (de la Bastie et al., 1990). When hemodynamic measurements are correlated with mRNA levels, no change in SERCA2 mRNA levels are noted in the case of compensated hypertrophy (Feldman et al., 1993). A decrease in SERCA2 mRNA levels can only be observed when heart failure occurs. In this study a good correlation between SERCA levels and systolic function is found. Comparable results are obtained with other models of cardiac hypertrophy induced by hemodynamic overload, that is, pulmonary artery banding (Rockman et al., 1994; Matsui et al., 1995) and volume overload (Shen et al., 1991).

The experimental models of hemodynamic overload are not necessarily comparable and often the hemodynamic effects on systolic and diastolic function are not even taken into account. Moreover, in some gene-expression studies, it is not explicitly known whether hypertrophied or failing hearts are included. Because SERCA2 expression correlates with hemodynamic function, hemodynamical evaluations should essentially complement the biochemical and molecular biology data.

B. SERCA Expression in Cardiac Failure

Changes in SERCA2 expression accompanying the progression from compensated hypertrophy to end-stage heart failure have been the subject of intensive experimental investigation both in animal models and humans. In humans, most studies report a decrease in SERCA2 mRNA levels, a decrease in SERCA2 protein levels and decreased Ca^{2+} uptake and Ca^{2+}-dependent ATPase activity in failing heart, as compared with nonfailing controls (Mercadier et al., 1990; Takahashi et al., 1992a; Arai et al., 1993; Hasenfuss et al., 1994). Decreased levels of SERCA2 expression are associated with decreased expression of phospholamban (Arai et al., 1993), cardiac ryanodine receptor (RYR2) (Brillantes et al., 1992; Vatner et al., 1994; Go et al., 1995) and dihydropyridine receptor (Takahashi et al., 1992b). While there are no changes in expression of the calsequestrin gene (Takahashi et al., 1992b), the expression levels of the Na^+/Ca^{2+} exchanger (Studer et al., 1994), the IP_3-receptor (Go et al., 1995) and ANF (Takahashi et al., 1992a) are increased. However, this decrease in SERCA2 expression level and in Ca^{2+} uptake in end-stage heart failure was not observed by Movsesian et al. (1992; 1994). On the whole, the data obtained by different independent groups suggest a decrease in SERCA2-expression levels in human end-stage heart failure. Evidence that the protein levels of SERCA2 are closely related to systolic contractile reserve as determined by frequency potentiation of contractile force, suggests that the level of SERCA2 expression determines the loading of intracellular Ca^{2+} stores, and hence determines the amount of Ca^{2+} available for contraction.

The human heart failure data are confirmed in different animal models. Two strains of cardiomyopathic Syrian hamster were used: the hypertrophic strain and the dilated strain. In the hypertrophic strain, SERCA2 expression level declines before any signs of disease are observed (Kuo et al., 1992). In the dilated strain, no expression studies were performed, but Ca^{2+}-uptake experiments suggest a decrease in expression with increasing age (Whitmer et al., 1988). It therefore seems likely that an abnormal Ca^{2+}-transport function can contribute to the diastolic Ca^{2+} overload and the decreased Ca^{2+} release during systole, both of which are characteristic of cardiac failure. Another frequently used model is heart failure induced by chronic rapid ventricular pacing. This approach allows the study of the relationship between mechanical properties and Ca^{2+} handling. The Ca^{2+} transient in these chronically paced animals shows an increased time-to-peak $[Ca^{2+}]_i$ during stimulation and a prolonged decline in free cytosolic Ca^{2+} levels (Perreault et al., 1992). This was reported to be associated with decreased SERCA activity (Cory et al., 1994; O'Brien et al., 1995). However, SERCA2 mRNA levels in the endomyocardium were not statistically different in the chronically stimulated animals (Williams et al., 1994). By contrast, a significant decrease in phospholamban mRNA levels is found in these conditions. This is quite remarkable since a strongly positive correlation exists between SERCA2 and phospholamban levels in the human heart-failure studies (Arai et al., 1993). Possibly various pathogenic mechanisms are involved in the different models of heart failure.

C. SERCA Expression in Arterial Hypertension

Arterial hypertension is a third condition in which the role of Ca^{2+}-pump expression levels has been extensively studied. Blood vessels from hypertensive animals are characterized by an increased wall thickness and altered contractile properties. The increased vascular tone in hypertensive animals is associated with an increased concentration of free cytosolic Ca^{2+} in the vascular smooth muscle cells. Several lines of evidence indicate that diminished Ca^{2+}-pump activity might contribute to the elevation in cytoplasmic Ca^{2+} levels in hypertension (Kwan, 1985). However, data are also available (Levitsky et al., 1993) which show increased Ca^{2+}-uptake in homogenates obtained from spontaneous hypertensive rats (SHR), as compared with homogenates from normal rats. Furthermore, the increase in SR function is accompanied by an increased expression of the SERCA2 gene without any alterations in the SERCA2 isoform expression pattern. The fact that the method used for preparation of the membranes was different may explain the divergent results obtained by this group. However, their findings are consistent and suggest that the increased Ca^{2+} levels in hypertensive vascular smooth muscle cells, cannot simply be explained by reduced SR Ca^{2+}-uptake activity. Alternatively, increased Ca^{2+}-store loading may be the result of increased Ca^{2+}-pump activity, which together with an increased sensitivity to circulating hormones, could explain the increased vascular tone. Further study of intracellular Ca^{2+} handling in hypertensive animals is required if these contradictory results are to be resolved.

D. SERCA2 Expression in Other Cardiac Disorders

The expression of Ca^{2+}-transport ATPases has been studied in other disorders. Decreased Ca^{2+}-uptake activity in the SR might well be involved in reversible contractile dysfunction which occurs in cardiac muscle after a short period of ischemia (myocardial stunning). However, several lines of evidence indicate an increase in SERCA activity after reperfusion (Kaplan et al., 1992; Lamers et al., 1993) and that this is associated with increased transcription of the SERCA2 gene (Frass et al., 1993; Knöll et al., 1994).

E. SERCA Expression in Skeletal Muscle Pathology

Little is known of the possible involvement of an abnormal Ca^{2+}-transport ATPase in the pathogenesis of skeletal muscle disorders. Only in Brody's disease, a rare inherited disorder of skeletal-muscle function presenting as pseudomyotonia, is a deficiency in the Ca^{2+}-transport ATPase activity found (Benders et al., 1994). Some patients with the disorder show normal levels of the SERCA1 protein in the *musculus quadriceps* samples, but the SERCA activity is decreased by about 50%. In contrast, Danon et al. (1988) could not immunologically detect SERCA1 protein in the fast-twitch fibers of Brody patients, implying that the ATPase is either absent from the fibers or in a form with altered antigenicity. The actual nature of the defect leading to this impaired muscle function remains therefore unknown. No mutations were detected in the coding or splice junction regions of the SERCA1 gene of three Brody patients, thus eliminating SERCA1 as a candidate gene for the disease in these patients (Zhang et al., 1995). In a more recent study, however, the same group documented three mutations in the SERCA1 gene, which lead to a different defect in the gene and consequently to a functional knockout of the Ca^{2+}-pump protein and a Brody phenotype (Odermatt et al.,1996). Clearly, Brody disease is a genetically heterogenous disorder. A reduced activity of the SERCA1 protein is also demonstrated in samples obtained from patients suffering from myotonic dystrophy (Benders et al., 1993).

IX. SUMMARY

Both in excitable and inexcitable cells intracellular Ca^{2+} stores play a central role in cellular Ca^{2+} signaling, and in the maturation and targeting of secreted and membrane proteins. Multiple isoforms of Ca^{2+}-transport ATPases accumulate Ca^{2+} in these stores. The first class of these enzymes is represented by the SERCA family. In vertebrates three genes and seven different splice variants, each with a different cellular/developmental expression pattern have been recognized: SERCAla (adult fast-twitch skeletal muscle), SERCAlb (neonatal fast-twitch skeletal muscle), SERCA2a (slow-twitch skeletal-/cardiac muscle), SERCA2b (nonmuscle/housekeeping isoform) and SERCA3a/b/c (platelets, myeloid-, lymphoid-, endothelial cells). Vertebrates also express another non-SERCA class of putative Ca^{2+}-transport ATPases related to a Golgi-type Ca^{2+} pump in yeast. This chapter discusses the different P-type Ca^{2+}-transport ATPases in non-vertebrate species. The mechanisms controlling alternative splicing underlying SERCA2a/b isoform diversity and the functional meaning of this diversity are more extensively explored. In addition, this chapter addresses the modulation of SERCA activity and SERCA expression in health and disease.

NOTE

1. The SERCA nomenclature introduced by Burk et al. (1989) will be used throughout this review.

ACKNOWLEDGEMENTS

JE is a Research Associate, LVDB a Postdoctoral Fellow and LM a Research Assistant supported by the National Fund for Scientific Research (NFWO, Belgium).

REFERENCES

Anderson, D. J., Mostov, K. E., & Blobel, G. (1983). Mechanisms of integration of *de novo*-synthesized polypeptides into membranes: Signal-recognition particle is required for integration into microsomal membranes of calcium ATPase and of lens MP26 but not of cytochrome b_5. Proc. Natl. Acad. Sci. USA. 80, 7249-7253.

Andersen, J. P. (1995). Dissection of the functional domains of the sarcoplasmic reticulum Ca^{2+}-ATPase by site-directed mutagenesis. Bioscience Rep. 15, 243-261.

Andres, D. A., Rhodes, J. D., Meisel, R. L., & Dixon, J. E. (1991). Characterization of the carboxyl-terminal sequences responsible for protein retention in the endoplasmic reticulum. J. Biol. Chem. 266, 14277-14282.

Anger, M., Samuel, J. L., Marotte, F., Wuytack, F., Rappaport, L., & Lompré, A.-M. (1993). The sarco(endo)plasmic reticulum Ca^{2+}-ATPase mRNA isoform, SERCA3, is expressed in endothelial and epithelial cells in various organs. FEBS Lett. 334, 45-48.

Antebi, A., & Fink, G. R. (1992). The yeast Ca^{2+}-ATPase homologue, PMR1, is required for normal Golgi function and localizes in a novel Golgi-like distribution. Mol. Biol. Cell 3, 633-654

Arai, M., Otsu, K., MacLennan, D. H., Alpert, N. R., & Periasamy, M. (1991). Effect of thyroid hormone on the expression of mRNA encoding sarcoplasmic reticulum proteins. Circ. Res. 69, 266-276.

Arai, M., Otsu, K., MacLennan, D. H., & Periasamy, M. (1992). Regulation of sarcoplasmic reticulum gene expression during cardiac and skeletal muscle development. Am. J. Physiol. 262, C614-620.

Arai, M., Alpert, N. R., MacLennan, D. H., Barton, P., & Periasamy, M. (1993). Alterations in sarcoplasmic reticulum gene expression in human heart failure. A possible mechanism for alterations in systolic and diastolic properties of the failing myocardium. Circ. Res. 72, 463-469.

Arai, M., Matsui, H., & Periasamy, M. (1994). Sarcoplasmic reticulum gene expression in cardiac hypertrophy and heart failure. Circ. Res. 74, 555-564.

Baba-Aissa, F., Raeymackers, L., Wuytack, F., Callewaert, G., Dode, L., Missiaen, L., & Casteels, R. (1996). Purkinje neurons express the SERCA3 isoform of the organellar type Ca^{2+}-transport ATPase. Mol. Brain Res. 41, 169-174.

Bayle, D., Weeks, D., & Sachs, G. (1995). The membrane topology of rat sarcoplasmic and endoplasmic reticulum calcium ATPases by in vitro translation scanning. J. Biol. Chem. 270, 25678-25684.

Balkman, C., Ojamaa, K., & Klein, I. (1992). Time course of the *in vivo* effects of thyroid hormone on cardiac gene expression. Endocrinology 130, 2001-2006.

Beekman, R. E., van Hardeveld, C., & Simonides, W. S. (1989). On the mechanism of the reduction by thyroid hormone of beta-adrenergic relaxation rate stimulation in rat heart. Biochem. J. 259, 229-236.

Benders, A. A., Veerkamp, J. H., Oosterhof, A., Jongen, P. J., Bindels, R. J., Smit, L. M., Busch, H. F., & Wevers, R. A. (1994). Ca^{2+} homeostasis in Brody's disease. A study in skeletal muscle and cultured muscle cells and the effects of dantrolene and verapamil. J. Clin. Invest. 94, 741-748.

Benders, A. A. G. M., Timmermans, J. A. H., Oosterhof, A., Ter Laak, H. J., Van Kuppevelt, T. H. M. S. M., Wevers, R. A., & Veerkamp, J. H. (1993). Deficiency of Na^+/K^+-ATPase and sarcoplasmic reticulum Ca^{2+}-ATPase in skeletal muscle and cultured muscle cells of myotonic dystrophy patients. Biochem. J. 293, 269-274.

Berridge, M. J. (1995). Capacitative calcium entry. Biochem. J. 312, 1-11.

Berridge, M. J., & Irvine, R. F. (1989). Inositol phosphates and cell signalling. Nature 341, 197-205.

Bobe, R., Bredoux, R., Wuytack, F., Quarck, R., Kovács, T., Papp, B., Corvazier, E., Magnier, C., & Enouf, J. (1994). The rat platelet 97-kDa Ca^{2+} ATPase isoform is the sarco/endoplasmic reticulum Ca^{2+} ATPase 3 protein. J. Biol. Chem. 269, 1417-1424.

Bokkala, S., El-Daher, S. S., Kakkar, V. V., Wuytack, F., & Authi, K. S. (1995). Localisation and identification of Ca^{2+} ATPases in highly purified human platelet plasma and intracellular membranes. Evidence that the monoclonal antibody PL/IM 430 recognises the SERCA3 Ca^{2+} ATPase in human platelets. Biochem. J. 306, 837-842.

Bonifacino, J. S., & Lippincott-Schwartz, J. (1991). Degradation of proteins within the endoplasmic reticulum. Curr. Opin. Cell. Biol. 3, 592-600

Brandl, C. J., deLeon, S., Martin, D. R., & MacLennan, D. H. (1987). Adult forms of the Ca^{2+} ATPase of sarcoplasmic reticulum. Expression in developing skeletal muscle. J. Biol. Chem. 262, 3768-3774.

Brandl, C.J., Green, N. M., Korczak, B., & MacLennan, D. H. (1986). Two Ca^{2+} ATPase genes: homologies and mechanistic implications of deduced amino acid sequences. Cell 44, 597-607.

Briggs, F. N., Lee, K. F., Feher, J. J., Wechsler, A. S., Ohlendieck, K., & Campbell, K. (1990). Ca-ATPase isozyme expression in sarcoplasmic reticulum is altered by chronic stimulation of skeletal muscle. FEBS Lett. 259, 269-272.

Brillantes, A. M., Allen, P., Takahashi, T., Izumo, S., & Marks, A. R. (1992). Differences in cardiac calcium release channel (ryanodine receptor) expression in myocardium from patients with end-stage heart failure caused by ischemic versus dilated cardiomyopathy. Circ. Res. 71, 18-26.

Brostrom, C. O., & Brostrom, M. A. (1990). Calcium-dependent regulation of protein synthesis in intact mammalian cells. Ann. Rev. Physiol. 52, 577-590.

Bull, P. C., & Cox, D. W. (1994). Wilson disease and Menkes disease: new handles on heavy-metal transport. TIG 10, 246-252.

Burk, S., & Shull, G. E. (1992). Structure of the rat plasma membrane Ca^{2+}-ATPase isoform 3 gene and characterization of alternative splicing and transcription products. Skeletal muscle specific splicing results in a plasma membrane Ca^{2+} ATPase with a novel calmodulin-binding domain. J. Biol. Chem. 267, 19683-19690

Burk, S. E., Lytton, J., MacLennan, D. H., & Shull, G. E. (1989). cDNA cloning, functional expression, and mRNA tissue distribution of a third organellar Ca^{2+} pump. J. Biol. Chem. 264, 18561-18568.

Callen, D. F., Baker, E., Lane, S., Nancarrow, J., Thompson, A., Whitmore, S. A., MacLennan, D. H., Berger, R., Cherif, D., Jarvela, I., Peltonen, L., Sutherland, G. R., & Gardiner, R. M. (1991). Regional mapping of the Batten disease locus (CLN3) to human chromosome 16pl2. Am. J. Hum. Genet. 49, 1372-1377.

Catty, P., & Goffeau, A. (1996). Identification and phylogenetic classification of eleven putative P-type calcium transport ATPase genes in the yeasts *Saccharomyces cerevisiae* and *Schizosaccharomyces pombe*. Bioscience Reports, 16, 75-85

Campbell, A. M., Kessler, P. D., Sagara, Y., Inesi, G., & Fambrough, D. M. (1991). Nucleotide sequences of avian cardiac and brain SR/ER Ca^{2+}-ATPases and functional comparisons with fast twitch Ca^{2+}-ATPase. Calcium affinities and inhibitor effects. J. Biol. Chem. 266, 16050-16055.

Campbell, A. M., Kessler, P. D., & Fambrough, D. M. (1992). The alternative carboxyl termini of avian cardiac and brain sarcoplasmic reticulum/endoplasmic reticulum Ca^{2+}-ATPases are on opposite sides of the membrane. J. Biol. Chem. 267, 9321-9325.

Campbell, A. M., Wuytack, F., & Fambrough, D. M. (1993). Differential distribution of the alternative forms of the sarcoplasmic/endoplasmic reticulum Ca^{2+}-ATPase, SERCA2b and SERCA2a in the avian brain. Brain Research 605, 67-76.

Cantilina, T., Sagara, Y., Inesi, G., & Jones, L. R. (1993). Comparative studies of cardiac and skeletal sarcoplasmic reticulum ATPases. Effect of a phospholamban antibody on enzyme activation by Ca^{2+}. J. Biol. Chem. 268, 17018-17025.

Carafoli, E. (1994). Biogenesis: plasma membrane calcium ATPase: 15 years of work on the purified enzyme. FASEB J. 8, 993-1002.

Chen, Q., & van Breemen, C. (1992). Function of smooth muscle sarcoplasmic reticulum. Adv. Second Messenger Phosphoprotein Res. 26, 335-350.

Chyn, T. L., Martonosi, A. N., Morimoto, T., & Sabatini, D. D. (1979). *In vitro* synthesis of the Ca^{2+} transport ATPase by ribosomes bound to sarcoplasmic reticulum membranes. Proc. Natl. Acad. Sci. USA 76, 1241-1245.

Clarke, D. M., Loo, T. W., Inesi, G., & MacLennan, D. H. (1989). Location of high-affinity Ca^{2+} binding sites within the predicted transmembrane domain of the sarcoplasmic reticulum Ca^{2+}-ATPase. Nature, 339, 476-478.

Cory, C. R., Shen, H., & O'Brien, P. J. (1994). Compensatory asymmetry in down-regulation and inhibition of the myocardial Ca^{2+} cycle in congestive heart failure produced in dogs by idiopathic dilated cardiomyopathy and rapid ventricular pacing. J. Mol. Cell. Cardiol. 26, 173-184.

Cunningham, K. W., & Fink, G. R. (1994). Calcineurin-dependent growth control in *Saccharomyces cerevisiae* mutants lacking PMC 1, a homolog of the plasma membrane Ca^{2+}-ATPases. J. Cell Biol. 124, 351-363.

Danon, M. J., Karpati, G., Charuk, J., & Holland, P. (1988). Sarcoplasmic reticulum adenosine triphosphatase deficiency with probable autosomal dominant inheritance. Neurology 38, 812-815.

Davis, B. A., Schwartz, A., Samaha, F. J., & Kranias, E. G. (1983). Regulation of cardiac sarcoplasmic reticulum calcium transport by calcium-calmodulin-dependent phosphorylation. J. Biol. Chem. 258, 13587-13591.

Dawson, A. P. (1990). Regulation of intracellular Ca^{2+}. Essays Biochem. 25, 1-37

De Jaegere, S., Wuytack, F., De Smedt, H., Van Den Bosch, L., & Casteels, R. (1993). Alternative processing of the gene transcripts encoding a plasma-membrane and a sarco/endoplasmic reticulum Ca^{2+} pump during differentiation of BC_3H1 muscle cells. Biochim. Biophys. Acta 1173, 188-194.

de la Bastie, D., Wisnewsky, C., Schwartz, K., & Lompre, A.-M. (1988). (Ca^{2+}-Mg^{2+})-dependent ATPase mRNA from smooth muscle sarcoplasmic reticulum differs from that in cardiac and fast skeletal muscles. FEBS Lett. 229, 45-48.

de la Bastie, D., Levitsky, D., Rappaport, L., Mercadier, J. J., Marotte, F., Wisnewsky, C., Brovkovich, V., Schwartz, K., & Lompré, A.-M. (1990). Function of the sarcoplasmic reticulum and expression of its Ca^{2+}-ATPase gene in pressure overload-induced cardiac hypertrophy in the rat. Circ. Res. 66, 554-564

de Mendonça, R. L., Beck, E., Rumjanek, F.D., & Goffeau, A. (1995). Cloning and characterization of a putative calcium-transporting ATPase gene from *Schistosoma mansoni.* Molec. Biochem. Parasitol. 72, 129-139.

De Smedt, H., Eggermont, J. A., Wuytack, F., Parys, J. B., Van Den Bosch, L., Missiaen, L., Verbist, J., & Casteels, R. (1991). Isoform switching of the sarco(endo)plasmic reticulum Ca^{2+} pump during differentiation of BC_3H1 myoblasts. J. Biol. Chem. 266, 7092-7095.

Dode, L., Wuytack, F., Kools, P. F. J., Baba-Aissa, F., Raeymaekers, L., Briké, F., Van De Ven, W. J. M., & Casteels, R. (1996). cDNA cloning, expression and chromosomal localization of the human sarco/endoplasmic reticulum Ca^{2+}-ATPase 3 gene. Biochem. J. 318, 689-699.

Dode L., Mountian, I., De Greef, C., Casteels, R., & Wuytack, F. (1997a). Three functional isoforms are generated by the alternative processing of the sarco/endoplasmic reticulum Ca^{2+}-ATPase 3 gene primary transcript. Submitted for publication.

Dode, L., De Greef, C., Attard, M., Town, M. M., Casteels, R., & Wuytack, F. (1997b). Structure of the human sarco/endoplasmic reticulum Ca^{2+}-ATPase 3 gene and functional characterization of its promoter. Submitted for publication.

Dormer, R. L., Capurro, D. E., Morris, R., & Wegg, R. (1993). Demonstration of two isoforms of the SERCA 2b type Ca^{2+}, Mg^{2+}-ATPase in pancreatic endoplasmic reticulum. Biochim. Biophys. Acta 1152, 225-230.

Dux, L. (1993). Muscle relaxation and sarcoplasmic reticulum function in different muscle types. Rev. Physiol. Biochem. Pharmacol. 122, 69-147

Dyer, M., Jackson, M., McWhinney, C., Zhao, G., & Mikkelsen, R. (1996). Analysis of a cation-transporting ATPase of *Plasmodium falciparum.* Molec. Biochem. Parasitol. 78, 1-12

Eggermont, J. A., Wuytack, F., & Casteels, R. (1991). Characterization of the 3' end of the pig sarcoplasmic/endoplasmic-reticulum Ca^{2+} pump gene 2. Biochim. Biophys. Acta. 1088, 448-451.

Eggermont, J. A., Wuytack, F., De Jaegere, S., Nelles, L., & Casteels, R. (1989). Evidence for two isoforms of the endoplasmic-reticulum Ca^{2+} pump in pig smooth muscle. Biochem. J. 260, 757-761.

Eggermont, J. A., Wuytack, F., Verbist, J., & Casteels, R. (1990a). Expression of endoplasmic-reticulum Ca^{2+}-pump isoforms and of phospholamban in pig smooth-muscle tissues. Biochem. J. 271, 649-653.

Eggermont, J. A., Wuytack, F., & Casteels, R. (1990b). Characterization of the mRNAs encoding the gene 2 sarcoplasmic-reticulum Ca^{2+} pump in pig smooth muscle. Biochem. J. 266, 901-907.

Escalante, R., & Sastre, L. (1993). Similar alternative splicing events generate two sarcoplasmic or endoplasmic reticulum Ca-ATPase isoforms in the crustacean *Artemia franciscana* and in vertebrates. J. Biol. Chem. 268, 14090-14095.

Escalante, R., & Sastre, L. (1994a). Structure of *Artemia franciscana* sarco/endoplasmic reticulum Ca-ATPase gene. J. Biol. Chem., 269, 13005-13012.

Escalante, R., & Sastre, L. (1994b). Identification of an *Artemia franciscana* retropseudogene containing part of the last exons of the sarco/endoplasmic Ca-ATPase-encoding SERCA gene. Gene 149, 377-378.

Escalante, R., & Sastre, L. (1995). Tissue-specific alternative promotors regulate the expression of the two sarco/endoplasmic reticulum Ca-ATPase isoforms from *Artemia franciscana*. DNA Cell Biol. 14, 893-900.

Evans, D. E., Briars, S., & Williams, L. E. (1991). Active transport by plant cell membranes. J. Exp. Botany 42, 285-303.

Fambrough, D. M., Lemas, M. V., Hamrick, M., Emerick, M., Renaud, K. J., Inman, E. M., Hwang, B., & Takeyasu, K. (1994). Analysis of subunit assembly of the Na K ATPase. Am. J. Physiol. 266, C579-589.

Feldman, A. M, Weinberg, E. O., Ray, P. E., & Lorell, B. H. (1993). Selective changes in cardiac gene expression during compensated hypertrophy and the transition to cardiac decompensation in rats with chronic aortic banding. Circ. Res. 73, 184-192.

Ferrol, N., & Bennett, A. B. (1996). A single gene may encode differentially localized Ca^{2+} ATPases in tomato. The Plant Cell, 8, 1159-1169.

Fisher, S. A., Buttrick, P. M., Sukovich, D., & Periasamy, M. (1993). Characterization of promoter elements of the rabbit cardiac sarcoplasmic reticulum Ca^{2+}-ATPase gene required for expression in cardiac muscle cells. Circ. Res. 73, 622-628.

Frass, O., Sharma, H. S., Knoll, R., Duncker, D. J., McFalls, E. O., Verdouw, P. D., & Schaper, W. (1993). Enhanced gene expression of calcium regulatory proteins in stunned porcine myocardium. Cardiovasc. Res. 27, 2037-2043.

Fujii, J., Ueno, A., Kitano, K., Tanaka, S., Kadoma, M., & Tada, M. (1987). Complete complementary DNA-derived amino acid sequence of canine cardiac phospholamban. J. Clin. Invest. 79, 301-304.

Fujii, J., Maruyama, K., Tada, M., & MacLennan, D. H. (1989). Expression and site-specific mutagenesis of phospholamban. Studies of residues involved in phosphorylation and pentamer formation. J. Biol. Chem. 264, 12950-12955.

Fujii, J., Maruyama, K., Tada, M., & MacLennan, D. H. (1990). Co-expression of slow-twitch/cardiac muscle Ca^{2+}-ATPase (SERCA2) and phospholamban. FEBS Lett. 273, 232-234.

Fujii, J., Zarain-Herzberg, A., Willard, H. F., Tada, M., & MacLennan, D. H. (1991). Structure of the rabbit phospholamban gene, cloning of the human cDNA, and assignment of the gene to human chromosome 6. J. Biol. Chem. 266, 11669- 11675.

Gabathuler, R., & Kvist, S. (1990). The endoplasmic reticulum retention signal of the E3.19K protein of adenovirus type 2 consists of three separate amino acid segments at the carboxy terminus. J. Cell Biol. 111, 1803- 1810.

Gambel, A. M., Gallien, T. N., Dantzler-Whithworth, T., Bowes, M., & Menick, D. R. (1992). Sequence of feline cardiac sarcoplasmic reticulum Ca^{2+}-ATPase. Biochim. Biophys. Acta 1131, 203-206.

Geisler, M., Richter, J., & Schumann, J. (1993). Molecular cloning of a P-type ATPase gene from the cyanobacterium *Synechocystis* sp. PCC6803. Homology to eukaryotic Ca^{2+}-ATPases. J. Mol. Biol. 234, 1284-1289.

Ghislain, M., Goffeau, A., Halachmi, D., & Eilam, Y. (1990). Calcium homeostasis and transport are affected by disruption of cta3, a novel gene encoding Ca^{2+}-ATPase in *Schizosaccharomyces pombe*. J. Biol. Chem. 265, 18400-18407.

Go, L. O., Moschella, M. C., Watras, J., Handa, K. K., Fyfe, B. S., & Marks, A. R. (1995). Differential regulation of two types of intracellular calcium release channels during end-stage heart failure. J. Clin. Invest. 95, 888-894.

Greber, U. F., & Gerace, L. (1995). Depletion of calcium from the lumen of endoplasmic reticulum reversibly inhibits passive diffusion and signal-mediated transport into the nucleus. J. Cell. Biol., 128, 5-14.

Green, N. M. (1992). Evolutionary relationships within the family of P-type cation pumps. Ann. N.Y. Acad. Sci. 671,104-112.

Greenway, D. C., & MacLennan, D. H. (1978). Assembly of the sarcoplasmic reticulum. Synthesis of calsequestrin and the Ca^{2+}+Mg^{2+}-adenosine triphosphatase on membrane-bound polyribosomes. Can. J. Biochem. 56, 452-456.

Grover, A. K., & Khan, I. (1992). Calcium pump isoforms: diversity, selectivity and plasticity. Cell Calcium 13, 9- 17.

Gunteski-Hamblin, A.-M., Greeb, J., & Shull, G. E. (1988). A novel Ca^{2+} pump expressed in brain, kidney, and stomach is encoded by an alternative transcript of the slow-twitch muscle sarcoplasmic reticulum Ca^{2+}-ATPase gene. Identification of cDNAs encoding Ca^{2+} and other cation-transporting ATPases using an oligonucleotide probe derived from the ATP-binding site. J. Biol. Chem. 263, 15032- 15040.

Gunteski-Hamblin, A.-M., Clarke, D. M., & Shull, G. E. (1992). Molecular cloning and tissue distribution of alternatively spliced mRNAs encoding possible mammalian homologues of the yeast secretory pathway calcium pump. Biochemistry 31, 7600-7608.

Hack, N., Wilkinson, J. M., & Crawford, N. (1988). A monoclonal antibody (PL/IM430) to platelet intracellular membranes which inhibits the uptake of Ca^{2+} without affecting the Ca^{2+} Mg^{2+} ATPase. Biochem. J. 250, 355-361.

Halachmi, D., Ghislain, M., & Eilam, Y. (1992). An intracellular ATP-dependent calcium pump within the yeast *Schizosaccharomyces pombe*, encoded by the gene cta3. Eur. J. Biochem. 207, 1003-1008.

Halachmi, D., & Eilam, Y. (1996). Elevated cytosolic free Ca^{2+} concentrations and massive Ca^{2+} accumulation within vacuoles in yeast mutant lacking PMR1, a homolog of Ca^{2+} ATPase. FEBS Lett., 392, 194-200.

Hartong, R., Wang, N., Kurokawa, R., Lazar, M. A., Glass, C. K., Apriletti, J. W., & Dillmann, W. H. (1994). Delineation of three different thyroid hormone-response elements in promoter of rat sarcoplasmic reticulum Ca^{2+} ATPase gene. Demonstration that retinoid X receptor binds 5' to thyroid hormone receptor in response element 1. J. Biol. Chem. 269, 13021-13029.

Hasenfuss, G., Reinecke, H., Studer, R., Meyer, M., Pieske, B., Holtz, J., Holubarsch, C., Posival, H., Just, H., & Drexler, H. (1994). Relation between myocardial function and expression of sarcoplasmic reticulum Ca^{2+}-ATPase in failing and nonfailing human myocardium. Circ. Res. 75, 434-442.

Hawkins, C., Xu, A., & Narayanan, N. (1994). Sarcoplasmic reticulum calcium pump in cardiac and slow-twitch skeletal muscle but not fast-twitch skeletal muscle undergoes phosphorylation by endogenous and exogenous Ca^{2+}/calmodulin-dependent protein kinase. Characterization of optimal conditions for calcium pump phosphorylation. J. Biol. Chem. 269, 31198-31206.

Heim, R., Iwata, T., Zvarich, E., Adamo, H. P., Rutishauser, B., Strehler, E. E., Guerini, D., & Carafoli, E. (1992). Expression, purification and properties of the plasma membrane Ca^{2+} pump and of its N-terminally truncated 105 kDa fragment. J. Biol. Chem. 267, 24476-24484.

Hilfiker, H., Strehler-Page, M.-A., Stauffer, T. P., Carafoli, E., & Strehler, E. E. (1993). Structure of the gene encoding the human plasma membrane calcium pump isoform 1. J. Biol. Chem. 268, 19717-19725.

Hu, P., Yin, C., Zhang, K.-M., Wright, L. D., Nixon, T. E., Wechsler, A. S., Spratt, J. A., & Briggs, N. (1995). Transcriptional regulation of phospholamban gene and translational regulation of SERCA2 gene produces coordinate expression of these two sarcoplasmic reticulum proteins during skeletal muscle phenotype switching. J. Biol. Chem. 270, 11619-11622.

Huang, L., Berkelman, T., Franklin, A. E., & Hoffman, N. E. (1993). Characterization of a gene encoding a Ca^{2+}-ATPase-like protein in the plastid envelope [published erratum appears in Proc. Natl. Acad. Sci. USA (1994) 91, 9664]. Proc. Natl. Acad. Sci. USA 90, 10066-10070.

Jackson, M. R., Nilsson, T., & Peterson, P. A. (1990). Identification of a consensus motif for retention of transmembrane proteins in the endoplasmic reticulum. EMBO J. 9, 3153-3162.

James, P., Inui, M., Tada, M., Chiesi, M., & Carafoli, E. (1989). Nature and site of phospholamban regulation of the Ca^{2+} pump of sarcoplasmic reticulum. Nature 342, 90-92.

Johnson, R. J., Pyun, H. Y., Lytton, J., & Fine, R. E. (1993). Differences in the subcellular localization of calreticulin and organellar Ca^{2+}-ATPase in neurons. Molec. Brain Res. 17, 9-16.

Kandarian, S. C., Peters, D. G., Taylor, J. A., & Williams, J. H. (1994). Skeletal muscle overload upregulates the sarcoplasmic reticulum slow calcium pump gene. Am. J. Physiol. 266, C1190-1197.

Kaplan, P., Hendrikx, M., Mattheussen, M., Mubagwa, K., & Flameng, W. (1992). Effect of ischemia and reperfusion on sarcoplasmic reticulum calcium uptake. Circ. Res. 71, 1123-1130.

Kaprielian, Z., Bandman, E., & Fambrough, D. M. (1991). Expression of Ca^{2+}-ATPase isoforms in denervated, regenerating, and dystrophic chicken skeletal muscle. Dev. Biol. 144, 199-211.

Kaprielian, Z., & Fambrough, D. M. (1987). Expression of fast and slow isoforms of the Ca^{2+}-ATPase in developing chick skeletal muscle. Develop. Biol. 124, 490-503.

Karin, N. J., Kaprielian, Z., & Fambrough, D. M. (1989). Expression of avian Ca^{2+}-ATPase in cultured mouse myogenic cells. Molec. Cell. Biol. 9, 1978-1986.

Karin, N. J., & Settle, V. J. (1992). The sarcoplasmic reticulum Ca^{2+}-ATPase, SERCAla, contains endoplasmic reticulum targeting information. Biochem. Biophys. Res. Commun. 186, 219-227.

Khan, I., Spencer, G. G., Samson, S. E., Crine, P., Boileau, G., & Grover, A. K. (1990). Abundance of sarcoplasmic reticulum calcium pump isoforms in stomach and cardiac muscles. Biochem. J. 268, 415-419.

Kimura, M., Yamaguchi, Y., Takada, S., & Tanabe, K. (1993). Cloning of a Ca^{2+}-ATPase gene of *Plasmodium falciparum* and comparison with vertebrate Ca^{2+}-ATPases. J. Cell Sciences 104, 1129-1136.

Kimura, Y., Otsu, K., Nishida, K., Kuzuya, T., & Tada, M. (1994). Thyroid hormone enhances Ca^{2+} pumping activity of the cardiac sarcoplasmic reticulum by increasing Ca^{2+} ATPase and decreasing phospholamban expression. J. Mol. Cell. Cardiol. 26, 1145-1154.

Kimura, Y., Kurzydlowski, K., Tada, M., & MacLennan, D.H. (1996). Phospholamban regulates the Ca^{2+}-ATPase through intramembrane interactions. J. Biol. Chem. 271, 21726-21731.

Kimura, Y., Kurzydlowski, K., Tada, M., & MacLennan, D. H. (1997). Phospholamban inhibitory function is activated by depolymerization. J. Biol. Chem. 272, 15061-15064.

Kirchberger, M. A., & Antonetz, T. (1982). Phospholamban: dissociation of the 22,000 molecular weight protein of cardiac sarcoplasmic reticulum into 11,000 and 5,500 molecular weight forms. Biochem. Biophys. Res. Commun. 105, 152-156.

Kirchberger, M. A., Borchman, D., & Kasinathan, C. (1986). Proteolytic activation of the canine cardiac sarcoplasmic reticulum calcium pump. Biochemistry 25, 5484-5492.

Kiss, E., Jakab, G., Kranias, E. G., & Edes, I. (1994). Thyroid hormone-induced alterations in phospholamban protein expression. Regulatory effects on sarcoplasmic reticulum Ca^{2+} transport and myocardial relaxation. Circ. Res. 75, 245-251.

Klausner, R.D., & Sitia, R. (1990). Protein degradation in the endoplasmic reticulum. Cell 62,611-614.

Knöll, R., Arras, M., Zimmerman, R., Schaper, J., & Schaper, W. (1994). Changes in gene expression following short coronary occlusions studied in porcine hearts with run-on assays. Cardiovasc. Res. 28, 1062-1069.

Komuro, I., Kurabayashi, M., Shibazaki, Y., Takaku, F., & Yazaki, Y. (1989). Molecular cloning anu characterization of a Ca^{2+} + Mg^{2+} -dependent adenosine triphosphatase from rat cardiac sarcoplasmic reticulum. J. Clin. Invest. 83, 1102-1108.

Korczak, B., Zarain-Herzberg, A., Brandl, C. J., Ingles, C. J., Green, N. M., & MacLennan, D. H. (1988). Structure of the rabbit fast-twitch skeletal muscle Ca^{2+}-ATPase gene. J. Biol. Chem. 263, 4813-4819.

Kovács, T., Corvazier, E., Papp, B., Magnier, C., Bredoux, R., Enyedi, A., Sarkadi, B., & Enouf, J. (1994). Controlled proteolysis of Ca^{2+}-ATPases in human platelet and nonmuscle cell membrane vesicles. Evidence for a multi-sarco/endoplasmic reticulum Ca^{2+}-ATPase system. J. Biol. Chem. 269, 6177-6184

Krause, K. H. (1991). Ca^{2+}-storage organelles. FEBS Lett. 285, 225-229.

Kuo, T. H., Tsang, W., Wang, K. K. W., & Carlock, L. (1992). Simultaneous reduction of the sarcolemmal and SR calcium ATPase activities and gene expression in cardiomyopathic hamster. Biochim. Biophys. Acta 1138, 343-349.

Kwan, C. Y. (1985). Dysfunction of calcium handling by smooth muscle in hypertension. Can. J. Physiol. Pharmacol. 63, 366-374.

Lamers, M. J., Duncker, D. J., Bezstarosti, K., McFalls, E. O., Sassen, L. M. A., & Verdouw, P. D. (1993). Increased activity of the sarcoplasmic reticular calcium pump in porcine stunned myocardium. Cardiovasc. Res. 27, 520-524.

Launay, S., Bobe, R., Lacabaratz-Porret, C., Bredoux, R., Kovacs, T., Enouf, J., & Papp, B. (1997). Modulation of endoplasmic reticulum calcium pump expression during T lymphocyte activation. J. Biol. Chem. 272, 10746-10750.

Le Peuch, C. J., Haiech, J., & Demaille, J. G. (1979). Concerted regulation of cardiac sarcoplasmic reticulum calcium transport by cyclic adenosine monophosphate dependent and calcium-calmodulin-dependent phosphorylations. Biochemistry 18, 5150-5157.

Liang, F., Cunningham, K. W., Harper, J. F., & Sze, H. (1997). ECA1 complements yeast mutants defective in Ca^{2+} pumps and encodes an endoplasmic reticulum-type Ca^{2+}-ATPase in *Arabidopsis thaliana*. Proc. Natl. Acad. Sci. USA 94, 8579-8584.

Leberer, E., Hartner, K. T., Brandl, C. J., Fujii, J., Tada, M., MacLennan, D. H., & Pette, D. (1989). Slow/cardiac sarcoplasmic reticulum Ca^{2+}-ATPase and phospholamban mRNAs are expressed in chronically stimulated rabbit fast-twitch muscle. Eur. J. Biochem. 185, 51-54.

Leberer, E., Seedorf, U., & Pette, D. (1986). Neural control of gene expression in skeletal muscle. Calcium-sequestering proteins in developing and chronically stimulated rabbit skeletal muscles. Biochem. J. 239, 295-300.

Levitsky, D..O., Clergue, M., Lambert, F., Souponitskaya, M. V, Le Jemtel, T. H., Lecarpentier, Y., & Lompré, A.-M. (1993). Sarcoplasmic reticulum calcium transport and Ca^{2+}-ATPase gene expression in thoracic and abdominal aortas of normotensive and spontaneously hypertensive rats. J. Biol. Chem. 268, 8325-8331.

Lompré, A.-M., Anger, M., & Levitsky, D. (1994). Sarco(endo)plasmic reticulum Calcium pumps in the cardiovascular system: function and gene expression. J. Mol. Cell. Cardiol. 26, 1109-1121

Luo, W., Grupp, I. L., Harrer, J., Ponniah, S., Grupp, G., Duffy, J. J., Doetschman, T., & Kranias, E. G. (1994). Targeted ablation of the phospholamban gene is associated with markedly enhanced myocardial contractility and loss of β-agonist stimulation. Circ. Res. 75, 401-409.

Lytton, J., & MacLennan, D. H. (1988). Molecular cloning of cDNAs from human kidney coding for two alternatively spliced products of the cardiac Ca^{2+}-ATPase gene. J. Biol. Chem. 263, 15024-15031.

Lytton, J., Zarain-Herzberg., Periasamy, M., & MacLennan, D. H. (1989). Molecular cloning of the mammalian smooth muscle sarco(endo)plasmic reticulum Ca^{2+} ATPase. J. Biol. Chem. 264, 7059-7065.

Lytton, J., Westlin, M., & Hanley, M. R. (1991). Thapsigargin inhibits the sarcoplasmic or endoplasmic reticulum Ca-ATPase family of calcium pumps. J. Biol. Chem. 266, 22912-22918.

Lytton, J., Westlin, M., Burk, S. E., Shull, G. E., & MacLennan, D. H. (1992). Functional comparisons between isoforms of the sarcoplasmic or endoplasmic reticulum family of calcium pumps. J. Biol. Chem. 267, 14483-14489.

MacLennan, D. H., Brandl, C. J., Champaneria, S., Holland, P. C., Powers, V. E., & Willard, H. F. (1987). Fast twitch and slow twitch/cardiac Ca^{2+} ATPase genes map to chromosomes 16 and 12. Somatic Cell Mol. Genet. 13, 341-346.

MacLennan, D. H., Rice, W. J., & Green, N. M. (1997). The mechanism of Ca^{2+} transport by sarco(endo)plasmic reticulum Ca^{2+}-ATPases. J. Biol. Chem. 272, 28815-28818.

Maeda, M., Ashiman, K.-I., Tamura, S., & Futai, M. (1990). Human gastric (H^+ + K^+)-ATPase gene. Similarity to (Na^+ + K^+)-ATPase genes in exon/intron organization but difference in control region. J. Biol. Chem., 265, 9027-9032.

Magnier, C., Papp, B., Corvazier, E., Bredoux, R., Wuytack, F., Eggermont, J., Maclouf, J., & Enouf, J. (1992). Regulation of sarco-endoplasmic reticulum Ca^{2+}-ATPases during platelet-derived growth factor-induced smooth muscle cell proliferation. J. Biol. Chem. 267, 15808-15815.

Magnier, C., Bredoux, R., Kovács, Quarck, B., Papp, B., Corvazier, E., de Gunzburg, J., & Enouf, J. (1994). Correlated expression of the 97 kDa sarcoendoplasmic reticulum Ca^{2+} ATPase and RapIB in platelets and various cell lines. Biochem. J. 297, 343-350.

Magyar, A., & Váradi, A. (1990). Molecular cloning and chromosomal localization of a sarco/endoplasmic reticulum-type- Ca^{2+}-ATPase of *Drosophila melanogaster*. Biochem. Biophys. Res. Commun. 173, 872-877.

Magyar, A., Bakos, E., & Váradi, A. (1995). Structure and tissue-specific expression of the *Drosophila melanogaster* organellar-type Ca^{2+}-ATPase gene. Biochem. J. 310, 757-763

Malmstrom, S., Askerlund, P., & Palmgren, M. G. (1997). A calmodulin-stimulated Ca^{2+}-ATPase from plant vacuolar membranes with a putative regulatory domain at its N-terminus. FEBS Lett. 400, 324-328.

Maruyama, K., & MacLennan, D. H. (1988). Mutation of aspartic acid-351, lysine-352, and lysine-515 alters the Ca^{2+} transport activity of the Ca^{2+}-ATPase expressed in COS-1 cells. Proc. Natl. Acad. Sci. USA 85, 3314-3318

Matsui, H., MacLennan, D. H., Alpert, N. R., & Periasamy, M. (1995). Sarcoplasmic reticulum gene expression in pressure overload-induced cardiac hypertrophy in rabbit. Am. J. Physiol. 268, C252-258.

Mercadier, J. J., Lompré, A.-M., Duc, P., Boheler, K. R., Fraysse, J. B., Wisnewsky, C., Allen, P. D., Komajda, M., & Schwartz, K. (1990). Altered sarcoplasmic reticulum Ca^{2+}-ATPase gene expression in the human ventricle during end-stage heart failure. J. Clin. Invest. 85, 305-309.

Mertens, L., Van Den Bosch, L., Verboomen, H., Wuytack, F., De Smedt, H., & Eggermont, J. (1995). Sequence and spatial requirements for regulated muscl-specific processing of the sarco/endoplasmic reticulum Ca^{2+}-ATPase 2 gene (SERCA2) transcript. J. Biol. Chem. 270, 11004-11011.

Michelangeli, F., Di Virgilio, F., Villa, A., Podini, P., Meldolesi, J., & Pozzan, T. (1991). Identification, kinetic properties and intracellular localization of the (Ca^{2+}-Mg^{2+})-ATPase from intracellular stores of chicken cerebellum. Biochem. J. 275, 555-561.

Miller, K. K., Verma, A., Snyder, S. H., & Ross, C. A. (1991). Localization of an endoplasmic reticulum calcium ATPase mRNA in rat brain by *in situ* hybridization. Neuroscience 43, 1-9.

Mintz, G., Pizzarello, R., & Klein, I. (1991). Enhanced left ventricular diastolic function in hyperthyroidism: noninvasive assessment and response to treatment. J. Clin. Endocrinol. Metab. 73, 146-150.

Møller, J. V., Juul, B., & le Maire, M. (1996). Structural organization, ion transport and energy transduction of P-type ATPases. Biochim. Biophys. Acta 1286, 1-51.

Moniakis, J., Coukell, M. B., & Forer, A. (1995). Molecular cloning of an intracellular P-type ATPase from *Dictyostelium* that is up-regulated in calcium-adapted cells. J. Biol. Chem. 270, 28276-28281.

Morgan, J. P. (1991). Abnormal intracellular modulation of calcium as a major cause of cardiac contractile dysfunction. N. Engl. J. Med. 325, 625-632.

Morris, G. L., Cheng, H. C., Colyer, J., & Wang, J. H. (1991). Phospholamban regulation of cardiac sarcoplasmic reticulum (Ca^{2+}-Mg^{2+})-ATPase. Mechanism of regulation and site of monoclonal antibody interaction. J. Biol. Chem. 266, 11270-11275.

Mostov, K. E., DeFoor, P., Fleischer, S., & Blobel, G. (1981). Co-translational membrane integration of calcium pump protein without signal sequence cleavage. Nature 292, 87-88

Movsesian, M. A., Karimi, M., Green, K., & Jones, L. R. (1994). Ca^{2+}-transporting ATPase, phospholamban, and calsequestrin levels in nonfailing and failing human myocardium. Circulation 90, 653-657.

Movsesian, M. A., Morris, G. L., Wang, J. H., & Krall, J. (1992). Phospholamban modulated Ca^{2+} transport in cardiac and slow twitch skeletal muscle sarcoplasmic reticulum. Second Messengers Phosphoproteins 14, 151-161.

Muller, A., van der Linden, G. C., Zuidwijk, M. J., Simonides, W. S., van der Laarse, W. J., & van Hardeveld, C. (1994). Differential effects of thyroid hormone on the expression of sarcoplasmic reticulum Ca^{2+}-ATPase isoforms in rat skeletal muscle fibers. Biochem. Biophys. Res. Commun. 203, 1035-1042.

Muller, A., van Hardeveld, C., Simonides, W. S., & van Rijn, J. (1991). The elevation of sarcoplasmic reticulum Ca^{2+}-ATPase levels by thyroid hormone in the L6 muscle cell line is potentiated by insulin-like growth factor-I. Biochem. J. 275, 35-40.

Munro, S., & Pelham, H. R. B. (1987). A C-terminal signal prevents secretion of luminal ER proteins. Cell 48, 899-907.

Murakami, K., Tanabe, K., & Takada, S. (1990). Structure of a *Pasmodium yoelii* gene-encoded protein homologous to the Ca^{2+}-ATPase of rabbit skeletal muscle sarcoplasmic reticulum. J. Cell Science. 97, 487-495.

Nagai, R., Zarain-Herzberg, A., Brandl, C. J., Fujii, J., Tada, M., MacLennan, D. H., Alpert, N. R., & Periasamy, M. (1989). Regulation of myocardial Ca^{2+}-ATPase and phospholamban mRNA expression in response to pressure overload and thyroid hormone. Proc. Natl. Acad. Sci. USA 86, 2966-2970.

Nolan, D. P., Revelard, P., & Pays, E. (1994). Overexpression and characterization of a gene for a Ca^{2+}-ATPase of the endoplasmic reticulum in *Trypanosoma brucei*. J. Biol. Chem. 269, 26045-26051.

Nørregaard, A., Vilsen, B., & Andersen, J. P. (1994). Transmembrane segment M3 is essential to thapsigargin sensitivity of the sarcoplasmic reticulum Ca^{2+}-ATPase. J. Biol. Chem. 269, 26598-26601.

O'Brien, P. J., Duke, A. L., Shen, H., & Shohet, R. V. (1995). Myocardial mRNA content and stability, and enzyme activities of Ca-cycling and aerobic metabolism in canine dilated cardiomyopathies. Mol. Cell Biochem. 142, 139-150.

Odermatt, A., Kurzydlowski, K., & MacLennan, D. H. (1996). The vmax of the Ca^{2+}-ATPase of cardiac sarcoplasmic reticulum (SERCA2a) is not altered by Ca^{2+}/calmodulin-dependent phosphorylation or by interaction with phospholamban. J. Biol. Chem. 271, 14206-14213.

Ojamaa, K., Samarel, A. M., Kupfer, J. M., Hong, C., & Klein, I. (1992). Thyroid hormone effects on cardiac gene expression independent of cardiac growth and protein synthesis. Am. J. Physiol. 263, E534-540.

Okorokov, L. A., Tanner, W., & Lehle, L. (1993). A novel primary Ca^{2+}-transport system from *Saccharomyces cerevisiae*. Eur. J. Biochem. 216, 573-577.

Oldershaw, K. A., & Taylor, C. W. (1990). 2,5-di-(tert-butyl)-1,4-benzohydroquinone mobilizes inositol 1,4,5-trisphosphate sensitive and insensitive Ca^{2+}-stores. FEBS Lett. 274, 214-216.

Olson, S., Wang, M. G., Carafoli, E., Strehler, E. E., & McBride, O. W. (1991). Localization of two genes encoding plasma membrane Ca^{2+}-transporting ATPases to human chromosomes lq25-32 and 12q21-23. Genomics 9, 629-641.

Otsu, K., Fujii, J., Periasamy, M., Difilippantonio, M., Uppender, M., Ward, D. C., & MacLennan, D. H. (1993). Chromosome mapping of five human cardiac and skeletal muscle sarcoplasmic reticulum protein genes. Genomics 17, 507-509.

Palmero, I., & Sastre, L. (1989). Complementary DNA cloning of a protein highly homologous to mammalian sarcoplasmic reticulum Ca-ATPase from the crustacean *Artemia*. J. Mol. Biol. 210, 737-748.

Papp, B., Corvazier, E., Magnier, C., Kovács, T., Bourdeau, N., Lévy-Tolédano, S., Bredoux, R., Lévy, B., Poitevin, P., Lompré, A.-M., Wuytack, F., & Enouf, J. (1993). Spontaneously hypertensive rats and platelet Ca^{2+}-ATPases: specific up-regulation of the 97 kDa isoform. Biochem. J. 295, 685-690.

Papp, B., Enyedi, A., Kovács, T., Sarkadi, B., Wuytack, F., Thastrup, O., Gardos, G., Bredoux, R., Lévy-Tolédano, S., & Enouf, J. (1991). Demonstration of two forms of calcium pumps by thapsigargin inhibition and radioimmunoblotting in platelet membrane vesicles. J. Biol. Chem. 266, 14593-14596.

Papp, B., Enyedi, A., Pászty, K., Kovács, T., Sarkadi, B., Gardos, G., Magnier, C., Wuytack, F., & Enouf, J. (1992). Simultaneous presence of two distinct endoplasmic-reticulum-type calcium-pump isoforms in human cells. Characterization by radio-immunoblotting and inhibition by 2,5-di-(t-butyl)-1,4-benzohydroquinone. Biochem. J. 288, 297-302.

Pedersen, P. L., & Carafoli, E. (1987). Ion motive ATPases. I. Ubiquity, properties and significance to cell function. TIBS 12, 146-150.

Peiffer, W., Desrosiers, M. G., & Menick, D. R. (1996). Cloning and expression of the unique Ca^{2+}-ATPase from *Flavobacterium odoratum*. J. Biol. Chem. 271, 5095-5100.

Perez-Prat, E., Narasimhan, M. L., Binzel, L., Botella, M. A., Chen, Z., Valpuesta, V., Bressan, R. A., & Hasegawa, P.M. (1992). Induction of a putative Ca^{2+}-ATPase mRNA in NaCl-adapted cells. Plant Physiol. 100, 1471-1478.

Perreault, C. L., Shannon, R. P., Komamura, K., Vatner, S. F., & Morgan, J. P. (1992). Abnormalities in intracellular calcium regulation and contractile function in myocardium from dogs with pacing-induced heart failure. J. Clin. Invest. 89, 932-938.

Perreault, C. L., Williams, C. P., & Morgan, J. P. (1993). Cytoplasmic calcium modulation and systolic versus diastolic dysfunction in myocardial hypertrophy and failure. Circulation 87 supplement VII, VII31-37.

Plessers, L., Eggermont, J. A., Wuytack, F., & Casteels, R. (1991). A study of the organellar Ca^{2+}-transport ATPase isozymes in pig cerebellar Purkinje neurons. J. Neurosc. 11, 650-656.

Pozzan, T., Rizzuto, R., Volpe, P., & Meldolesi, J. (1994). Molecular and cellular physiology of intracellular calcium stores. Physiol. Rev. 74, 595-636.

Putney, J. W. Jr. (1993). Excitement about calcium signalling in inexcitable cells. Science 262, 676-678.

Raeymaekers, L., Verbist, J., Wuytack, F., Plessers, L., & Casteels, R. (1993). Expression of Ca^{2+} binding proteins of the sarcoplasmic reticulum of striated muscle in the endoplasmic reticulum of pig smooth muscles. Cell Calcium 14, 581-589.

Raschke, B. C., & Wolf, A. H. (1996). Molecular cloning of a P-type Ca^{2+}-ATPase from the halotolerant alga *Dunaliella bioculata*. Planta 200, 78-84.

Reddy, L. G., Jones, L. R., Pace, R. C., & Stokes, D. L. (1996). Purified, reconstituted cardiac Ca^{2+}-ATPase is regulated by phospholamban but not by direct phosphorylation with Ca^{2+}/calmodulin-dependent protein kinase. J. Biol. Chem. 271, 14964-14970.

Reithmeier, R. A. F., de Leon, S., & MacLennan, D. H. (1980). Assembly of the sarcoplasmic reticulum cell free system of the Ca^{2+}+Mg^{2+}-adenosinetriphosphatase and calsequestrin. J. Biol. Chem. 255, 11839-11-846.

Revelard, P. & Pays, E. (1991). Structure and transcription of a P-ATPase gene from *Trypanosoma brucei*. Molec. Biochem. Parasitol. 46, 241-252.

Rockman, H. A., Ono, S., Ross, R. S., Jones, L. R., Karimi, M., Bhargava, V., Ross, J., & Chien, K. R. (1994). Molecular and physiological alterations in murine ventricular dysfunction. Proc. Natl. Acad. Sci. USA 91, 2694-2698.

Rohrer, D. K., & Dillmann, W. H. (1988). Thyroid hormone markedly increases the mRNA coding for sarcoplasmic reticulum calcium ATPase in the rat heart. J. Biol. Chem. 263, 6941-6944.

Rohrer, D. K., Hartong, R., & Dillmann, W. H. (1991). Influence of thyroid hormone and retinoic acid on slow sarcoplasmic reticulum Ca^{2+} ATPase and myosin heavy chain alpha gene expression in cardiac myocytes. Delineation of cis-active DNA elements that confer responsiveness to thyroid hormone but not to retinoic acid. J. Biol. Chem. 266, 8638-8646.

Rooney, E. K., Gross, J. D., & Satre, M. (1994). Characterisation of an intracellular Ca^{2+} pump in *Dictyostelium*. Cell Calcium 16, 509-522.

Rudolph, H. K., Antebi, A., Fink, G. R., Buckley, C. M., Dorman, T. E., LeVitre, J., Davidow, L. S., Mao, J., & Moir, D. T. (1989). The yeast secretory pathway is perturbed by mutations in *PMRI*, a member of a Ca^{2+} ATPase family. Cell 58, 133-145.

Salvatori, S., Damiani, E., Zorzato, F., Volpe, P., Pietrobon, S., Quaglino, D. Jr., Salviati, G., & Margreth, A. (1988). Denervation-induced proliferative changes of triads in rabbit skeletal muscle. Muscle Nerve 11, 1246-1259.

Sambrook, J. F. (1990). Involvement of calcium in transport of secretory proteins from the endoplasmic reticulum. Cell 61, 197- 199.

Sayen, M. R., Rohrer, D. K., & Dillmann, W. H. (1992). Thyroid hormone response of slow and fast sarcoplasmic reticulum Ca^{2+} ATPase mRNA in striated muscle. Mol. Cell. Endocrinol. 87, 87-93.

Schulte, L. M., Navarro, J., & Kandarian, S. C. (1993). Regulation of sarcoplasmic reticulum calcium pump gene expression by hindlimb unweighting. Am. J. Physiol. 264, C1308-1315.

Schulte, L., Peters, D., Taylor, J., Navarro, J., & Kandarian, S. (1994). Sarcoplasmic reticulum Ca^{2+} pump expression in denervated skeletal muscle. Am. J. Physiol. 267, C617-622.

Seidler, N. W., Jona, I., Vegh, M., & Martonosi, A. (1989). Cyclopiazonic acid is a specific inhibitor of the Ca^{2+} ATPase of the sarcoplasmic reticulum. J. Biol. Chem. 264, 17816-17823.

Shen, H., Mirsalimi, S. M., Weiler, J. E., Julian, R. J., & O'Brien, P. J. (1991). Effects of mild cardiac hypertrophy, induced by volume overload in turkeys, on myocardial sarcoplasmic reticulum calcium-pump and calcium-channel activities and on the creatine kinase system. J. Vet. Res. 52, 1527-1530.

Shull, M. M., Pugh, D. G., & Lingrel, J. B. (1989). Characterization of the human Na,K-ATPase $\alpha 2$ gene and identification of intragenic restriction fragment length polymorphisms. J. Biol. Chem. 264, 17532-17543.

Simmerman, H. K., Collins, J. H., Theibert, J. L., Wegener, A. D., & Jones, L. R. (1986). Sequence analysis of phospholamban. Identification of phosphorylation sites and two major structural domains. J. Biol. Chem. 261, 13333-13341.

Simmerman, H. K., Lovelace, D. E., & Jones, L. R. (1989). Secondary structure of detergent-solubilized phospholamban, a phosphorylatable, oligomeric protein of cardiac sarcoplasmic reticulum. Biochim. Biophys. Acta 997, 322-329.

Simmerman, H. K., Kobayashi, Y. M., Autry, J. M., & Jones, L. R. (1996). A leucine zipper stabilizes the pentameric membrane domain of phospholamban and forms a coiled-coil pore structure. J. Biol. Chem. 271, 5941-5946.

Simonides, W. S., van der Linden, G. C., & van Hardeveld, C. (1990). Thyroid hormone differentially affects mRNA levels of Ca-ATPase isozymes of sarcoplasmic reticulum in fast and slow skeletal muscle. FEBS Lett. 274, 73-76.

Sitia, R., & Meldolesi, J. (1992). Endoplasmic reticulum: a dynamic patchwork of specialized subregions. Molec. Biol. Cell 3, 1067-1072.

Skerjanc, I. S., Clarke, D. M., Loo, T. W.~& MacLennan, D. H. (1993). Deletion of NH_2- and COOH-terminal sequences destroys function of the Ca^{2+} ATPase of rabbit fast-twitch skeletal muscle sarcoplasmic reticulum. FEBS Lett. 336, 168170.

Song, Y., & Fambrough, D. (1994). Molecular evolution of the calcium-transporting ATPases analyzed by the maximum parsimony method. In: Molecular Evolution of Physiological Processes. (Fambrough, D. M., Ed.) Soc. Gen. Physiol. Ser. 49, 271-283.

Sorin, A., Rosas, G., & Rao, R. (1997). PMR1, a Ca^{2+} ATPase in yeast Golgi, has properties distinct from sarco/endoplasmic reticulum and plasma membrane calcium pumps. J. Biol. Chem. 272, 9895-9901.

Stendahl, O., Krause, K.-H., Krischer, J., Jerstrom, P., Theler, J.-M., Clark, R. A., Carpentier, J.-L., & Lew, D. P. (1994). Redistribution of intracellular Ca^{2+} stores during phagocytosis in human neutrophils. Science 265, 1439-1441.

Strehler, E. E. (1991). Recent advances in the molecular characterization of plasma membrane Ca^{2+} pumps. J. Memb. Biol. 120, 1-15.

Studer, R., Reinecke, H., Bilger, J., Eschenhagen, T., Bohm, M., Hasenfuss, G., Just, H., Holtz, J., & Drexler, H. (1994). Gene expression of the cardiac Na^{+}/Ca^{2+} exchanger in end-stage human heart failure. Circ. Res. 75, 443-453.

Sukovich, D. A., Shabbeer, J., & Periasamy, M. (1993). Analysis of the rabbit cardiac/slow twitch muscle sarcoplasmic reticulum calcium ATPase (SERCA2) gene promoter. Nucleic Acids Res. 21, 2723-2728.

Suzuki, C. K., Bonifacino, J. G., Lin, A. Y., Davis, M. M., & Klausner, R. D. (1991). Regulation of the retention of T-cell receptor α chain variants within the endoplasmic reticulum: Ca^{2+} dependent association with Bip. J. Cell Biol. 114, 189-205.

Tada, M., Ohmori, F., Yamada, M., & Abe, H. (1979). Mechanism of stimulation of Ca^{2+}-dependent ATPase of cardiac sarcoplasmic reticulum by adenosine 3':5'-monophosphate-dependent protein kinase. Role of the 22000-Dalton protein. J. Biol. Chem. 254, 319-326.

Tada, M., & Katz, A. M. (1982). Phosphorylation of the sarcoplasmic reticulum and sarcolemma. Ann. Rev. Physiol. 44, 401-423.

Tada, M., Kadoma, M., Inui, M., & Fujii, J. (1988). Regulation of Ca^{2+}-pump from cardiac sarcoplasmic reticulum. Methods Enzymol. 157, 107-154.

Takahashi, T., Allen, P. D., & Izumo, S. (1992a). Expression of A-, B-, and C-type natriuretic peptide genes in failing and developing human ventricles. Correlation with expression of Ca^{2+}-ATPase gene. Circ. Res. 71, 9-17.

Takahashi, T., Allen, P. D., Lacro, R. V., Marks, A. R., Dennis, A. R., Schoen, F. J., Grossman, W., Marsh, J. D., & Izumo, S. (1992b). Expression of dibydropyridine receptor (Ca^{2+} channel) and calsequestrin genes in the myocardium of patients with end-stage heart failure. J. Clin. Invest. 90, 927-935.

Takei, K., Stukenbrok, H., Metcalf, A., Mignery, G. A., Sudhof, T. C., Volpe, P., & De Camilli, P. (1992). Ca^{2+} stores in Purkinje neurons: Endoplasmic reticulum subcompartments demonstrated by the heterogeneous distribution of the InsP3 receptor, Ca^{2+}-ATPase, and calsequestrin. J. Neurosc. 12, 489-505.

Taubman, M. B., Smith, C. W. J., Isumo, S., Grant, J. W., Endo, T., Andreadis, A., & Nadal-Ginard, B. (1989). The expression of sarcomeric muscle-specific contractile protein genes in BC_3H1 cells resemble skeletal myoblasts that are defective for commitment to terminal differentiation. J. Cell. Biol. 198, 1799-1806.

Thastrup, O., Cullen, P. J., Drobak, B. K., Hanley, M. K., & Dawson, A. P. (1990). Thapsigargin, a tumor promotor, discharges intracellular Ca^{2+} stores by specific inhibition of the endoplasmic reticulum Ca^{2+} ATPase. Proc. Natl. Acad. Sci. USA 87, 2466-2470.

Thelen, M. H., Muller, A., Zuidwijk, M. J., van der Linden, G. C., Simonides, W. S., & van Hardeveld, C. (1994). Differential regulation of the expression of fast-type sarcoplasmic-reticulum Ca^{2+}-ATPase by thyroid hormone and insulin-like growth factor-I in the L6 muscle cell line. Biochem. J. 303, 467-474.

Toyofuku, T., Kurzydlowski, K., Lytton, J., MacLennan, D. H. (1992). The nucleotide binding hinge domain plays a crucial role in determining isoform-specific Ca^{2+} dependence of organellar Ca^{2+}-ATPases. J. Biol. Chem. 267, 14490-14496.

Toyofuku, T., Kurzydlowski, K., Tada, M., & MacLennan, D. H. (1993). Identification of regions in the Ca^{2+}-ATPase of sarcoplasmic reticulum that affect functional association with phospholamban. J. Biol. Chem. 268, 2809-2815.

Toyofuku, T., Kurzydlowski, K., Tada, M., & MacLennan, D. H. (1994a). Amino acids Glu2 to Ilel8 in the cytoplasmic domain of phospholamban are essential for functional association with the Ca^{2+}-ATPase of sarcoplasmic reticulum. J. Biol. Chem. 269, 3088-3094.

Toyofuku, T., Kurzydlowski, K., Tada, M., & MacLennan, D. H. (1994b). Amino acids Lys-Asp-Asp-Lys-Pro-Val402 in the Ca^{2+}-ATPase of cardiac sarcoplasmic reticulum are critical for functional association with phospholamban. J. Biol. Chem. 269, 22929-22932.

Toyofuku, T., Kurzydlowski, K., Narayanan, K., & MacLennan, D. H. (1994c). Identification of Ser38 as the site in cardiac sarcoplasmic reticulum Ca^{2+}-ATPase that is phosphorylated by Ca^{2+}/calmodulin-dependent protein kinase. J. Biol. Chem. 269, 26492-26496.

Trottein, F., Thompson, J., & Cowman, A. F. (1995). Cloning of a new cation ATPase from *Plasmodium falciparum:* conservation of critical amino acids involved in calcium binding in mammalian organellar Ca^{2+}-ATPases. Gene 158, 133-137.

van Breemen, C., & Saida, K. (1989). Cellular mechanisms regulating $[Ca^{2+}]_i$ in smooth muscle. Ann. Rev. Physiol. 51, 315-329.

Van Den Bosch, L., Eggermont, J., De Smedt, H., Mertens, L., Wuytack, F., & Casteels, R. (1994). Regulation of splicing is responsible for the expression of the muscle-specific 2a isoform of the sarco/endoplasmic-reticulum Ca^{2+}-ATPase. Biochem. J. 302, 559-566.

van der Linden, G. C., Simonides, W. S., & van Hardeveld, C. (1992). Thyroid hormone regulates Ca^{2+}-ATPase mRNA levels of sarcoplasmic reticulum during neonatal development of fast skeletal muscle. Mol. Cell. Endocrinol. 90, 125-131.

Váradi, A., Gilmore-Heber, M., & Benz, E. J. (1989). Amplification of the phosphorylation site-ATP-binding site cDNA fragment of the Na^{+},K^{+}-ATPase and the Ca^{2+}-ATPase of *Drosophila melanogaster* by polymerase chain reaction. FEBS Lett. 258, 203-207.

Váradi, A. Molnar, E., Ostenson, C. G., & Ashcroft, S. J. (1996). Isoforms of endoplasmic reticulum Ca^{2+}-ATPase are differentially expressed in normal and diabetic islets of Langerhans. Biochem. J. 319, 521-527.

Vatner, D. E., Sato, N., Kiuchi, K., Shannon, R. P., & Vatner, S. F. (1994). Decrease in myocardial ryanodine receptors and altered excitation-contraction coupling early in the development of heart failure. Circulation 90, 1423-1430.

Verboomen, H., Wuytack, F., De Smedt, H., Himpens, B., & Casteels, R. (1992). Functional difference between SERCA2a and SERCA2b Ca^{2+} pumps and their modulation by phospholamban. Biochem. J. 286, 591-596.

Verboomen, H., Wuytack, F., Van Den Bosch, L., Mertens, L., & Casteels, R. (1994). The functional importance of the extreme C-terminal tail in the gene 2 organellar Ca^{2+}-transport ATPase (SERCA2a/b). Biochem. J. 303, 979-984.

Vilsen, B., & Andersen, J.-P. (1992). Deduced amino acid sequence and E1=E2 equilibrium of the sarcoplasmic reticulum Ca^{2+}-ATPase of frog skeletal muscle. Comparison with the Ca^{2+}-ATPase of rabbit fast twitch muscle. FEBS Lett. 306, 213-218.

Webb, R., & Dormer, R. L. (1995). Photoaffinity labelling of the ATP-binding sites of two Ca^{2+}, Mg^{2+}-ATPase isoforms in pancreatic endoplasmic reticulum. Biochim. Biophys. Acta, 1233, 1-6.

Wegener, A. D., Simmerman, H. K., Lindemann, J. P., & Jones, L. R. (1989). Phospholamban phosphorylation in intact ventricles. Phosphorylation of serine 16 and threonine 17 in response to beta-adrenergic stimulation. J. Biol. Chem. 264, 11468-11474.

Whitmer, J. T., Kumar, P., & Solaro, R. J. (1988). Calcium transport properties of cardiac sarcoplasmic reticulum from cardiomyopathic Syrian hamsters (BIO 53.58 and 14.6): evidence for a quantitative defect in dilated myopathic hearts not evident in hypertrophic hearts. Circ. Res. 62, 81-85.

Wileman, T., Kane, T. P., Carson, G. R., & Terhorst, C. (1991). Depletion of cellular calcium accelerates protein degradation in the endoplasmic reticulum. J. Biol. Chem. 266, 4500-4507

Williams, R. E., Kass, D. A., Kawagoe, Y., Pak, P., Tunin, R. S., Shah, R., Hwang, A., & Feldman, A. M. (1994). Endomyocardial gene expression during development of pacing tachycardia-induced heart failure in the dog. Circ. Res. 75, 615-623.

Wimmers, L. E., Ewing, N. N., & Bennett, A. B. (1992). Higher plant Ca^{2+}-ATPase: primary structure and regulation of mRNA abundance by salt. Proc. Natl. Acad. Sci. USA 89, 9205-9209.

Wu, K. D., & Lytton, J. (1993). Molecular cloning and quantification of sarcoplasmic reticulum Ca^{2+}-ATPase isoforms in rat muscles. Am. J. Physiol. 264, C333-341.

Wu, K. D., Lee, W.-S., Wey, J., Bungard, D., & Lytton, J. (1995). Localization and quantification of endoplasmic reticulum Ca^{2+}-ATPase isoform transcripts. Am. J. Physiol. 269, C775-784.

Wuytack, F., Dode, L., Baba-Aissa, F., & Raeymaekers, L. (1995). The SERCA3-type of organellar Ca^{2+} pumps. Bioscience Rep. 15, 299-306.

Wuytack, F., Eggermont, J. A., Raeymaekers, L., Plessers L., & Casteels, R. (1989). Antibodies against the non-muscle isoform of the endoplasmic reticulum Ca^{2+}-transport ATPase. Biochem. J. 264, 765-769.

Wuytack, F., Raeymaekers, L., Verbist, J., Jones, L. R., & Casteels, R. (1987). Smooth-muscle endoplasmic reticulum contains a cardiac-like form of calsequestrin. Biochim. Biophys Acta. 899, 151-158.

Wuytack, F., Papp, B., Verboomen, H., Raeymaekers, L., Dode, L., Bobe, R., Enouf, J., Bokkala, S., Authi, K. S., & Casteels, R. (1994). A sarco/endoplasmic reticulum Ca^{2+}-ATP ase 3 -type Ca^{2+} pump is expressed in platelets, in lymphoid cells, and in mast cells. J. Biol. Chem. 269, 1410-1416.

Wuytack, F., Raeymaekers, L., De Smedt, H., Eggermont, J. A., Missiaen, L., Van Den Bosch, L., De Jaegere, S., Verboomen, H., Plessers, L., & Casteels, R. (1992). Ca^{2+} transport ATPases and their regulation in muscle and brain. Ann. N.Y. Acad. Sciences 671, 82-91.

Wuytack, F., Raeymaekers, L., Verbist, J., De Smedt, H., & Casteels, R. (1984). Evidence for the presence in smooth muscle of two types of Ca^{2+}-transport ATPase. Biochem. J. 224, 445-451.

Xu, A., Hawkins, C., & Narayanan, N. (1993). Phosphorylation and activation of the Ca^{2+}-pumping ATPase of cardiac sarcoplasmic reticulum by Ca^{2+}/calmodulin-dependent protein kinase. J. Biol. Chem. 268, 8394-8397.

Zádor, E., Mendler, L., Ver Heyen, M., Dux, L., & Wuytack, F. (1996) Changes of mRNA levels of the sarco/endoplasmic reticulum Ca^{2+}-ATPase isoforms in rat soleus muscle regenerating from notexin-induced necrosis. Biochem. J. 320, 107-113.

Zarain-Herzberg, A., MacLennan, D. H., & Periasamy, M. (1990). Characterization of rabbit cardiac sarco(endo)plasmic reticulum Ca^{2+}-ATPase gene. J. Biol. Chem. 265, 4670-4677.

Zarain-Herzberg, A., Marques, J., Sukovich, D., & Periasamy, M. (1994). Thyroid hormone receptor modulates the expression of the rabbit cardiac sarco (endo) plasmic reticulum Ca^{2+}-ATPase gene. J. Biol. Chem. 269, 1460-1467.

Zhang, Y., Fujii, J., Phillips, M. S., Chen H.-S., Karpati, G., Yee W.-C., Schrank, B., Cornblath, D. R., Boylan, K. B., & MacLennan, D. H. (1995) Characterization of cDNA and genomic DNA encoding SERCAl, the Ca^{2+}-ATPase of human fast-twitch skeletal muscle sarcoplasmic reticulum, and its elimination as a candidate gene for Brody disease. Genomics 30, 415-424.

Printed and bound by CPI Group (UK) Ltd, Croydon, CR0 4YY
03/10/2024
01040433-0012